Helmut Werner
Peter Schreier
(Eds.)

**Selective Reactions of
Metal-Activated Molecules**

Third Symposium

W0246182

__ **Chemistry** _____

G. Helmchen with J. Dibo, D. Flubacher, B. Wiese (Eds.)
Organic Synthesis via Organometallics (OSM 5)

P. Schreier, M. Herderich, H.-U. Humpf, W. Schwab
Natural Product Analysis
Chromatography – Spectroscopy – Biological Testing

A. Müller, A. Dress, F. Vögtle (Eds.)
From Simplicity to Complexity in Chemistry
– and Beyond (Part I)

K. Mainzer, A. Müller, W. G. Saltzer (Eds.)
From Simplicity to Complexity (Part II)

H. Engelhardt, W. Beck, Th. Schmitt
Capillary Electrophoresis

Ch. Heller (Ed.)
Analysis of Nucleic Acids by Capillary Electrophoresis
CHROMATOGRAPHIA CE-Series, Vol. I

CHROMATOGRAPHIA
An International Journal for Rapid Communications
in Chromatography, Electrophoresis and
Associated Techniques

__ **Vieweg** _____

Helmut Werner
Peter Schreier
(Eds.)

Selective Reactions of Metal-Activated Molecules

Proceedings of the Third Symposium
held in Würzburg,
September 17–19, 1997

vieweg

Produced by Hubert & Co., Göttingen

ISBN 978-3-662-00977-2   ISBN 978-3-662-00975-8 (eBook)
DOI 10.1007/978-3-662-00975-8

Preface

The rapid and impressive development which occurred and still occurs in all parts of science would not have been possible without interdisciplinary research. In order to help the scientists to cross the boundaries of their own field and to collaborate with research groups of neighbouring disciplines, the Deutsche Forschungsgemeinschaft (DFG) initiated interdisciplinary research units (Sonderforschungsbereiche) already in the 1960s. This program has been very successful and nowadays it is one of the big wheels to spur first-class research at German universities.

The Sonderforschungsbereich (SFB) 347 was established at the University of Würzburg in 1990. It is entitled "Selektive Reaktionen Metall-aktivierter Moleküle" and should promote efficient cooperation between the departments of inorganic, organic and physical chemistry, nutrition chemistry and experimental physics. Organometallic chemistry in all its different facets provides the frame of the project and the metal and its influence on interacting molecules stands in its center. The following important roles played by the metal are considered:

- activation of relatively unreactive molecules (e.g. oxygen, hydrocarbons, CO etc.)
- stabilization of short-lived reaction intermediates (oxenes, nitrenes, carbenes, sulfenes etc.)
- coordination of substrates in highly-ordered templates (e.g. prochiral biaryllactones)
- modification of reactive substrates (e.g. peroxides or phosphorus ylides) with concomitant achievement of chemo-, regio- and stereoselectivity.

Catalysis in general and by metalloenzymes in particular is also becoming an important part of the program. This aspect as well as spectroscopical and theoretical studies on metal-ligand interactions have been taken into account in several contributions to this volume.

An essential ingredient of this interdisciplinary research project is the periodic holding of meetings in order to present current results and discuss modern trends in organometallic chemistry and related areas. The third such symposium was held at Würzburg from September 17 to 19, 1997 with many guests from other universities and industry attending. The scientific program consisted of 36 poster contributions from research groups inside and outside the Sonderforschungsbereich, and 12 plenary lectures from renowned scientists, which covered a broad spectrum of metal catalysis, organometallic synthesis, bioorganic reactions, physicochemical and theoretical investigations on organometallic compounds and intermediates. We are grateful to all the participants for their high-quality contributions to this proceedings book. We are convinced that the results collected in this volume present an useful (admittedly selected) survey of the trends and concepts which will lead to further progress in this stimulating and important area of modern chemistry.

The editors also thank Dr. Angelika Schulz (Vieweg-Verlag) for the excellent cooperation.

Würzburg, January 10, 1998　　　　　　　　　　　Helmut Werner (Chairman, SFB 347)
　　　　　　　　　　　　　　　　　　　　　　　　　　Peter Schreier

Contents

I Selective Coupling and Cleavage of Covalent Bonds by Metal Complexes and Metallo Enzymes

C-O Bonds

C-C and C-H Bonds

C-E and E-E Bonds (E = Heteroatom)

II Spectroscopical and Theoretical Studies on the Structure and Dynamics of Metal Complexes

Chapter I

Selective Coupling and Cleavage of Covalent Bonds by Metal Complexes and Metallo Enzymes

C-O Bonds

Enantioselective Metal Catalysis

Carsten Bolm

Institut für Organische Chemie der RWTH Aachen, Professor-Pirlet-Str. 1, D-52056
Aachen, Germany, Fax (int.) + 241 8888 391, E-mail: Carsten.Bolm@RWTH- Aachen.de

Efficient asymmetric metal-catalyzed processes require catalysts with excellent properties in many respects. Their substrate range should be broad giving products in very good yield with high enantiomeric excess (*ee*). Functional groups must be tolerated and should not have negative effects on the various steps of the chirality transfer. In addition, the catalyst is expected to function rapidly in a predictable manner under convenient reaction conditions. High turnover numbers indicate catalyst stability.

Only very few enantioselective catalysts fulfil most of these requirements [1]. For example, in the area of asymmetric hydrogenations appropriate metal-ligand combinations allow the reduction of C–C double bonds and carbonyl groups with excellent enantio-selectivities [2]. Recently, exceedingly high turnover numbers have been achieved in such processes [3]. In this article we describe various asymmetric metal-catalyzed trans-formations - mainly oxidations - with special focus on our attempts to increase the efficiency of these catalyses.

In 1994 we [4] and others [5] reported on the first asymmetric Baeyer-Villiger reactions catalyzed by chiral metal complexes. In the presence of 1 mol% of (*S,S*)-**3**, racemic 2-phenylcyclohexanone (**1**) was oxidized to the corresponding lactone (*R*)-**2** in 41% yield having 65% *ee*. A combination of dioxygen and pivaldehyde served as oxidant.

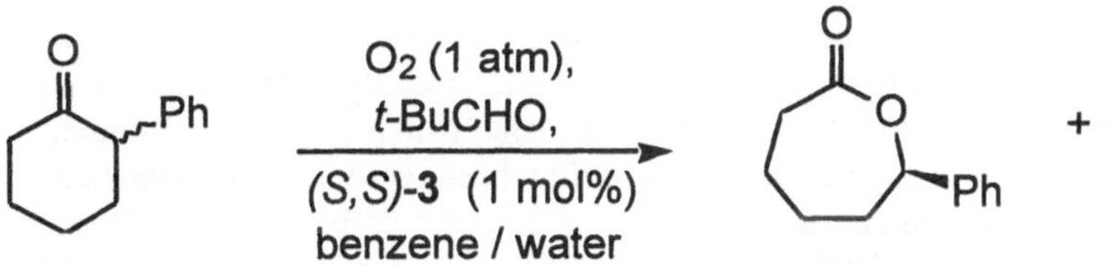

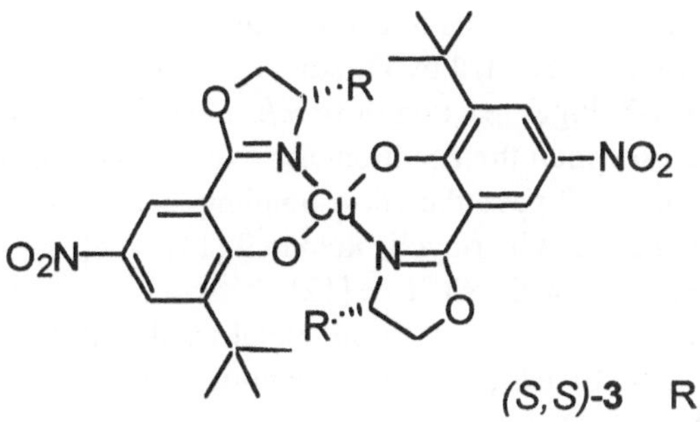

Favorable factors in this catalysis were the mild reaction conditions and the ease of accessibility of the copper catalyst [6]. Compared to many other asymmetric oxidations the catalyst amount was relatively small. Problems, however, remained. In particular, the substrate range was very limited, and in the series of 6-membered cycloalkanones only 2-aryl substituted derivatives were reactive enough to give the desired optically active lactones [4].

Next, we investigated transformations of cyclobutanone derivatives such as **4**. Compounds of this type are readily available, and the corresponding lactones are valuable intermediates in organic synthesis. To our surprise we found that from racemic **4** *two* isomeric lactones were formed under the standard reaction conditions (1 mol% of catalyst and pivaldehyde in benzene solution at ambient temperature under an atmosphere of dioxygen) [7].

O_2 (1 atm),
t-BuCHO,
(*S*,*S*)-**3** (1 mol%)
benzene

rac-**4**   67% ee   92% ee

(*R*,*R*)-**5**   (*S*,*R*)-**6**

These two isomeric lactones, **5** and **6**, differ in three major aspects: 1. They result from oxygen insertion at either side of the carbonyl group. Whereas **5** is the expected product of a regular Baeyer-Villiger reaction [8], **6** is an unusual regioisomer. Control experiments with **4** and MCPBA as oxidant gave **5** with great preference, and **6** was only detected in traces. 2. The enantiomeric excesses of the two products are different. Only the unexpected regioisomer **6** is obtained with high enantioselectivity (92% *ee*). Presumably due to a parallel uncatalyzed pathway giving racemic product, the *ee* of lactone **5** was only 67%. 3. Lactones **5** and **6** result from enantiomeric ketones. Thus, the reaction proceeds in an enantiodivergent manner giving regioisomeric products which differ in absolute stereochemistry. Thus **5** has *R*-configuration at the bridgehead carbon atom attached to the oxygen, whereas **6** has *S*-configuration at the carbon α to the carbonyl group. Processes of this kind are rare, and the very few examples have been summarized by Kagan [9]. In its overall behavior in Baeyer-Villiger reactions copper complex (*S*,*S*)-**3** resembles its biocatalytic counterpart [10]: enzymatic oxidations of compounds of type **4** also proceed via enantiodivergent pathways, as previously revealed by Furstoss and coworkers [11].

Attempts to use (*S*,*S*)-**3** in Baeyer-Villiger reaction of *prochiral* cyclobutanones had only limited success [12, 13]. With one exception the enantiomeric excesses remained moderate. Thus, 3-monosubstituted cyclobutanones **7** gave the corresponding lactones **8** with enantioselectivities in the range of 23-47% *ee*. Kelly's tricyclic ketone **9** [14] was the only substrate which afforded an optically active lactone with >90% *ee* [12].

Studies to further improve the efficiency of these metal-catalyzed Baeyer-Villiger reactions are in progress, and we hope to achieve higher enantioselectivities and improved turnover numbers by modifying the metal catalyst and changing the reaction conditions.

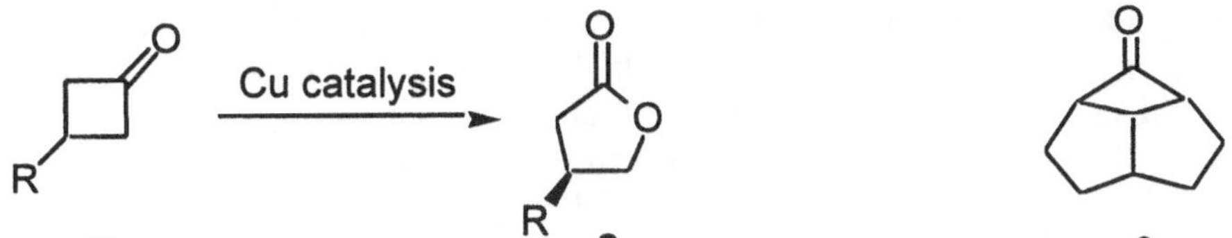

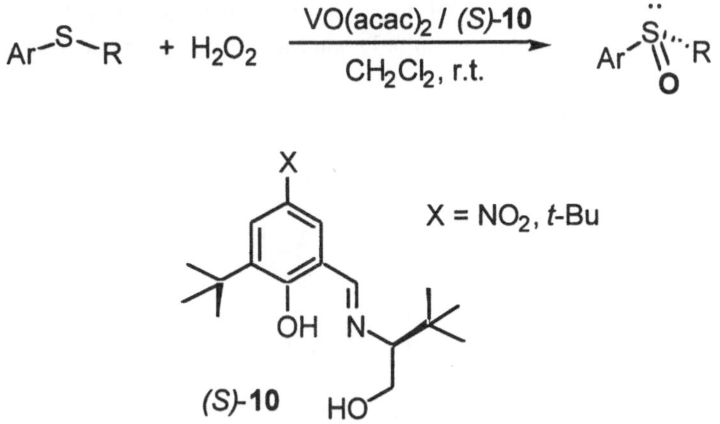

Whereas the Baeyer-Villiger reaction is catalyzed by a chiral copper complex using a combination of dioxygen and aldehyde as oxidant, an asymmetric sulfide oxidation with hydrogen peroxide has been achieved with the help of a vanadium complex [15].

In this oxidation, five aspects are particularly noteworthy. Firstly, inexpensive and environmentally friendly aqueous hydrogen peroxide (30%) is used as the oxidant. With 1.1 equiv. of H_2O_2 optically active sulfoxides are formed in good yield with up to 85% *ee* [16]. Secondly, the vanadium source [VO(acac)$_2$] is commercially available, and the ligands are easily prepared from salicyl aldehydes and *tert*-leucinol. Thirdly, the catalyst loading is exceptionally low, usually 21 mol%. Even with 0.01 mol% of catalyst optically active sulfoxides were obtained. Fourthly, the reaction conditions are simple and do not require exclusion of air or moisture. Fifthly, the process is 'ligand-accelerated'. This phenomenon was first described by Sharpless [17] and with respect to the catalysis described here, it means that achiral vanadium species, which may also be present in the reaction mixture, are less effective catalysts for the sulfide oxidation under these conditions than optically active ones. In the presence of the ligand 'chiral pathways' are 'turned on' and begin to dominate giving rise to the formation of products with good enantiomeric excess.

The sulfide oxidation described above benefits from an inherent property of this particular vanadium catalysis - the ligand acceleration. We expect other asymmetric catalyses to show the same (or related) phenomena, and we look forward to discovering and to using them to improve catalyst performance.

The ligands in the Baeyer-Villiger reaction and the sulfide oxidation are inexpensive and readily available. In cases, however, where highly valuable compounds are employed as ligands and in particular in large-scale applications, ligand recovery becomes of major importance. This aspect must also be considered when catalyst efficiency is discussed [18]. We chose the Sharpless asymmetric dihydroxylation of olefins (AD) [19] as model reaction to demonstrated the principle of ligand recovery by simple solvent addition followed by filtration [20-22].

Bis(9-O-dihydroquinidinyl)bis(biaryl)pyrazinopyridazine **11** [21] was readily prepared in analogy to procedures describing the syntheses of the parent AD ligands [23]. Attachment of the MeO-PEG chains was accomplished by either esterification or nucleophilic substitution. MALDI mass spectrometry [24] revealed that both methods predominantly yielded products with only one MeO-PEG tether (ester **12** and ether **13**).

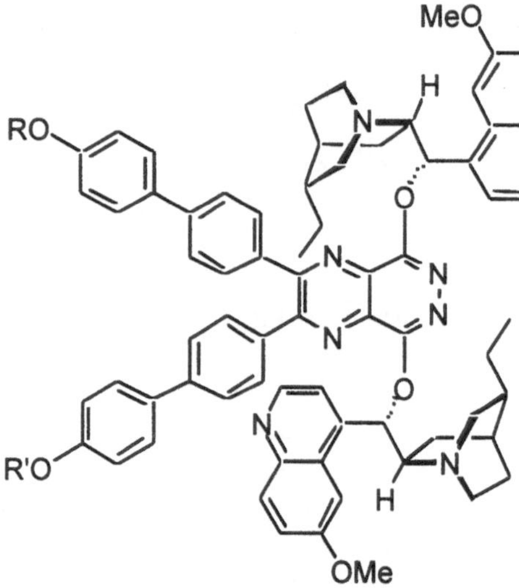

11: R = R' = H
12: R/R' = H or OC(O)(CH$_2$)$_2$C(O)O-PEG-OMe
13: R/R' = H or O-PEG-OMe

AD reactions under the standard conditions using 1 mol% of K$_2$OsO$_2$(OH)$_4$, 1.2 mol% of ligand and a K$_3$[Fe(CN)$_6$]/K$_2$CO$_3$ reoxidation system in *tert*-butanol/water mixtures gave diols with very high enantiomeric excess. The *ee*-values (Table 1) were in good accordance with those reported by Sharpless for the dihydroxylation with the original ligands [19,23].

After the catalyses, **12** and **13** were easily recovered (>98%) by addition of an ethereal solvent (such as MTBE) to the reaction mixture followed by filtration of the precipitated alkaloids. Consecutive use of **12** in the AD of styrene revealed a slight drop of the enantioselectivity after several runs (run 1: 98% *ee*; run 6: 96% *ee*). No such decrease in *ee* was observed in consecutive reactions with **13**. The enantioselectivity remained constantly high (98% *ee*), and even when the recycled ligand was used 6 times, the activity and selectivity of the catalyst system did not change [25].

8

Table 1. AD with MeO-PEG-bound alkaloids.

Olefin	Ester **12**	Ether **13**
Ph⁀⁀Ph	99% *ee*	99% *ee*
Ph⁀⁀	98% *ee*	98% *ee*

The AD reactions with MeO-PEG-bound ligands occur under *homogeneous* conditions and therefore benefit from all advantages of such type of catalysis (e. g. the ligand acceleration [17]). Ligand recovery is easy and almost complete, but suffers from the fact that large quantities of ethereal solvents are required to precipitate the ligand after completion of the reaction. Various studies have been directed towards the use of *heterogenized* alkaloids in the AD [26]. Usually, lower enantioselectivities and reduced activity were observed and only very recently, significant progress has been achieved [26g,h]. Of particular interest is the attachment of alkaloids to silica gel because ligands with high mechanical and thermal stability can be obtained [27]. We therefore investigated the anchoring of ligands of type **11** to functionalized silica and studied the use of such heterogenized alkaloids in the AD. Again, various linkers connecting the solid support with the alkaloid were tested, and ether **14** proved to be the most efficient compound [28]. In the ADs of stilbene and styrene the products were obtained with 99 and 98% *ee*, respectively, indicating that the silica did not have negative effects on the enantioselectivity.

We also investigated the use of **14** in consecutive AD reactions. No loss of enantio-selectivity and activity was observed, even when the ligand was employed in a sequence of 7 runs. Clearly, silica-supported **14** is an excellent ligand for the *heterogeneous* AD, and ligand recovery by filtration is particularly simple.

All catalytic reactions described so far were performed in a batch mode. We also wondered if chiral catalysts could be developed for the synthesis of optically active products in continuously operating systems. As test reaction we chose the asymmetric borane reduction of prochiral ketones. In previous studies we had demonstrated that optically active sulfoximines of type **15** were good catalysts for this transformation giving products with up to 93% *ee* (batch mode in toluene at ambient temperature; 10 mol% of **15** with Ar = Ph) [29].

Furthermore, we confirmed that despite of the molecular size of **16** and the significant steric crowding around the sulfoximine unit, **16** was still able to catalyze enantioselective borane reductions. In a batch reaction with **17** as substrate, the presence of 10 mol% of **16** led to the formation of **18** with 87% *ee* [32].

Finally, we were also able to verify the overall concept [30]. In a continuously operating membrane reactor acetophenone was reduced over a period of more than 40 residence times. Under the present conditions the enantiomeric excess was significantly lowered [from 76% *ee* (batch) to ca. 43% *ee* in the continuous mode]. However, based on previous studies we are very confident that a detailed optimization of the process will allow us to produce *continuously* large quantities of a variety of optically active alcohols with much higher enantiomeric excess. Studies directed towards this goal are in progress.

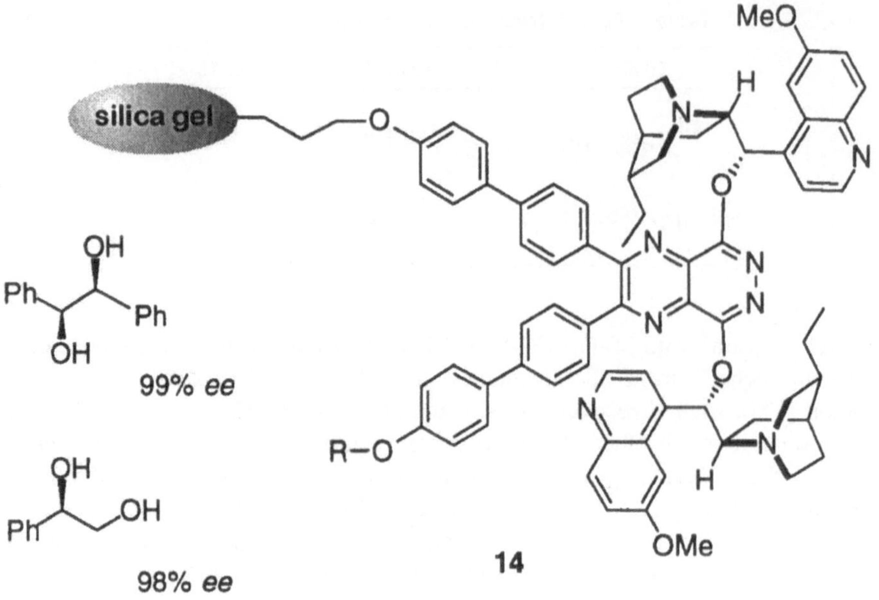

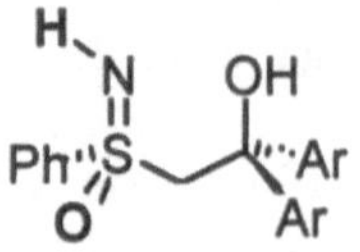

silica gel

OH
Ph   Ph
OH
99% ee

OH
Ph   OH
98% ee

14

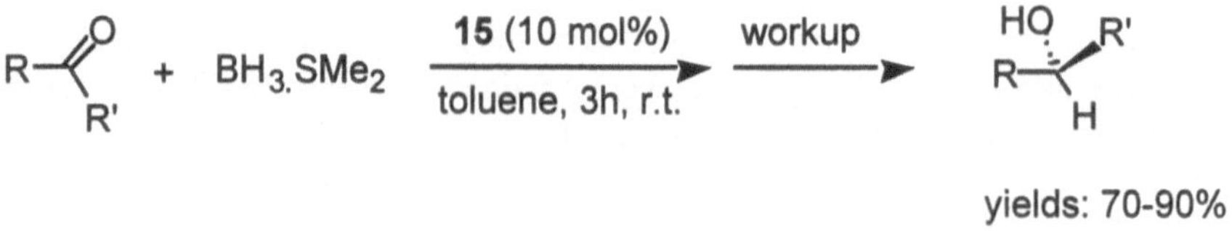

R—C(=O)—R' + BH$_3$·SMe$_2$ →[**15** (10 mol%), toluene, 3h, r.t.] →[workup] R—CH(OH)—R'

yields: 70-90%

H—N=S(=O)Ph ... OH ... Ar / Ar   **15**: Ar = Ph

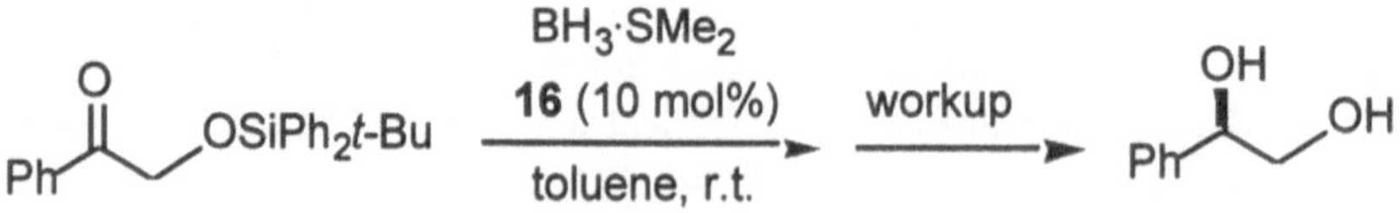

Ph—C(=O)—CH$_2$—OSiPh$_2$t-Bu

17

→[BH$_3$·SMe$_2$, **16** (10 mol%), toluene, r.t.] →[workup] Ph—CH(OH)—CH$_2$—OH

18

Acknowledgment

I wish to express my sincere thanks to all my coworkers who participated in this research. Financial support was kindly provided by the *Deutsche Forschungsgemeinschaft* (Schwerpunktprogramm: Sauerstofftransfer/Peroxidchemie), the *BMBF* and the *Fonds der Chemischen Industrie*. I also thank our *BMBF* partners for their collaboration.

10

References and Notes

[1] a) R. Noyori, *Asymmetric Catalysis in Organic Synthesis,* Wiley, New York, **1994**; b) *Catalytic Asymmetric Synthesis* (Ed.: I. Ojima), VCH, Weinheim, **1993**.

[2] Reviews: a) H. Takaya, T. Ohta, R. Noyori in Ref. [1b], p. 1; b) H. Brunner in *Methods of Organic Chemistry (Houben Weyl) 4th ed.* (Eds.: G. Helmchen, R. W. Hoffmann, J. Mulzer, E. Schaumann), Vol. E21d, chapter 2.3.1, Thieme, Stuttgart, **1995**.

[3] H. -U. Blaser, F. Spindler, *Chimia* **1997**, *51*, 297.

[4] C. Bolm, G. Schlingloff, K. Weickhardt, *Angew. Chem.* **1994**, *106*, 1944; *Angew. Chem. Int. Ed. Engl.* **1994**, *33*, 1848.

[5] For a Pt-catalyzed asymmetric Baeyer-Villiger reaction with hydrogen peroxide: a) A. Gusso, C. Baccin, F. Pinna, G. Strukul, *Organometallics* **1994**, *13*, 3442; b) G. Strukul, A. Varagnolo, F. Pinna, *J. Mol. Cat.* **1997**, *117*, 413.

[6] a) C. Bolm, K. Weickhardt, M. Zehnder, T. Ranff, *Chem. Ber.* **1991**, *124*, 1173; b) C. Bolm, K. Weickhardt, M. Zehnder, D. Glasmacher, *Helv. Chim. Acta* **1991**, *74*, 717; c) C. Bolm, T. K. K. Luong, K. Harms, *Chem. Ber./Recueil* **1997**, *130*, 887.

[7] a) C. Bolm, G. Schlingloff, *J. Chem. Soc. Chem. Commun.* **1995**, 1247; (b) C. Bolm, G. Schlingloff, F. Bienewald, *J. Mol. Cat.* **1997**, *117*, 347.

[8] Reviews: a) C. H. Hassal, *Org. React.* **1957**, *9*, 73; b) G. R. Krow, *Org. React.* **1993**, *43*, 251; c) C. Bolm in *Advances in Catalytic Processes* (Ed.: M. P. Doyle), JAI Press, Greenwich, in press.

[9] H. B. Kagan, *Croatica Chem. Acta* **1996**, *69*, 669.

[10] Reviews: a) V. Alphand, R. Furstoss in *Enzyme Catalysis in Organic Synthesis* (Eds.: K. Drauz, H. Waldmann), VCH, Weinheim, **1995**, p. 745; b) K. Faber, *Biotransformations in Organic Chemistry*, Springer, Berlin, **1995**, p. 203; c) R. Azerad, *Bull. Chem. Soc. Fr.* **1995**, *132*, 17.

[11] V. Alphand, A. Archelas, R. Furstoss, *Tetrahedron Lett.* **1989**, *30*, 3663.

[12] C. Bolm, G. Schlingloff, T. K. K. Luong, *Synlett* in press.

[13] For another approach using the Sharpless epoxidation catalyst see: M. Lopp, A. Paju, T. Kanger, T. Pehk, *Tetrahedron Lett.* **1996**, *37*, 7583.

[14] a) D. R. Kelly, C. J. Knowles, J. G. Mahdi, I. N. Taylor, M. A. Wright, *J. Chem. Soc. Chem. Commun.* **1995**, 729; b) D. R. Kelly, C. J. Knowles, J. G. Mahdi, M. A. Wright, I. N. Taylor, D. E. Hibbs, M. B. Hursthouse, A. K. Mish'al, S. M. Roberts, P. W. H. Wan, G. Grogan, A. J. Willetts, *J. Chem. Soc. Perkin Trans 1*, **1995**, 2057.

[15] C. Bolm, F. Bienewald, *Angew. Chem.* **1995**, *107*, 2883; *Angew. Chem. Int. Ed. Engl.* **1995**, *34*, 2640.

[16] Recently, the enantioselectivity of this vanadium catalysis could be improved by further modifying the ligand. Prof. Dr. J. Skarzewski, Wroclaw, Poland, private communication.

[17] Review: D. J. Berrisford, C. Bolm, K. B. Sharpless, *Angew. Chem.* **1995**, *107*, 1159; *Angew. Chem. Int. Engl.* **1995**, *34*, 1050.

[18] For a general discussion see in: *Applied Homogeneous Catalysis with Organometallic Compounds* (Eds.: B. Cornils, W. A. Herrmann), VCH, Weinheim, **1996**, chapter 3.1.

[19] Reviews: a) R. A. Johnson, K. B. Sharpless in Ref. [1b], p. 227; b) H. C. Kolb, M. S. VanNieuwenhze, K. B. Sharpless, *Chem. Rev.* **1994**, *94*, 2483.

[20] The general concept is based on some earlier work by Bayer and Schurig: a) E. Bayer, V. Schurig, *Angew. Chem.* **1975**, *87*, 484; *Angew. Chem. Int. Ed. Engl.* **1975**, *14*, 493; b) E. Bayer, V. Schurig, *Chemtech* **1976**, *6*, 212; c) For an extension of this work see: B. Koppenhoefer, Ph. D. Dissertation, Univ. of Tübingen, **1976**; d) For a recent related review: D. J. Gravert, K. D. Janda, *Chem. Rev.* **1997**, *97*, 489.

[21] C. Bolm, A. Gerlach, *Angew. Chem.* **1997**, *109*, 773; *Angew. Chem. Int. Ed. Engl.* **1997**, *36*, 773.

[22] For different MeO-PEG-bound ligands used in AD reactions see: a) H. Han, K. D. Janda, *J. Am. Chem. Soc.* **1996**, *118*, 7632; b) H. Han, K. D. Janda, *Tetrahedron Lett.* **1997**, *38*, 1527; c) H. Han, K. D. Janda, *Angew. Chem.* **1997**, *109*, 1835; *Angew. Chem. Int. Ed. Engl.* **1997**, *36*, 0000.

[23] a) H. Becker, K. B. Sharpless, *Angew. Chem.* **1996**, *108*, 447; *Angew. Chem. Int. Ed. Engl.* **1996**, *35*, 448; b) G. A. Crispino, K. -S. Jeong, H. C. Kolb, Z. -M. Wang, D. Xu, K. B. Sharpless, *J. Org. Chem.* **1993**, *58*, 3785.

[24] We thank Prof. Dr. M. -R. Kula, Jülich, for the opportunity to use her MALDI spectrometer.

[25] C. Bolm, A. Gerlach, submitted for publication.

[26] a) B. M. Kim, K. B. Sharpless, *Tetrahedron Lett.* **1990**, *31*, 3003; b) D. Pini, A. Petri, A. Nardi, C. Rosini, P. Salvadori, *ibid.* **1991**, *32*, 5175; c) B. B. Lohray, E. Nandanan V. Bushan, *ibid.* **1994**, *35*, 6559; d) C. E. Song, E. J. Roth, S. -g. Lee, I. O. Kim, *Tetrahedron: Asymmetry* **1995**, *6*, 2687; e) D. Pini, A. Petri, P. Salvadori, *Tetrahedron* **1994**, *50*, 11321; f) A. Petri, D. Pini, S. Rapaccini, P. Salvadori, *Chirality* **1995**, *7*, 580; g) C. E. Song, J. W. Yang, H. J. Ha, S. -g. Lee, *Tetrahedron: Asymmetry* **1996**, *7*, 645; h) P. Salvadori, D. Pini, A. Petri, *J. Am. Chem. Soc.* **1997**, *119*, 6929.

[27] a) B. B. Lohray, E. Nandanan, V. Bhushan, *Tetrahedron: Asymmetry* **1996**, *7*, 2805; b) C. E. Song, J. W. Yang, H.-J. Ha, *ibid.* **1997**, *8*, 841.

[28] C. Bolm, A. Maischack, A. Gerlach, submitted for publication.

[29] a) C. Bolm, M. Felder, *Tetrahedron Lett.* **1993**, *34*, 6041; b) C. Bolm, A. Seger, M. Felder, *ibid.* **1993**, *34*, 8079; c) C. Bolm, M. Felder, *Synlett* **1994**, 655.

[30] These investigations were done in collaboration with Prof. Dr. C. Wandrey and Dr. U. Kragl, Jülich.

[31] a) Review: U. Kragl, C. Dreisbach, C. Wandrey in *Applied Homogeneous Catalysis with Organometallic Compounds* (Eds.: B. Cornils, W. A. Herrmann), VCH, Weinheim, **1996**, chapter 3.2.2, p. 832; b) For the use of a membrane reactor in the addition of diethyl zinc to aldehydes see: U. Kragl, C. Dreisbach, *Angew. Chem.* **1996**, *108*, 684; *Angew. Chem. Int. Ed. Engl.* **1996**, *35*, 642.

[32] C. Bolm, N. Derrien, to be published.

Direct Asymmetric Epoxidation of Aldehydes Using Catalytic Amounts of Enantiomerically Pure Sulfides

Varinder K. Aggarwal,[a*] J. Gair Ford,[a] Alison Thompson,[a] Ray V. H. Jones,[b] and Mike C.H. Standen[b]

(a) Department of Chemistry, University of Sheffield, Sheffield S3 7HF, England.
(b) Zeneca FCMO, Process Technology Department, Earls Road, Grangemouth, Stirling-shire FK3 8XG, UK.

Introduction

The development of catalytic methods for the synthesis of non-racemic epoxides has been a long standing goal in asymmetric synthesis[1]. Most attention has focused on the asymmetric oxidation of alkenes and good enantioselectivities are now beginning to emerge for an increasing range of substrates.[2] Direct epoxidation of carbonyl compounds using sulfur ylides[3] has also been studied but the process usually requires stoichiometric amounts of sulfides/sulfur ylides[4] and often only gives moderate enantioselectivities.[5] We recently described a *catalytic* process for epoxidation involving sulfur ylides which overcomes the former limitation[6,7] (Scheme 1) and also described the use of sulfide 1 for the preparation of non-racemic epoxides.[6] The levels of enantioselectivity were poor and we now describe significant improvements in asymmetric induction using easily accessible chiral sulfides.[8]

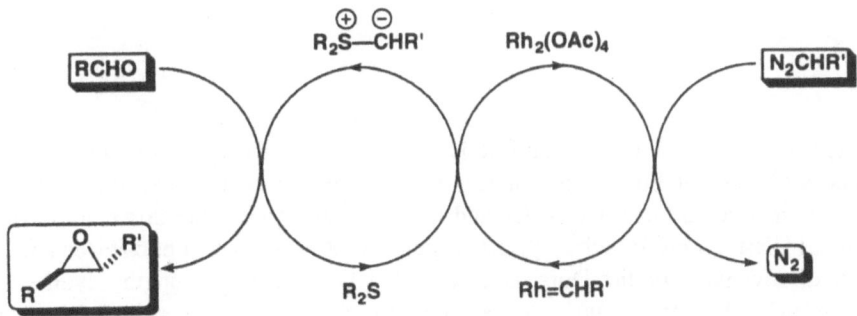

Scheme 1

13

Results

In our first attempts at improving enantioselectivity we studied a more substituted analogue of **1** as it had been previously shown by Durst that the benzyl sulfur ylide derived from sulfide **2** reacted with aldehydes to give epoxides with very high enantioselectivity.[5k,9] Sulfide **2** was prepared and tested in the catalytic cycle but no epoxide was obtained, only stilbene.[10] However, using Cu(acac)$_2$ in place of Rh$_2$(OAc)$_4$ and employing a stoichiometric amount of the Durst sulfide **2**, we were delighted to find that epoxidation was the dominant process again (Scheme 2). The significant difference in epoxide yield using Cu(acac)$_2$ and Rh$_2$(OAc)$_4$ is a reflection of the difference in rate of reaction of the metal carbenoid with either the sulfide (to give ylide) or diazocompound (to give stilbene). Evidently, the copper carbenoid is less sterically hindered than the rhodium carbenoid and can therefore react with relatively hindered sulfides.[7] However, the enantiomeric excess was still only moderate[11] and so new chiral sulfides were sought.

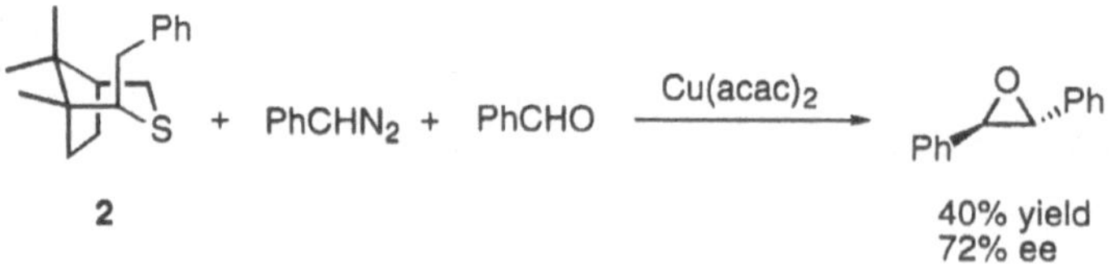

Scheme 2

A positive feature of the Durst sulfide is that only one of the two diastereomeric lone pairs reacts with the metallocarbene resulting in the formation of a single sulfur ylide. In the design of alternative sulfides it was deemed important to incorporate this feature to avoid formation of diastereomeric sulfur ylides which could react with opposite enantioselectivity.[5f] A disadvantage of the Durst sulfide is that because of its lengthy synthesis it is difficult to tune the steric and/or electronic environment of the sulfur to maximize enantioselectivity. Sulfide **3** was therefore designed as it possesses only one reactive sulfur lone pair and, being a thioacetal, the R groups are readily amenable to 'tuning'.

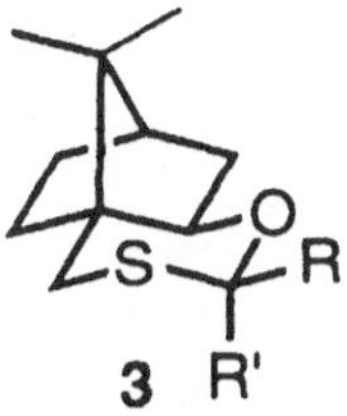

3a R' = H, R = H
3b R' = H, R = Me
3c R' = H, R = *i*-Pr
3d R' = H, R = *t*-Bu
3e R' = H, R = CH₂Ph

3f R' = H, R = CH₂OPh
3g R' = H, R = CH₂OMe
3h R' = R = Me
3i R' = R = c-C₃H₆

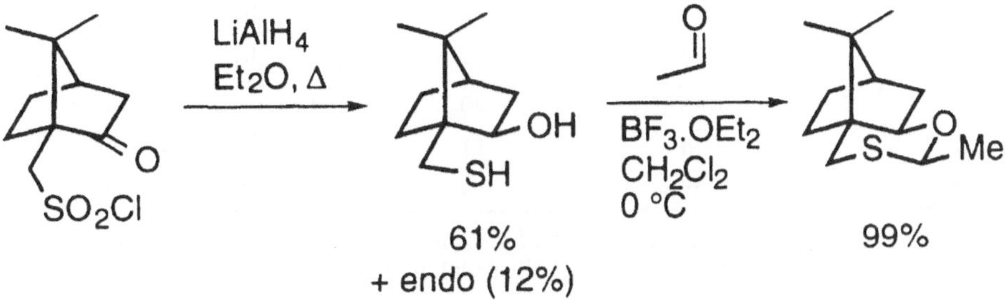

3   R'

LiAlH₄
Et₂O, Δ

61%
+ endo (12%)

BF₃.OEt₂
CH₂Cl₂
0 °C

99%

Scheme 3

Sulfides **3a-g** were prepared as shown in Scheme 3[12] (**3b** shown) and incorporated in the catalytic cycle with benzaldehyde (Table 1). It was found that high enantioselectivity could be obtained provided that the thioacetal was substituted at the 2 position (entries 2-7). Sterically hindered (entries 2, 3, 4 and 5) or electron withdrawing groups (entries 6 and 7) resulted in lower yields in the epoxidation process. The optimum sulfide in terms of yield (73%) and enantioselectivity (93%) was **3b** (entry 2).[13,14] This is the highest enantioselectivity yet reported for trans-stilbene oxide formation by any method and uses an easily accessible sulfide, employed in only catalytic quantities.

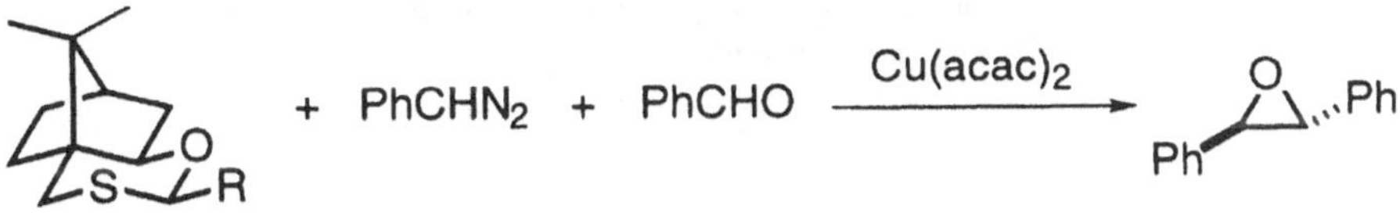

+ PhCHN₂ + PhCHO

Cu(acac)₂

Sulfide **3b** was tested with a range of aldehydes and the results are summarized in Table 2. It was found that high enantioselectivity was maintained with both aromatic and aliphatic aldehydes. Aliphatic aldehydes gave lower yields compared to aromatic aldehydes and gave a mixture of *trans* and *cis* epoxides whereas aromatic aldehydes only gave trans epoxides.[8] This contrasts with sulfides **1** and **2** in which mixtures of *trans:cis* epoxides were obtained with benzaldehyde.[5k]

Table 1: Yields, Enantioselectivities and Ratios of Stilbene Oxide Formed From Benzaldehyde Using 0.2 eq of Sulfides **3a-i**.

entry	sulfide	yield %	ee %[a]	trans: cis
1	3a	83	41 (R,R)[b]	>98:2
2	3b	73	93 (R,R)[b]	>98:2
3	3c	45	93 (R,R)[b]	>98:2
4	3d	0	-	-
5	3e	56	88 (R,R)[b]	>98:2
6	3f	43	83 (R,R)[b]	>98:2
7	3g	70	92 (R,R)[b]	>98:2
8	3h	11	70 (R,R)[b]	>98:2
9	3i	18	89 (R,R)[b]	>98:2

(a) Enantiomeric excess determined by chiral HPLC using a Chiralcel OD column. (b) Absolute configuration determined by comparison of $[\alpha]_D$ values with literature values.[5b]

Table 2: Yields, Enantioselectivities and Ratios of Epoxides Formed from Aldehydes Using 0.2 eq. of Sulfide **3**.

entry	aldehyde	yield %	ee %[a]	trans:cis
1	benzaldehyde	73	93 (R,R)[b]	>98:2
2	p-chlorobenzaldehyde	72	92(R,R)[b]	>98:2
3	p-tolualdehyde	64	92(R,R)[b]	>98:2
4	cinnamaldehyde	73	89[c]	>98:2
5	valeraldehyde	35	68[c]	92:8
6	cyclohexanecarboxaldehyde	32	90[c]	70:30

(a) Enantiomeric excess determined by chiral HPLC using a Chiralcel OD column. (b) Absolute configuration determined by comparison of $[\alpha]_D$ values with literature values.[5b] (c) Absolute configuration (R,R) assumed by analogy with entries 1-3.

Origin of diastereoselectivity

In order to account for the origin of the enantioselectivity/diastereoselectivity, we needed to know whether the sulfur ylide reactions were under kinetic or thermodynamic control. From detailed kinetic and cross-over experiments we had found that the addition of benzyl-sulfonium ylide to aldehydes was remarkably finely balanced.[15] The *trans* epoxide was derived directly from *irreversible* formation of the *anti* betaine or indirectly from reversible formation of the *syn* betaine. The *cis* epoxide was derived from partial reversible formation of the *syn* betaine (Scheme 4). The higher *trans* selectivity observed in reactions with aromatic aldehydes compared to aliphatic aldehydes was due to greater reversibility in the formation of the *syn* betaine.

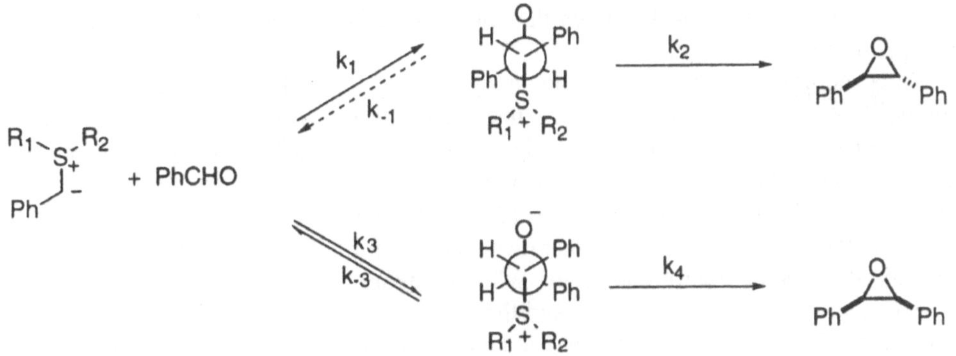

Scheme 4

Reactions of simple sulfides (Me$_2$S, tetrahydrothiophene) in the catalytic cycle with benzaldehyde give stilbene oxide as an 84:16 ratio of *trans:cis* epoxides. However, epoxidation using our camphor-derived 1,3 oxathiane **3b** only gave *trans* epoxides with a range of aromatic and unsaturated aldehydes. This higher selectivity must be due to an increase in k_{-3} relative to k_4. An increase in k_{-3} relative to k_4 would be expected for sulfides of increasing steric hindrance or in cases where the ylide had increased stability. In the case of the 1,3 oxathiane, we believe that the corresponding ylide shows increased stability relative to simple benzyl sulfonium ylides as a result of the anomeric effect (Scheme 5). The positive charge on sulfur can be delocalized over the oxygen and this will lead to increased stability and therefore greater reversibility, and therefore greater *trans* selectivity. The moderate increase in stability of the ylide is evidently not sufficient to promote reversibility in *syn* betaine formation with aliphatic aldehydes.

Scheme 5

Origin of Enantioselectivity

Having established that *trans* epoxides are derived from irreversible formation of *anti* betaines we needed information on the transition state leading to their formation in order to understand the origin of the enantioselectivity. As this was not possible, we focused on gaining information on the structure of the ylide. We believe a single sulfonium ylide is formed as we have previously shown that alkylation of the related oxathiane **3a** gave only the equatorial sulfonium salt.[5i]

Ylide conformation has been studied by X-ray,[16,17] NMR[18-22] and theoretical calculations.[23,24] All of these studies indicate that the preferred conformation of sulfur ylides is one in which the filled orbital on the ylide carbon is perpendicular to the lone pair on sulfur. The barrier to rotation around the C-S bond of the semi-stabilized ylide, dimethyl sulfonium fluorenide, has been found to be 42 ± 1.0 kJmol^{-1}.[25] This information implies that the ylide will adopt conformations **4a/b** and that these will be in rapid equilibrium at room temperature. Of these two, conformation **4b** will be favored as **4a** suffers from 1,3 diaxial interactions between the phenyl ring and the axial groups. The aldehyde can then approach either face of the ylide but the *Re* face is more accessible as the *Si* face is hindered by the equatorial methyl group (Scheme 6).

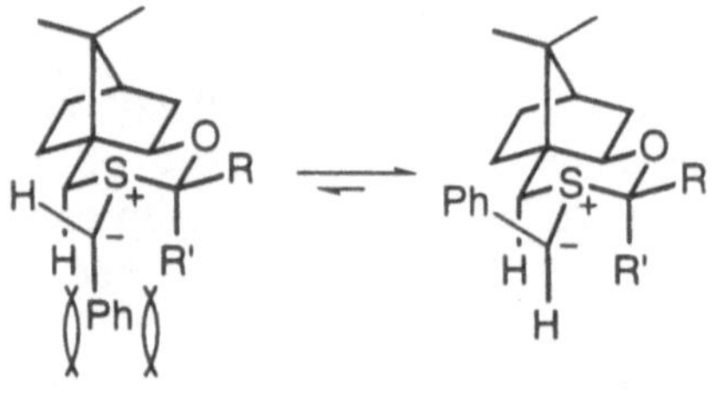

4a   4b

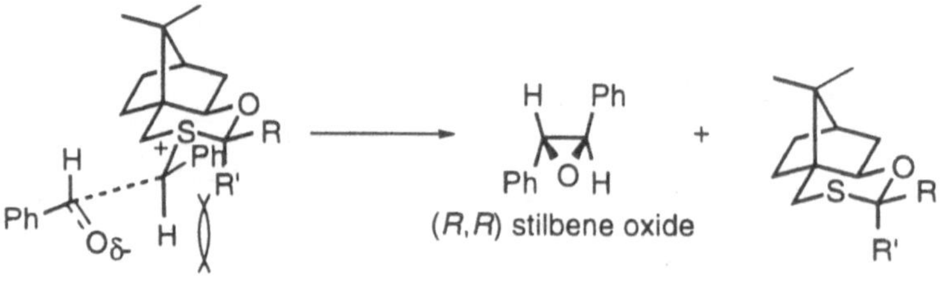

(R,R) stilbene oxide

Scheme 6

The aldehyde can react in an end-on or [2+2] mode but there is no evidence, experimental or theoretical, to indicate which is preferred. End-on and [2+2] transition states in the reaction of ylide **5** with benzaldehyde, leading to *trans* epoxides are shown. From analysis of molecular models of these transition states it is clear that they can all be accommodated. Thus, our own results do not provide any further evidence as to which mode (end-on *versus* [2+2]) is favored.

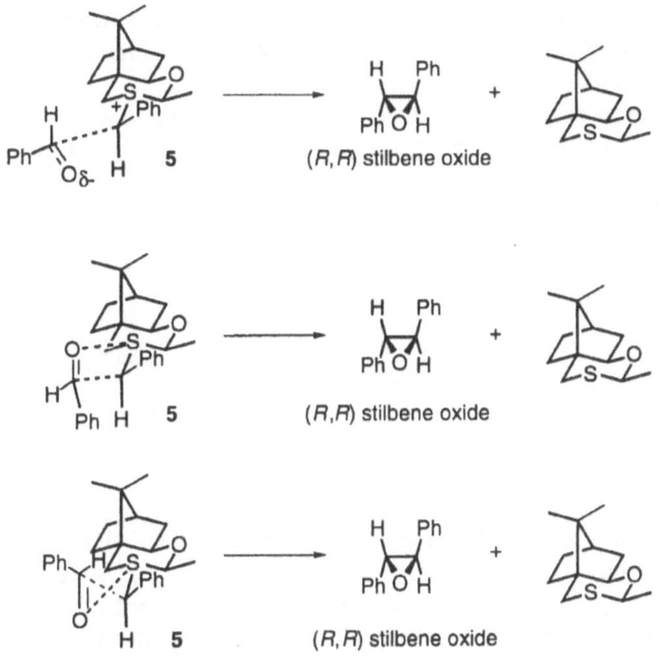

Scheme 7

The transition states shown in Scheme 7 account for the high enantioselectivities observed. We were curious as to the origin of the minor enantiomer and considered the possibility that it arose from *Si* face attack (Scheme 8). If this were the case, enantio-selectivity should be highly dependent upon the size of the equatorial substituent. However, increasing the size of this substituent from Me to Pr^i did not result in a concomitant increase in selectivity (Table 1, entries 2,3).

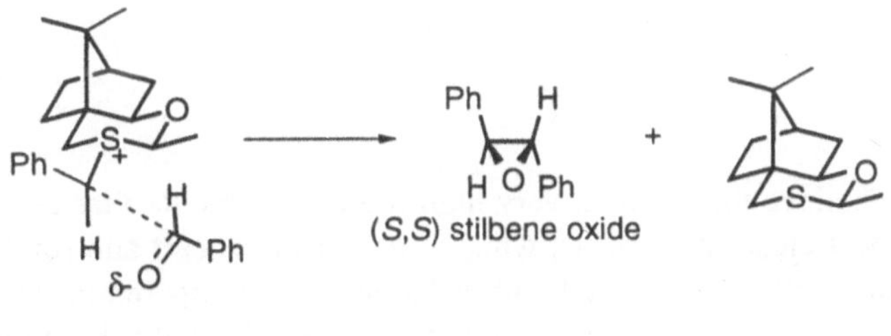

Scheme 8

Indeed, the enantioselectivity was essentially the same for a range of substituents (Table 1, entries 2,3,5,7) suggesting that the facial selectivity was essentially complete. Even, in the absence of a substituent (**3a**, Table 1, entry 1) good *Re* face selectivity was still observed.

19

This latter result in particular suggests that the oxygen of the oxathiane is affecting the facial selectivity of the ylide and we believe it is exerting this effect through a combination of the anomeric and Cieplak[26] effects.

A resonance form of the ylide is shown in Scheme 9. If there is a contribution from this resonance form to the ground state structure of the ylide, then the C_2–S bond should be more electron rich than the C_6–S bond.

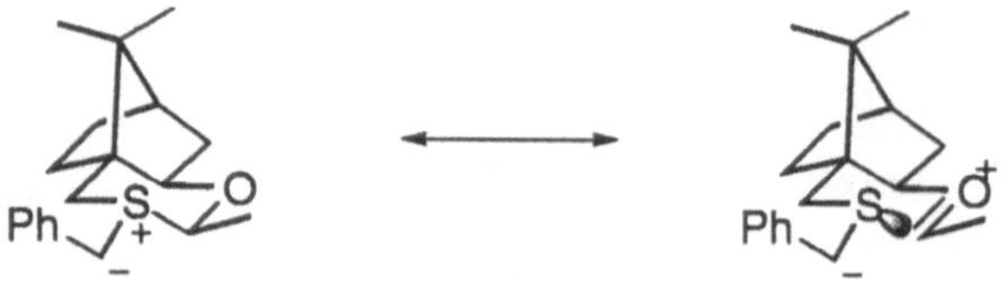

Scheme 9

According to the Cieplak effect reactions occur on the face antiperiplanar to the more electron rich σ bond.[26] The only other example in which an anomeric or anomeric-type effect controlled the outcome of a reaction in a similar manner to that described above is provided by Le Noble.[27] He studied the addition of nucleophiles to 5-azaadamantone and found a slight preference for addition of MeLi *anti* to nitrogen. NaBH$_4$ reduction in MeOH gave addition *syn* to nitrogen possibly because of hydrogen bonding with the nitrogen-lone pair (Scheme 10).

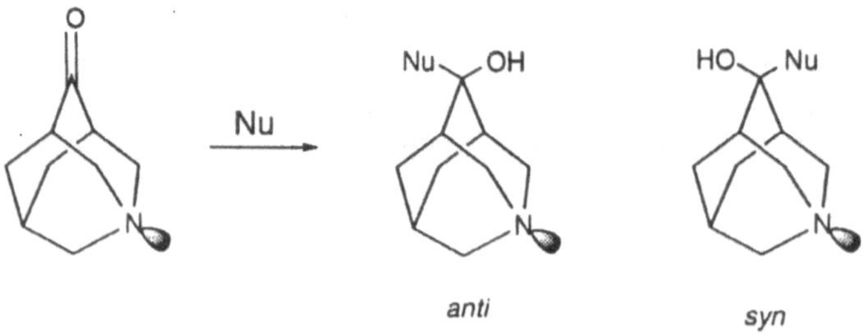

anti   *syn*

Scheme 10

We therefore believe that there is very high preference for *Re* face attack as a result of combined steric and electronic effects, which both act in concert and reinforce each other. The lower enantioselectivity observed with valeraldehyde compared to the other aldehydes could be due to its smaller size, which may allow some *Si* face attack on the ylide. In order to test this, the size of the equatorial substituent on the sulfide was increased. Using sulfide **3c** higher enantioselectivity was indeed obtained (Scheme 11).

This confirmed that facial selectivity in the reaction of sulfide **3b** with aldehydes in the catalytic cycle is dependent on the size of the aldehyde: α-branched, aromatic and unsaturated aldehydes react with essentially complete facial selectivity whilst unbranched aldehydes react with moderate facial selectivity.

20

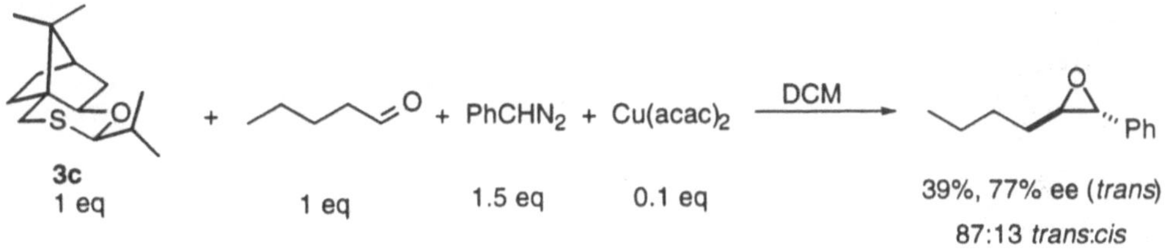

Scheme 11

As facial selectivity appears to be complete, the source of the minor enantiomer could be due to reaction of the minor conformation of the ylide **4a** (Scheme 6). In order to reduce the amount of this minor conformation we prepared thioacetals bearing axial substituents **3h** (R = Me) and **3i** (R = R^1 = *spiro*-cyclobutyl) to increase the 1,3-diaxial interactions. However, instead of increased selectivity, a decrease in enantioselectivity was observed (Table 1). A study of the conformation of these thioacetals revealed the origin of this reduction in selectivity. NOE experiments carried out on **3b** and **3h** revealed that **3h** existed in both chair and boat forms whilst **3b** adopted the chair form only. Reaction of the corresponding ylide from the boat conformer of **3h** should give the opposite enantiomer to that from the chair form (Scheme 12).

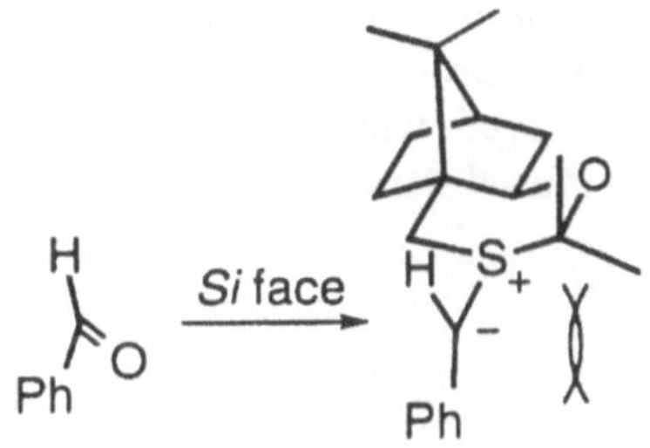

Scheme 12

An alternative way to increase the 1,3 diaxial interactions and favour ylide conformer **4b** over **4a** would be to increase the bulk of the aromatic ring. Thus mesityldiazomethane was prepared and used in the catalytic cycle with thioacetals **4b** and benzaldehyde (Scheme 13) and essentially complete selectivity was obtained.

It therefore seems likely that the minor enantiomer formed in the reactions involving phenyldiazomethane, arises from reaction of the minor conformer **4a**. Attempts to reduce the proportion of the minor conformer **4a** by lowering the temperature were unsuccessful. Reaction of *trans*-cinnamaldehyde[28] with thioacetal **3b**, phenyldiazomethane and Rh$_2$(OAc)$_4$[29] at 0 °C furnished the epoxide with similar enantioselectivity. Evidently, the small difference in temperature was not sufficient to make a significant difference in enantioselectivity. At lower temperatures the diazocompound did not decompose.

21

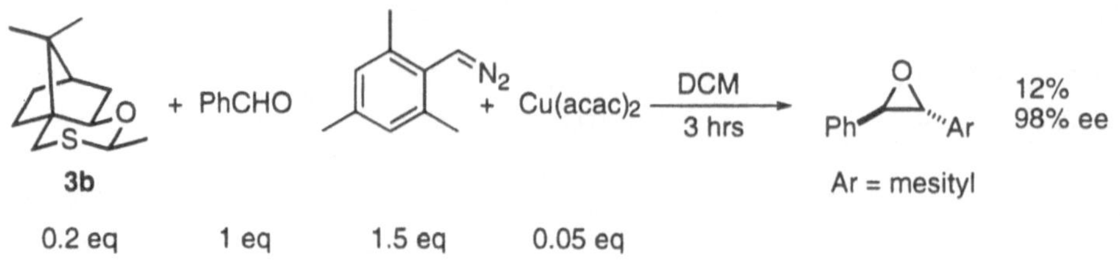

Scheme 13

Conclusion.

Oxathiane **3b** has been found to be a highly efficient sulfide for carbonyl epoxidation. It is easily prepared in two steps from camphor sulfonyl chloride (which is available in both enantiomeric forms) and can be used in substoichiometric amounts in our catalytic cycle. In reactions with aromatic and unsaturated aldehydes only *trans* epoxides are obtained, and this high selectivity is a result of irreversible formation of the *anti* betaine and greater reversibility in the formation of the *syn* betaine. This is believed to be a result of the enhanced stabilisation of the ylide due to an anomeric effect from the ring oxygen.

A single sulfonium ylide is formed which can adopt conformations **4b** (favoured) and **4a**. Facial selectivity of the ylide is essentially complete with aromatic, unsaturated and α-branched aldehydes and is a result of steric effects (the equatorial methyl group) and electronic effects (a combination of the anomeric and Cieplak effects) which reinforce each other. The minor enantiomer comes from reaction of the less favored conformer of the ylide **4a**. The lower selectivity observed with valeraldehyde is a result of reduced facial selectivity of the ylide due to reduced steric effects.

References

(1)   Besse, P.; Veschambre, H. *Tetrahedron* **1994**, *50*, 8885-8927.
(2)   (a) Jacobsen, E. N.; Zhang, W.; Muci, A. R.; Ecker, J. R.; Deng, L. *J. Am. Chem. Soc.* **1991**, *113*, 7063-7064. (b) Brandes, B. D.; Jacobsen, E. N. *J. Org. Chem.* **1994**, *59*, 4378-4380. (c) Chang, S.; Heid, R. M.; Jacobsen, E. N. *Tetrahedron. Lett.* **1994**, *35*, 669-672. (d) Chang, S. B.; Galvin, J. M.; Jacobsen, E. N. *J. Am. Chem. Soc.* **1994**, *116*, 6937-6938. (e) Palucki, M.; Pospisil, P. J.; Zhang, W.; Jacobsen, E. N. *J. Am. Chem. Soc.* **1994**, *116*, 9333-9334. (f) Irie, R.; Noda, K.; Ito, Y.; Matsumoto, N.; Katsuki, T. *Tetrahedron Asymm.* **1991**, *2*, 481-494. (g) Hamada, T.; Irie, R.; Katsuki, T. *Synlett* **1994**, 479-481. (h) Katsuki, T. *Coord. Chem. Rev.* **1995**, *140*, 189-214.
(3)   (a) Corey, E. J.; Chaykovsky, M. *J. Am. Chem. Soc.* **1965**, *87*, 1353-1364. (b) Trost, B. M.; Melvin, L. S. *Sulfur Ylides*; Academic Press: New York, 1975. (c) Robertson, G. In *Comprehensive Organic Synthesis*; B. M. Trost and I. Fleming, Eds.; Pergamon Press: Oxford, 1991; Vol. 3; pp 563-611.

(4)   In one exceptional case Furukawa has carried out epoxidation using 0.5 equivalents of sulfide in the presence of an alkyl halide, base and aldehyde to give epoxide. However, the yields of the epoxides obtained were very low (<23% based on aldehyde) and he acknowledges that this is a poor method for the preparation of epoxides.[5b] We have also tried this method and concur with Furukawa. However, Dai and Huang have recently described essentially the same process for epoxidation and obtained surprisingly high yields of epoxides (>90%).[5j] This method is only applicable to non-enolisable aldehydes and indeed has only been carried out using PhCHO, p-Me-C_6H_4CHO and p-Cl-C_6H_4CHO. Our catalytic method can be used for both aliphatic and aromatic aldehydes and can also be applied to base sensitive aldehydes due to the neutral reaction conditions employed[5h].

(5)   (a) Trost, B. M.; Hammen, R. F. *J. Am. Chem. Soc.* **1973**, *95*, 962-964.   (b) Furukawa, N.; Sugihara, Y.; Fujihara, H. *J. Org. Chem.* **1989**, *54*, 4222-4224.   (c) Breau, L.; Ogilvie, W. W.; Durst, T. *Tetrahedron. Lett.* **1990**, *31*, 35-38.   (d) Solladie-Cavallo, A.; Adib, A. *Tetrahedron* **1992**, *48*, 2453-2464.   (e) Solladie-Cavallo, A.; Adib, A.; Schmitt, M.; Fischer, J.; Decian, A. *Tetrahedron Asymm.* **1992**, *3*, 1597-1602.   (f) Aggarwal, V. K.; Kalomiri, M.; Thomas, A. P. *Tetrahedron Asymm.* **1994**, *5*, 723-730.   (g) Aggarwal, V. K.; Abdel-Rahman, H.; Jones, R. V. H.; Lee, H. Y.; Reid, B. D. *J. Am. Chem. Soc.* **1994**, *116*, 5973-5974.   (h) Aggarwal, V. K.; Abdel-Rahman, H.; Jones, R. V. H.; Standen, M. C. H. *Tetrahedron. Lett.* **1995**, *36*, 1731-1732.   (i) Aggarwal, V. K.; Thompson, A.; Jones, R. V. H.; Standen, M. *Tetrahedron Asymm.* **1995**, *6*, 2557-2564.   (j) Li, A. H.; Dai, L. X.; Hou, X. L.; Huang, Y. Z.; Li, F. W. *J. Org. Chem.* **1996**, *61*, 489-493.
In several cases >90% ee has been obtained: (k) Breau, L.; Durst, T. *Tetrahedron Asymm.* **1991**, *2*, 367-370.   (l) Solladie-Cavallo, A.; Diep-Vohuule, A. *J. Org. Chem.* **1995**, *60*, 3494-3498.   (m) Solladie-Cavallo, A.; Diep-Vohuule, A.; Sunjic, V.; Vinkovic, V. *Tetrahedron Asymm.* **1996**, *7*, 1783-1788.

(6)   Aggarwal, V. K.; Abdel-Rahman, H.; Jones, R. V. H.; Lee, H. Y.; Reid, B. D. *J. Am. Chem. Soc.* **1994**, *116*, 5973-5974.

(7)   Aggarwal, V. K.; Abdel-Rahman, H.; Li, F.; Jones, R. V. H.; Standen, M. *Chem. Eur. Jn.* **1996**, *2*, 212-218.

(8)   Aggarwal, V. K.; Ford, J. G.; Thompson, A.; Jones, R. V. H.; Standen, M. *J. Am. Chem. Soc.* **1996**, *118*, 7004-7005.

(9)   (a) Durst, T.; Breau, L.; Ben, R. N. *Phosphorus Sulfur* **1993**, *74*, 215-232.   (b) Ben, R. N.; Breau, L.; Bensimon, C.; Durst, T. *Tetrahedron* **1994**, *50*, 6061-6076.

(10)   Diazocompounds readily dimerise in the presence of metal catalysts. Shankar, B. K. R.; Shechter, H. *Tetrahedron. Lett.* **1982**, *23*, 2277-2280.

(11)   Durst reported that the benzylide of **2** reacted with benzaldehyde to give stilbene oxide with >96% ee. The enantioselectivity we obtained in our catalytic cycle was significantly lower (72% ee) and so we repeated Durst's original work but only obtained 71% ee. We have measured our ee's by chiral HPLC using a diode array detector to take a UV trace of the enantiomers as they elute. The UV traces of the two peaks were superimposable. Durt's ee's were determined by NMR using Eu shift reagents.

(12)   (a) Eliel, E. L.; Frazee, W. J. *J. Org. Chem.* **1979**, *44*, 3598-3599.   (b) de Lucchi, O.; Lucchini, V.; Marchioro, C.; Valle, G.; Modena, G. *J. Org. Chem.* **1986**, *51*, 1457.

23

(13)   (a) Using stoichiometric amounts of sulfide **3b**, high yield (90%) and high enantioselectivity (93% ee) was obtained with benzaldehyde but with 0.2 equivalents of sulfide a slightly lower yield (73%) but the same enantioselectivity was obtained. To obtain reasonable yields when using sub-stoichiometric amounts of sulfide it was found necessary to conduct reactions at the same effective concentration of sulfide. This presumably resulted in similar rates of ylide formation, and reaction with the aldehyde and therefore allowed the sulfide to be returned and recycled at the same rate as the stoichiometric process.

(14)   Whilst commercial Cu(acac)$_2$ worked well with sulfide **2**, no epoxide was obtained with sulfide **3b**. However, using Cu(acac)$_2$ prepared by adding saturated sodium carbonate to a solution of copper oxide and acetylacetone (see Bryant B. E.; Fernelius W. C. *Inorganic Synthesis* **1957**, *5*, 115) the yields reported in tables 1 and 2 were obtained. The following general procedure was used in all of the reactions: To a stirred solution of sulfide **3b** (0.2 mmol), Cu(acac)$_2$ (0.05 mmol) and the aldehyde (1 mmol) in dichloromethane (0.5 ml), under nitrogen was added a solution of phenyldiazomethane (1.5 mmol in 0.5 ml of dichloromethane) at room temperature over a period of 3 hours using a syringe pump. After stirring for a further 1 hour the solvent was removed *in vacuo* and the residue was chromatographed on silica gel.

(15)   Aggarwal, V. K.; Calamai, S.; Ford, J. G. *J. Chem. Soc., Perkin Trans. 1* **1997**, 593-599.

(16)   Christensen, A. T.; Witmore, W. G. *Acta. Cryst. B* **1969**, *25*, 73-78.

(17)   Christensen, A. T.; Thom, E. *Acta. Cryst. B* **1971**, *27*, 581-586.

(18)   Ratts, K. W. *Tetrahedron Lett.* **1966**, 4707-4712.

(19)   Cook, A. F.; Moffatt, J. G. *J. Am. Chem. Soc.* **1968**, *90*, 740-747.

(20)   Galloy, J.; Watson, W. H.; Craig, D.; Guidry, C.; Morgan, M.; McKellar, R.; Ternay, A. L.; Martin, G. *J. Heterocycl. Chem.* **1983**, *20*, 399-405.

(21)   Matsuyama, H.; Minato, H.; Kobayashi, M. *Bull. Chem. Soc. Jap.* **1977**, *50*, 3393-3396.

(22)   Johnson, A. W.; Amel, R. T. *J. Org. Chem.* **1969**, *34*, 1240-1247.

(23)   Bernardi, F.; Schlegel, H. B.; Whangbo, M.-Y.; Wolfe, S. *J. Am. Chem. Soc.* **1977**, *99*, 5633-5636.

(24)   Eades, R. A.; Gassman, P. G.; Dixon, D. A. *J. Am. Chem. Soc.* **1981**, *103*, 1066-1068.

(25)   Aggarwal, V. K.; Schade, S.; Taylor, B. *J. Chem. Soc., Perkin Trans. 1* **1997**, in press.

(26)   Cieplak, A. S. *J. Am. Chem. Soc.* **1981**, *103*, 4540-4552.

(27)   Hahn, J. M.; le Noble, W. J. *J. Am. Chem. Soc.* **1992**, *114*, 1916-1917.

(28)   This aldehyde was chosen as it gave a slightly lower enantioselectivity compared to other aldehydes, thereby allowing us to monitor changes more easily

(29)   Use of Cu(acac)$_2$ was unsuccessful. No epoxide, only stilbenes were formed.

Transition Metal Complexes as Models for Metallo Enzymes: Mechanistic Studies and Preparative Applications

Albrecht Berkessel

Institut für Organische Chemie der Universität zu Köln, Greinstr. 4, D-50939 Köln, Germany, e-mail: berkessel@uni-koeln.de

1 Introduction

Catalysis is one of the most important research areas in chemistry for various reasons, e.g.: Catalytic processes do generally not require (produce) stoichiometric amounts of auxiliaries (by-products). Novel catalysts are hoped to enable us to perform in one step transformations that to date require a number of steps, like e.g. the direct and selective oxygenation of hydrocarbons with oxygen. Within the last decade, quite impressive advances have been made in the field of asymmetric catalysis, i.e. the catalytic production of only one enantiomerically pure product in cases where two enantiomers are possible.

In principle, the "world of catalysts" may be devided into those that are purely man-designed and -made, and those that occur naturally, i.e. the enzymes. The principles leading to rate enhancement in general, and to the differentiation of reaction rates (i.e. in asymmetric catalysis) may be similar (e.g. acid-base catalysis) or different (e.g. effects of ligating heteroatoms or metal ions not naturally occuring).

Our research interest focuses on bioorganic aspects of catalysis in a threefold sense: (1) Based on biophysical and structural data, low-molecular weight models of active sites are synthesized, aiming at the elucidation of novel mechanisms of enzymatic catalysis. This approach also aims at the design and the synthesis of novel mechanism-based inhibitors. (2) Based on how enzymes effect catalysis, we try to develop low-molecular weight models for preparative applications. In this approach, our work is based on a natural system, but we do not restrict ourselves to the means nature uses.(3) Finally, we like to test newly discovered enzymes for their usefulness in synthetic organic chemistry. In the lecture, examples for these three lines of research are given. As it turned out, enzyme modelling always yielded quite interesting results, but not necessarily the desired model systems!

2 Enzyme Models as Catalysts

2.1 Models for [Ni,Fe]-Hydrogenases Catalyze the Reduction of Imines with Silanes

Hydrogenases[1] catalyze the reaction shown in equation 2.1.1. This simple process is of vital importance for all hydrogen-metabolizing and -producing organisms. Thus, hydrogenases have been developed to highly efficient catalysts, operating at physiogiocal H_2 partial pressures as low as $10^{-4} - 10^{-5}$ atm. In *in vitro*-experiments, hydrogenase activity is usually detected either by the reduction of viologens (equation 2.1.2) or by the so-called isotope exchange reaction shown in equation 2.1.3.

$$H_2 \xrightleftharpoons[\text{hydrogenase}]{} 2\,H^+ + 2\,e^- \quad (2.1.1)$$

$$2\,\text{viologen}^{2+} + H_2 \xrightleftharpoons[\text{hydrogenase}]{} 2\,\text{viologen}^{+\bullet} + 2\,H^+ \quad (2.1.2)$$

$$H^+_{(\text{solv})} + D_2\,{}_{(\text{gas})} \xrightleftharpoons[\text{hydrogenase}]{} D^+_{(\text{solv})} + HD_{(\text{gas})} \quad (2.1.3)$$

We hope that hydrogenases may provide guidelines for the design of new catalysts for hydrogenation, or substrate reduction in a more general sense. As revealed by the X-ray crystal structure[2,3] of the [Ni,Fe]-hydrogenase from *Desulfovibrio gigas*, the active site harbors two transition metal ions (Ni, Fe), the nickel ion being coordinated by four sulfur atoms derived from cysteines.

We have been trying to construct nickel complexes with hydrogenase activity for some time[4-6]. Usually, tripodal tridentate[4,6] and pentadentate[5] N,O,S-ligands were used: It was hoped that an H_2-molecule could be bound and heterolyzed at the vacant coordination site, e.g. as shown in Scheme 2.1.1 for a pentadentate ligand. As it turned out, we (and others) have not yet been able to achieve the desired heterolysis of the H_2-molecule with nickel model systems[7].

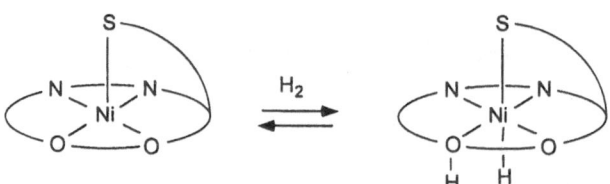

Scheme 2.1.1 A possible mode of hydrogen binding to a hydrogenase model.

In 1992, nickel complexes of the thiosemicarbazones of salicylic aldehydes were for the first time shown to activate the Si-H bond in hydrosilanes[9]: Compounds of the general formula **1** catalyze the alcoholysis of silanes[9]. We found that these easily accessible materials also catalyze the reduction of imines with silanes in a highly chemoselective fashion[10]: Unless the substrates are extremely sterically hindered, smooth reduction to the corresponding secondary amines occurs (Figure 2.1.1). This process is highly chemoselective: No competing reduction of CC multiple bonds, nitro or carbonyl groups (amide, ester, ketone or even aldehyde) occurs. The presence of e.g. bromine substituents (problematic with low-valent transition metal catalysts) is tolerated.

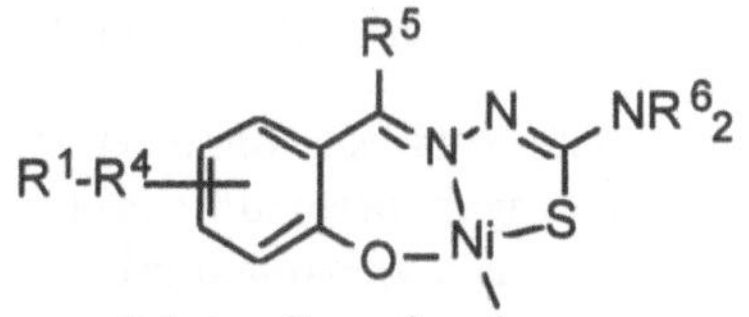

1, L: co-ligand   L

What is the mechanism of this reaction? It is not a "normal" hydrosilylation, N-silylated amines are not formed as primary products. Instead, one equivalent of water is required, and a silanol is the second product (equation 2.1.4). A number of X-ray crystal structures[11] proves binding of the substrate imine to the catalyst, as shown for example in Figure 2.1.2. Additional coordination of the silane and nucleophilic attack by water may give rise to a hydride complex, the precursor of an amide complex and the product amine, respectively. Work concerning asymmetric variants of this imine reduction is in progress.

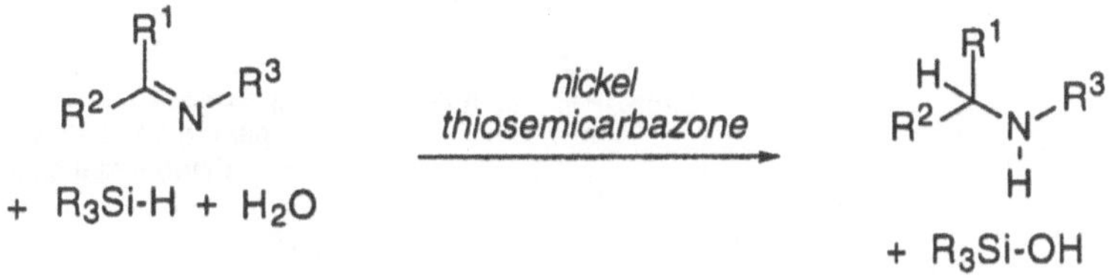

$$\text{(2.1.4)}$$

Figure 2.1.1 A selection of imine substrates that are quantitatively reduced to secondary amines.

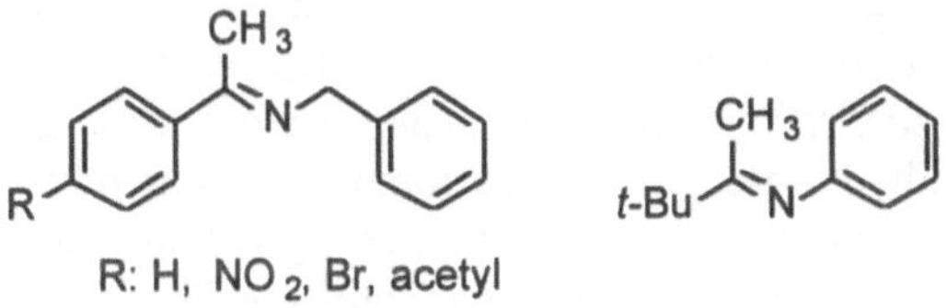

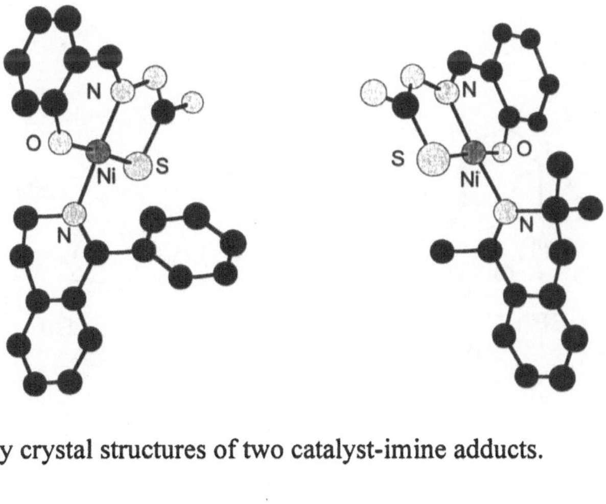

R: H, NO$_2$, Br, acetyl

Figure 2.1.2 X-ray crystal structures of two catalyst-imine adducts.

2.2 Peroxidase Models Catalyze the Asymmetric Epoxidation of Olefins with H_2O_2

As a part of our studies on biomimetic catalysts for selective oxyfunctionalizations[12-15], we addressed the question whether hydrogen peroxide could be employed as the terminal oxidant. Its advantages are obvious: It is a cheap and mild reagent, with only water being formed as waste product. In many peroxidases, i.e. hydrogen peroxide utilizing enzymes, the catalytically active iron center is coordinated by the four pyrrole nitrogen atoms of its heme ligand plus an axial imidazole donor. It appeared desirable to combine the features of a peroxidase-like coordination sphere and a (chiral) manganese(III) salen complex (**A**, Figure 2.2.1). In such an arrangement, a fifth, axial donor, preferably an imidazole group, should be covalently attached to a salen-type complex (**B**, Figure 2.2.1).

Figure 2.2.1 Pentacoordinated manganese(III) complexes as peroxidase models.

A
salen-type manganese chelates

B
peroxidase model based on pentadentate dihydrosalen manganese chelates

Formula **2** demonstrates how this design was put into practice. In principle, one C=N-double bond of a manganese(III) salen complex is modified by attaching an imidazolylmethyl side-chain to the carbon atom, and an alkyl substituent to the nitrogen atom. This dihydrosalen chelate harbors two centers of chirality. We synthesized a number of such pentadentate ligands and the corresponding manganese(III) complexes[13-15]. The X-ray crystal structure of one example is shown in Figure 2.2.2.

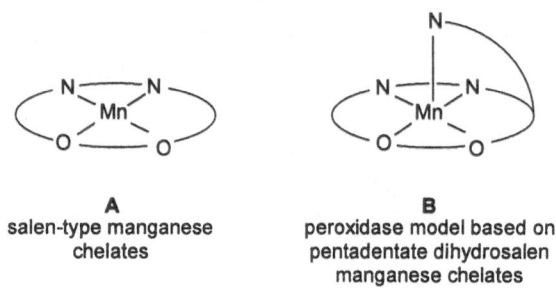

2

3

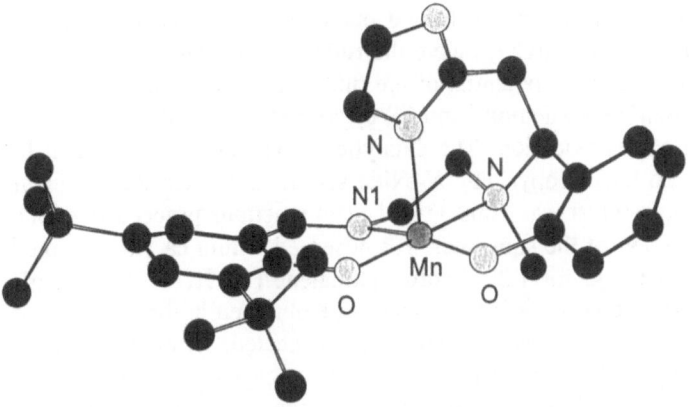

Figure 2.2.2 X-ray crystal structure of a pentacoordinated manganese(III) complex.

As expected, the pentadentate peroxidase models such as **2** catalyze the epoxidation of olefins with hydrogen peroxide - even at H_2O_2-concentrations as low as 1 %[13-15]. Significant asymmetric induction occurs: For example, when dihydronaphthaline is used as the substrate, the corresponding epoxide is obtained in 72 % yield and 64 % ee. It should finally be mentioned that the four-coordinate dihydrosalen complex **3** is not catalytically active.

2.3 Asymmetric Epoxidation and Cyclopropanation with a Chiral Ru-Porphyrin

In 1991, *Halterman* et al. described a synthesis for the chiral, D_4-symmetric porphyrin **4**[16]. Furthermore, the manganese(III) complex of the porphyrin **4** was shown to be an effective catalyst for the asymmetric epoxidation of olefins and sulfoxidation of thioethers using sodium hypochlorite as the terminal oxidant. For example, in the case of 1,2-dihydronaphthalene, an ee of 56 % at almost quantitative chemical yield was achieved.

Attempts by us (and by others, most likely) to convert the porphyrin **4** into its ruthenium complex **5** by the usual treatment with $Ru_3(CO)_{12}$ in standard solvents like benzene, toluene, decalin or diglyme failed completely, or gave at best poor yields (< 15 %) of the desired metalloporphyrin. Finally, phenol turned out to be the solvent of choice: Refluxing the porhyrin **4** with excess $Ru_3(CO)_{12}$ gave the ruthenium porphyrin **5** within 30 min in yields up to 90 %[17]. The catalytic potential of the ruthenium porphyrin **5** was probed with respect to the (i) asymmetric epoxidation[17] and (ii) cyclopropanation of olefins[18].

(i) Asymmetric epoxidation: The catalytic epoxidations were run in benzene at room temperature[17]. 2,6-Dichloropyridine *N*-oxide served as the terminal oxidant[19]. The ratio of substrate:oxidant:catalyst was 1000:1100:1. The reactions proceeded smoothly, and almost complete conversion of the substrates was observed within ca. 48 h. Quite high yields and ees were achieved, e.g. for 1,2-dihydronaphthalene (88 %, 77 % ee) and styrene (79 %, 70 % ee). In fact, the latter ee is the highest one obtained in the asymmetric epoxidation of styrene using a chiral porphyrin catalyst. As expected, terminal olefins like 1-octene or *trans*-1,2-disubstituted substrates like *trans*-stilbene gave only poor results.

(ii) Asymmetric cyclopropanation: In the presence of only 0.1 mol-% of the chiral ruthenium porphyrin **5**, styrene and ethyl diazoacetate reacted smoothly, affording the cyclopropanes **6**/*ent*-**6** and **7**/*ent*-**7** (equation 2.3.1). Quantitative yields were generally obtained,

$$
\begin{array}{ccc}
\text{styrene} & \xrightarrow[\text{2 h, r.t.}]{\substack{N_2CHCO_2Et \\ \text{0.1 mol-% 5 or } ent\text{-5}}} & \text{6, 7}
\end{array}
\qquad (2.3.1)
$$

6: $R^1 = CO_2Et$, $R^2 = H$; **7**: $R^1 = H$, $R^2 = CO_2Et$

at a *trans*- to *cis*-ratio of 96:4. Using ethyl diazoacetate as the carbene source, this diastereoselectivity is the highest one achieved so far[18]. The enantiomeric excess of the major (*trans*) product was 91 %. Furthermore, diethyl fumarate or maleate were at best formed in trace amounts.

3 Real Enzymes: How Metal-Free Haloperoxidases Activate H_2O_2

Haloperoxidases catalyze the formation of hypohalites from hydrogen peroxide and chloride, bromide or iodide (equation 3.1). The electrophiles thus formed are able to halogenate suitable organic substrates and can thus play an important role in the biosynthesis of halogenated natural products. Haloperoxidases can also catalyze the transfer of oxygen from hydrogen peroxide to organic substrates such as olefins or thioethers. Therefore, this class of enzymes has been intensively studied with respect to preparative transformations, for example, the asymmetric epoxidation of olefins and sulfoxidation of thioethers. So far, mainly the heme-containing haloperoxidases have proven suitable for preparative applications, in particular the chloroperoxidase from the fungus *Caldariomyces fumago*. However, a serious drawback of this enzyme is its relatively limited stability.

$$\text{H}_2\text{O}_2 + \text{Hal}^- \xrightleftharpoons[\text{Hal: I, Br, Cl}]{\textit{haloperoxidase}} \text{OHal}^- + \text{H}_2\text{O} \qquad (3.1)$$

Metal-free haloperoxidases from bacterial sources were first described by *van Pée* et al.: The chloroperoxidases from *Pseudomonas pyrrocinia* ("CPO-P")[20] and from *Streptomyces aureofaciens* (CPO-T)[21]. We were interested both in solving the problem of how a metal-free enzyme can activate hydrogen peroxide, and in probing the preparative potential of this novel class of enzymes. Interestingly, an X-ray crystal structure[22] of the closely related bromoperoxidase A2 from *Streptomyces aureofaciens* (ATCC 10762) revealed the presence of a "catalytic triad" (Asp-His-Ser) in the enzyme's active site! As it turned out, this "micro-machinery" typical for serin-esterases proved to be of central importance for the enzymatic mechanism.

The potential of the enzymes of preparative transformations turned out to be limited[23]: Olefins are not converted to epoxides, but in the presence of bromide, they yield bromohydrins in quantitative yield. Similarly, thioethers are converted to sulfoxides. However, in the case of prochiral substrates, the resulting products were found to be racemic. These results pointed to a diffusible, achiral oxidizing species. Since both CPO-P and CPO-T are only active in acetate or propionate buffer, it appeared reasonable to assume that these enzymes do in fact catalyze the equilibration shown in Scheme 3.1. Obviously, the "normal" esterase reaction (Scheme 3.1, top reaction) is very similar to the formation of peracids from H_2O_2 and carboxylic acids (Scheme 3.1, bottom reaction). We were indeed able to prove the specificity of CPO-T for the equilibration shown in Scheme 3.1 by high-resolution CI-MS. Phenylmethanesulfonyl fluoride ("PMSF"), a well-known inhibitor for serine esterases, inhibits the oxidizing activity of CPO-P and CPO-T. Furthermore, these enzymes showed esterase activity using *p*-nitrophenyl acetate as substrate. The latter activity is also inhibited by PMSF. We also tested a variety of serine esterases for activity in the H_2O_2-reaction. As it turned out, acetylcholin esterase from *Torpedo californica* showed an activity comparable to that of CPO-T, albeit, at much lower stability towards H_2O_2. Our mechanistic interpretation for the action CPO-P and CPO-T is summarized in Scheme 3.2[23]. Since the substrate oxidation does not take place in the enzyme's active site, the potential of CPO-P and CPO-T for catalytic stereoselctive transformations appears limited. On the other hand, these enzymes may prove useful for the catalytic activation of hydrogen peroxide for bleaching purposes[24].

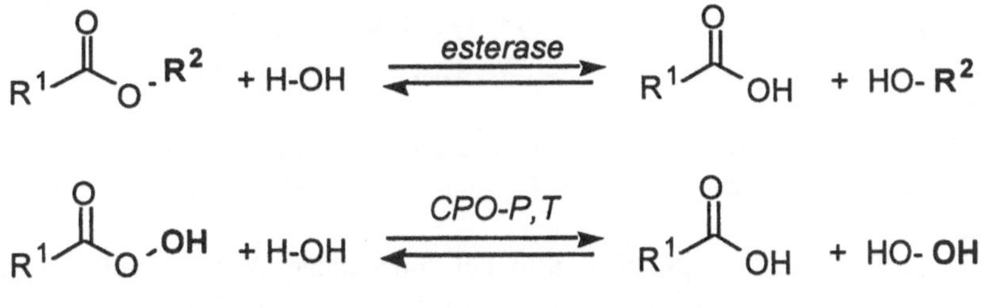

Scheme 3.1 Comparison of a "normal" esterase reaction with that of CPO-T, CPO-P.

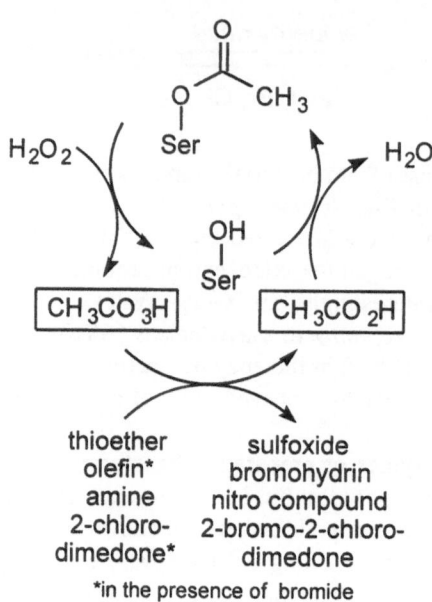

Scheme 3.2 Catalytic cycle of CPO-P,T.

Acknowledgements

The work presented here was supported by the *Bundesministerium für Forschung und Technologie* (grant no. 9342 A/H2-04-17-92), by the *Deutsche Forschungsgemeinschaft* (SFB 247, Universität Heidelberg, and grants nos. Be 998/6-1,2), by the *Fonds der Chemischen Industrie*, and by the *BASF AG*.

4 References and Notes

1. S. P. J. Albracht, *Biochim. Biophys. Acta* **1994**, *1188*, 167-204.
2. A. Volbeda, M.-H. Charon, C. Piras, E. C. Hatchikian, M. Frey, J. C. Fontecilla-Camps, *Nature* **1995**, 373, 580–587.
3. A. L. de Lacey, E. C. Hatchikian, A. Volbeda, M. Frey, J. C. Fontecilla-Camps, V. M. Fernandez, *J. Am. Chem. Soc.* **1997**, *119*, 7181-7189.
4. A. Berkessel, J. W. Bats, M. Hüber, W. Haase, T. Neumann, L. Seidel, *Chem. Ber.* **1995**, *128*, 125-129.
5. A. Berkessel, M. Bolte, T. Neumann, L. Seidel, *Chem. Ber.* **1996**, *129*, 1183-1189.
6. A. Berkessel, J. W. Bats, M. Bolte, T. Neumann, L. Seidel, *Chem. Ber./Recueil* **1997**, *130*, 891-897.
7. In the publication cited in ref. 8, a nickel thiosemicarbazone complex is claimed to catalyze the isotope exchange reaction shown in equation 2.1.3.

8.   R. H. Crabtree, M. Zimmer, G. Schulte, X.-L. Luo, *Angew. Chem.* **1991**, *103*, 205-207; *Angew. Chem. Int. Ed. Engl.* **1991**, *30*, 193-194.

9.   D. E. Braber, Z. Lu, T. Richardson, R. H. Crabtree, *Inorg. Chem.* **1992**, *31*, 4709-4711.

10.   A. H. Vetter, A. Berkessel, *Synthesis* **1995**, 419-422.

11.   A. Berkessel, G. Hermann, O.-T. Rauch, M. Büchner, A. Jacobi, G. Huttner, *Chem. Ber.* **1996**, *129*, 1421-1423.

12.   T. Schwenkreis, A. Berkessel, *Tetrahedron Lett.* **1993**, *34*, 4785-4788.

13.   A. Berkessel, M. Bolte, T. Schwenkreis, *J. Chem. Soc., Chem. Commun.* **1995**, 535-536.

14.   A. Berkessel, M. Frauenkron, T. Schwenkreis, A. Steinmetz, G. Baum,. D. Fenske, *J. Mol. Catal. A,* **1996**, *113*, 321-342.

15.   A. Berkessel, M. Frauenkron, T. Schwenkreis, A. Steinmetz, *J. Mol. Catal. A,* **1997**, *117*, 339-346.

16.   R. L. Halterman, S.-T. Jan, *J. Org. Chem.* **1991**, *56*, 5253-5254.

17.   A. Berkessel, M. Frauenkron, *J. Chem. Soc., Perkin Trans. 1* **1997**, 2265-2266.

18.   M. Frauenkron, A. Berkessel, *Tetrahedron Lett.* **1997**, *38*, in press.

19.   (a) M. Hirobe, T. Higuchi, H. Ohtake, *Tetrahedron Lett.* **1989**, *30*, 6545-6548; (b) M. Hirobe, T. Higuchi, H. Ohtake, *J. Am. Chem. Soc.* **1992**, *114*, 10660-10662; (c) M. Hirobe, T. Higuchi, H. Ohtake, *Heterocycles* **1995**, *40*, 867-903.

20.   W. Wiesner, K.-H. van Pée, F. Lingens, *J. Biol. Chem.* **1988**, *263*, 13725-13732.

21.   K.-H. van Pée, G. Sury, F. Lingens, *Biol. Chem. Hoppe-Seyler* **1987**, *368*, 1225-1232.

22.   H. J. Hecht, H. Sobek, T. Haag, O. Pfeifer, K.-H. van Pée, *Nat. Struct. Biol.* **1994**, *1*, 532-537.

23.   M. Picard, J. Gross, E. Lübbert, S. Tölzer, S. Krauss, K.-H. van Pée, A. Berkessel, *Angew. Chem.* **1997**, *109*, 1245-1248; *Angew. Chem. Int. Ed. Engl.* **1997**, *36*, 1196-1199.

24.   Patent DE 44 30 327 C 1.

The Potential of Selenium-Containing Peroxidases in Asymmetric Catalysis: Glutathione Peroxidase and Seleno-Subtilisin

Ellen Schüler, Dietmar Häring, Barbara Boss, Markus Herderich, Peter Schreier*

Lehrstuhl für Lebensmittelchemie, Universität Würzburg, Germany

Waldemar Adam, Cordula Mock-Knoblauch, Michael Renz, Chantu R. Saha-Möller and Oliver Weichold

Institut für Organische Chemie, Universität Würzburg, Am Hubland, D-97074 Würzburg, Germany

1 Introduction

Metalloenzymes have been successfully used as biocatalysts in asymmetric synthesis. Previously, we have reported the enantioselective reduction of racemic hydroperoxides by heme-containing horseradish peroxidase (HRP) in the presence of guaiacol [1]. HRP selectively recognized sterically uncumbered hydroperoxides to yield enantiomerically pure hydroperoxides and alcohols. However, hydroperoxides with branched aliphatic residues were poorly recognized by HRP, while tertiary hydroperoxides were not converted at all. In continuation of this work, we studied the potential of glutathione peroxidase (GSH-Px) and seleno-subtilisin, a semisynthetic glutathione peroxidase mimic, for kinetic resolution of racemic hydroperoxides.

2 Results

2.1 Glutathione Peroxidase

Glutathione peroxidase is a selenium-containing peroxidase in mammals which catalyzes the reduction of hydroperoxides in the presence of glutathione [2]. In comparison to heme-containing peroxidases, GSH-Px exhibits a broader substrate acceptance. The enzyme reduces achiral hydroperoxides such as hydrogen peroxide, ethyl hydroperoxide and tertiary butyl hydroperoxide; furthermore, hydroperoxides of fatty acids and steroids are converted [3].

$$ROOH + 2\,GSH \xrightarrow{\text{GSH-Px}} ROH + H_2O + GSSG$$

Studies of the enantioselective recognition of chiral hydroperoxides by GSH-Px have not been carried out to date. To assess the substrate specificity and the enantioselectivity, we investigated the GSH-Px-catalyzed reduction of a number of alkyl aryl hydroperoxides, whose structures are shown below. The enantiomer distribution of the alcohols and the unreacted hydroperoxides were determined by chiral-phase HPLC and multidimensional gas chromatography (MDGC) [4].

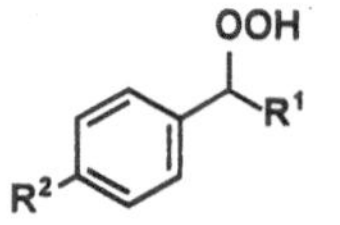

$R^1 = Me, Bu, CH_2OH$
$R^2 = H, Cl$

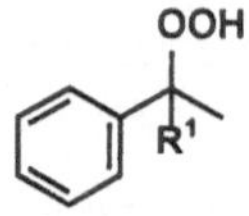

$R^1 = CH_2OH, C_6H_{11}$

In another series of experiments we studied racemic 13- and 9-hydroperoxy-octadecadienoic acids as substrates for GSH-Px.

The results on the reduction of a variety of hydroperoxides demonstrate that GSH-Px unspecifically reacts with these substrates at considerable rates to yield racemic mixtures. Thus, our studies emphasize the biological role of GSH-Px in living cells, in which hydroperoxides generated by enzymatic and non-enzymatic processes are effectively reduced by GSH-Px to protect cells against oxidative stress.

2.2 Seleno-Subtilisin

The design of artificial enzymes that mimic their natural archetypes, is a promising field for the development of novel biocatalysts. However, *de novo* synthesis of enzymes is quite cumbersome and the man-made catalysts are mostly inferior in terms of selectivity and catalytic efficiency compared to native enzymes [5]. As an alternative, semisynthetic enzymes are obtained by chemical modification of the active site of well-known enzymes by utilizing the optimized molecular framework of native enzymes; thus, new catalytic properties may developed [6]. Incorporation of selenocystein - the catalytic unit responsible for the oxido-reduction catalyzed by GSH-Px - into the active site of subtilisin Carlsberg, the former protease gains peroxidase activity.

By starting from subtilisin, we obtained the semisynthetic seleno-subtilisin in a three-step protocol according to Figure 1.

$$Enz\text{-}OH \xrightarrow{PMSF} Enz\text{-}OSO_2\text{-}CH_2Ph \xrightarrow{NaHSe} Enz\text{-}SeH \xrightarrow{H_2O_2} Enz\text{-}SeO_2H$$

Figure 1 Chemical modification of the active site in the protease subtilisin to afford the semisynthetic peroxidase seleno-subtilisin (Enz-OH: Ser221 of subtilisin; PMSF: phenylmethanesulfonyl fluoride) [7].

Our studies on seleno-subtilisin reveal for the first time that the application of semisynthetic enzymes in asymmetric catalysis is feasible (Table 1). Seleno-subtilisin catalyzes the kinetic resolution of racemic hydroperoxides to afford hydroperoxides and alcohols in good optical purity. Various acyclic, cyclic and aromatic hydroperoxides were examined as suitable substrates for this enantioselective reduction. The enantiomeric distribution of all products was determined by MDGC. The kinetic studies indicate seleno-subtilisin as the first semisynthetic enzyme with catalytic efficiency comparable to native enzymes [8].

Table 1 Kinetic parameters and enantiomer distribution of seleno-subtilisin-catalyzed kinetic resolution of racemic hydroperoxides

Hydroperoxide	Alcohol $(R):(S)$	Hydroperoxide $(R):(S)$	K_m [mM]	k_{cat} [min^{-1}]	k_{cat}/K_m [mM^{-1}min^{-1}]
(1-phenyl-2-hydroperoxy-ethanol structure, OOH/OH)	99 : 1	1 : 99	2.1	2443	1150
(1-phenylethyl hydroperoxide, OOH)	20 : 80	76 : 24	15.7	2125	135
(1-(4-chlorophenyl)ethyl hydroperoxide, OOH)	22 : 78	74 : 26	6.0	1723	287
(2-methyl-1-phenylpropyl hydroperoxide, OOH)	71 : 29	30 : 70	18.0	1745	97
(1-indanyl hydroperoxide, OOH)	62 : 38	47 : 53	3.9	699	172
(1-phenyl-propanediol hydroperoxide, OOH/OH)	49 : 51^a	53 : 47^a	9.3	905	96
(phenyl-cyclohexyl hydroperoxide, OOH)	48 : 52^a	52 : 48^a	1.8	33	19
(HOO / OH vinyl structure)	65 : 35^b	43 : 57^b	5.3	449	84
(HOO / OH vinyl structure)	31 : 69^b	61 : 39^b	12.2	643	53
(silyl cyclohexenyl hydroperoxide, OOH)	98 : 2^a	10 : 90^a	8.5	820	97
(cyclohexenyl hydroperoxide, OOH)	54 : 46^a	46 : 54^a	29.1	981	34

$^{a)}$ absolute configuration unknown $^{b)}$ *threo*, $(R, R) : (S, S)$

According to our studies, reduction of hydroperoxides by seleno-subtilisin provides enantioselectivies that are comparable to results obtained for the acylation and esterification reactions catalyzed by the native protease [9]. Since the tertiary structure of subtilisin and seleno-subtilin is identical, the knowledge of substrate-catalyst interactions of subtilisin can be transferred to the semisynthetic enzyme [10]. This fact will facilitate the proper selection of hydroperoxides for further studies.

We herewith contribute a first application of semisynthetic enzymes in asymmetric catalysis of considerable potential. The concept utilizes well defined apo-enzymes as molecular framework, thus minimizing the synthetic work required for obtaining the custom-designed catalyst. Established models of the substrate binding sites allow identification of structural units required for the efficiency and enantioselectivity in the catalysis.

Acknowledgment

The financial support by the Deutsche Forschungsgemeinschaft (Sonderforschungsbereich 347 „Selektive Reaktionen Metall-aktivierter Moleküle") is gratefully acknowledged.

References

[1]   (a) Adam, W.; Hoch, U.; Saha-Möller, C. R.; Schreier, P; Angew. Chem. Int. Ed. 1993, 105, 1737-1739. (b) Adam, W.; Hoch, U.; Lazarus, M.; Saha-Möller, C.R.; Schreier, P.; J. Am. Chem. Soc. 1995, 117, 11898-11901. (c) Adam W., Hoch, U.; Humpf, H.-U.; Saha-Möller, C.R.; Schreier, P.; *Chem. Commun.* **1996**, 2701-2702.

[2]   Wendel, A.; in Enzymatic Basis of Detoxification, Vol.1, (Ed.: Jakoby W.B.), Academic, New York, **1980**, 333-353.

[3]   (a) Flohé, L; in Free Radicals in Biology, Vol. 5, Academic, New York, 1982, 223-254. (b) Flohé, L; in Oxygen Free Radicals and Tissue Damage, CIBA Fdn. Symp. No.65, Excerpta Media, Amsterdam, **1979**, 95-122.

[4]   Schüler, E.; Doctoral thesis, in preparation.

[5]   Kirby, A. J.; *Angew. Chem Int. Ed.* **1996**, 35, 705-724.

[6]   Kaiser, E. T.; Lawrence, D.S.; *Science* **1984**, 226, 505-511.

[7]   Wu, Z.-P.; Hilvert D., *J. Am. Chem. Soc.* **1990**, 112, 5647-5648.

[8]   Häring, D.; Herderich, M.; Schüler, E.; Withopf, B; Schreier P.; *Tetrahedron: Asymmetry* **1997**, 8, 853-856.

[9]   (a) Fitzpatrick, P. A.; Klibanov A. M., *J. Am. Chem. Soc.* **1991**, 113, 3166-3171.(b) Kitaguchi, H.; Fitzpatrick, P. A.; Huber, J.E.; Klibanov, A. M.; *J. Am. Chem. Soc.* **1989**, 111, 3094-3095.

[10]   Syed R.; Wu Z.-P.; Hogle, J. M.; Hilvert D., *Biochemistry* **1993**, 32, 6157-6164.

Biocatalytic Synthesis of Optically Active Oxyfunctionalized Compounds

W. Adam*, M.T. Diaz, R.T. Fell, P. Groer, C. Mock-Knoblauch, C.R. Saha-Möller, O. Weichold
Institut für Organische Chemie, Universität Würzburg, D-97074 Würzburg, Germany

B. Boss, U. Hoch, H.-U. Humpf, M. Lazarus, Z. Lukacs, P. Schreier
Lehrstuhl für Lebensmittelchemie, Universität Würzburg, D-97074 Würzburg, Germany

Summary

High reactivity under mild reaction conditions, excellent selectivity, and good environmental tolerance establish enzymes and microorganisms as valuable biocatalysts in asymmetric synthesis. In the last years we have investigated intensively the biocatalytic synthesis of optically active hydroperoxides, alcohols, α-hydroxy carbonyl compounds and α-methylene β-lactones. A selection of highlights is presented below.

Highlights

Recently we have reported [1] that horseradish-peroxidase(HRP)-catalyzed kinetic resolution of alkyl aryl hydroperoxides affords the hydroperoxides and alcohols in high optical purities. This biocatalytic method was extended to diastereomeric hydroperoxy homoallylic alcohols and esters (Scheme 1) to produce the functionalized hydroperoxides and alcohols in high enantiomeric excess (up to 99%) [2]. Racemic hydroperoxides may also be resolved by lipases [3]. A comparative study of HRP- *versus* lipase-catalyzed kinetic resolution of hydroperoxy vinylsilanes (Scheme 2) revealed that the lipases are the enzymes of choice

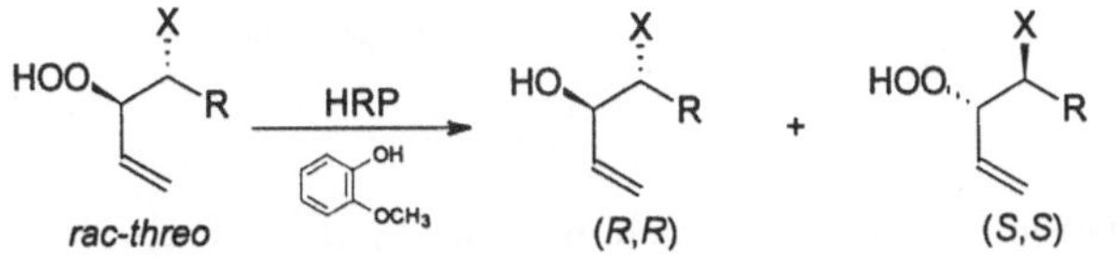

R	X	ee (%)	
		(R,R)	(S,S)
Et	OH	99	99
iPr	OH	99	99
Me	COOiPr	95	93

Scheme 1 HRP-catalyzed kinetic resolution of functionalized hydroperoxides [2].

for the transformation of sterically demanding substrates, which are not or only reluctantly converted by HRP [4]. The racemic alcohols of the corresponding hydroperoxy vinylsilanes were resolved in excellent enatioselectivities by lipase-catalyzed acetylation [5].

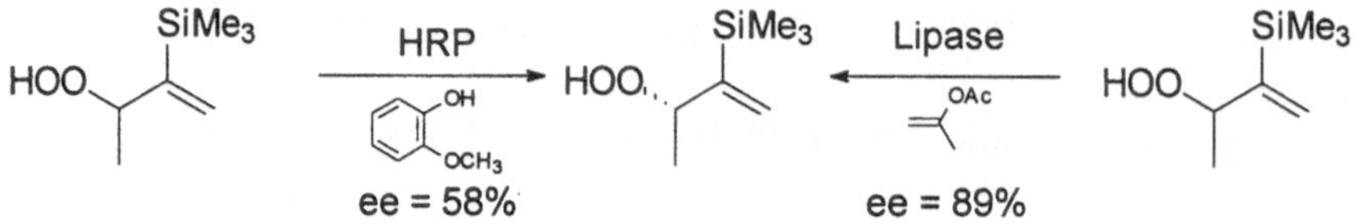

SiMe₃ ... (scheme)

Scheme 2 HRP- and lipase-catalyzed kinetic resolution of hydroperoxy vinylsilanes [4].

Our preliminary experiments have shown that microorganisms, which have been isolated from topsoil by a selective screening procedure, may be utilized as biocatalysts for the preparation of optically active hydroperoxides (Scheme 3).

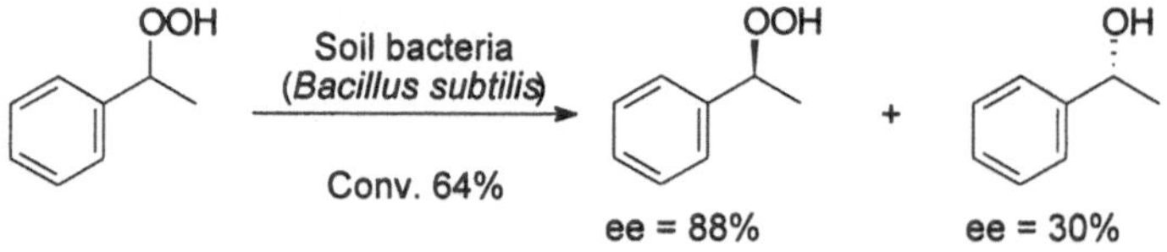

Scheme 3 Kinetic resolution of (1-phenyl) ethyl hydroperoxide with soil bacteria.

Optically active α-hydroxy carbonyl compounds are valuable building blocks for the asymmetric synthesis of natural products. Therefore, recently we have made efforts to develop biocatalytic methods for the synthesis of optically active α-hydroxy acids, esters and ketones. The direct α hydroxylation of saturated and unsaturated carboxylic acids with molecular oxygen by α oxidases of crude pea-leaf homogenate (Scheme 4) afforded the cor-

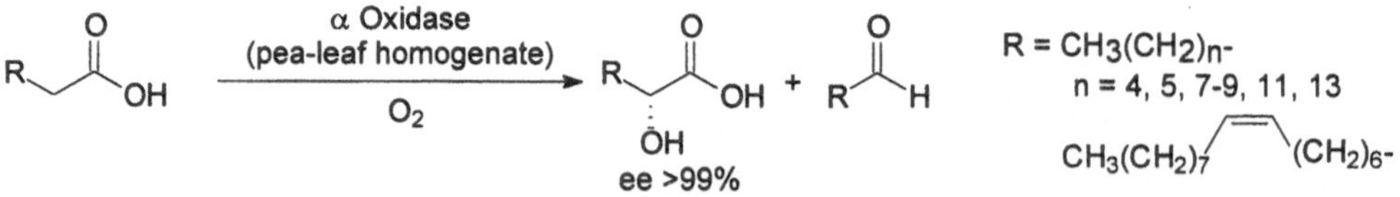

Scheme 4 α Oxidation of acids by crude homogenates of pea leaves with molecular oxygen [6].

responding α-hydroxy acids in enantiomerically pure form [6]. Furthermore, for the first time [7] we have achieved the efficient enzymatic resolution of 2-hydroxy acids by the enantioselective oxidation with molecular oxygen catalyzed by the glycolate oxidase from spinach (Scheme 5).

40

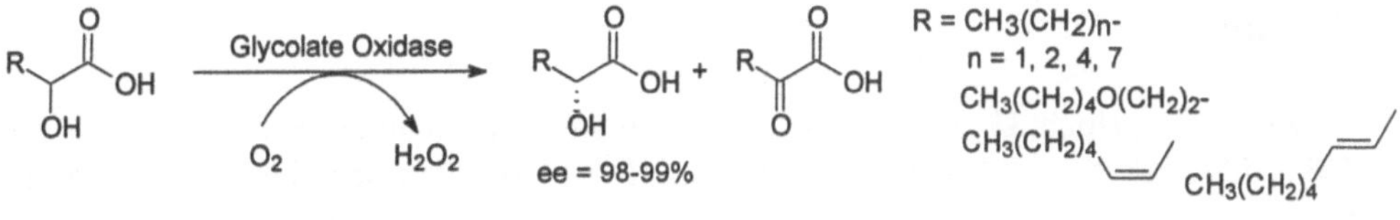

Scheme 5 Enzymatic resolution of 2-hydroxy acids by glycolate oxidase (*Spinacia oleracea*) [7].

Optically active α-hydroxy ketones were made available in high enantiomeric excess (up to 99%) [8] by the lipase-catalyzed enantioselective acetylation of racemic ketones with iso-propenyl acetate (Scheme 6).

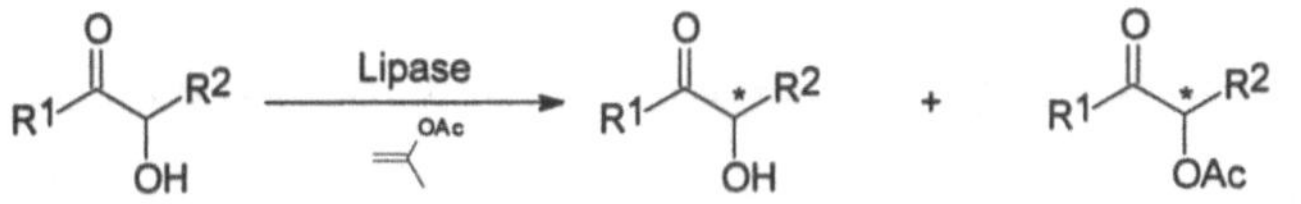

R^1	R^2	Conv. (%)	ee (%)	Config.	ee (%)	Config.
Ph	Me	52	99	*S*-(−)	90	*R*-(+)
Me	Ph	51	99	*R*-(−)	95	*S*-(+)

Scheme 6 Lipase-catalyzed kinetic resolution of α-hydroxy ketones [8].

Recently we have demonstrated [9] for the first time that the kinetic resolution of α-methylene β-lactones by lipase-catalyzed asymmetric transesterification with benzyl alcohol leads to the β-lactones and benzyl esters in enantiomerically pure form (Scheme 7).

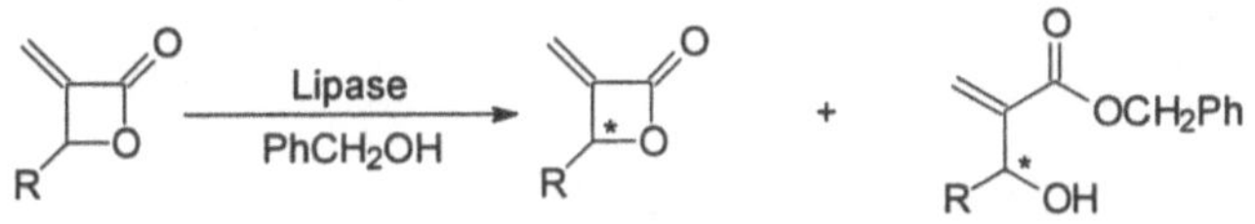

R	Conv. (%)	ee (%)	Config.	ee (%)	Config.
Et	50	99	*R*-(+)	99	*S*-(−)
*i*Pr	50	99	*S*-(+)	99	*R*-(−)

Scheme 7 Lipase-catalyzed kinetic resolution of racemic α-methylene β-lactones [9].

These results demonstrate that enzymes are useful and efficient biocatalysts for the preparation of enantiomerically pure building blocks in organic synthesis.

Acknowledgement

The generous financial support of the *Deutsche Forschungsgemeinschaft* (SFB 347 "Selektive Reaktionen Metall-aktivierter Moleküle") and the *Fonds der Chemischen Industrie* is gratefully appreciated. We thank Boehringer Mannheim GmbH for the generous gift of enzymes.

References

1. (a) Adam, W.; Hoch, U.; Lazarus, M.; Saha-Möller, C.R.; Schreier, P. *J. Am. Chem. Soc.* **1995**, *117*, 11898-11901. (b) Adam, W.; Fell, R.T.; Hoch, U.; Schreier, P.; Saha-Möller, C.R. *Tetrahedron: Asymmetry* **1995**, *6*, 1047-1050.
2. (a) Adam, W.; Hoch, U.; Humpf H.-U.; Saha-Möller, C.R.; Schreier, P. *J. Chem. Soc., Chem. Commun.* **1997**, 2701-2702. (b) Hoch, U.; Adam, W.; Fell, R.; Saha-Möller, C.R.; Schreier, P. *J. Mol. Cat. A: Chem.* **1997**, *117*, 321-328. (c) Hoch, U.; Humpf, H.-U.; Schreier, P.; Saha-Möller, C.R. *Chirality* **1997**, 69-74.
3. Höft, E.; Haman, H.-J.; Kunath, A.; Adam, W.; Hoch, U.; Saha-Möller, C.R.; Schreier, P. *Tetrahedron: Asymmetry* **1995**, *6*, 603-608.
4. Adam, W.; Mock-Knoblauch, C.; Saha-Möller, C.R. *Tetrahedron: Asymmetry* **1997**, *8*, 1947-1950.
5. Adam, W.; Mock-Knoblauch, C.; Saha-Möller, C.R. *Tetrahedron: Asymmetry* **1997**, *8*, 1441-1444.
6. Adam, W.; Lazarus, M.; Saha-Möller, C.R.; Schreier, P. *Tetrahedron: Asymmetry* **1996**, *7*, 2287-2292.
7. Adam, W.; Lazarus, M.; Boss, B.; Saha-Möller, C.R.; Humpf, H.-U.; Schreier, P. *J. Org. Chem.* **1997**, in press.
8. Adam, W.; Diaz, M.T.; Fell, R.T.; Saha-Möller, C.R. *Tetrahedron: Asymmetry* **1996**, *7*, 2207-2210.
9. Adam, W.; Groer, P.; Saha-Möller, C.R. *Tetrahedron: Asymmetry* **1997**, *8*, 833-836.

Selective Metal-Catalyzed Oxygen Transfer

Waldemar Adam*, Rainer T. Fell, Marion N. Korb, Catherine M. Mitchell,
Chantu R. Saha-Möller
Institut für Organische Chemie, Universität Würzburg, Am Hubland
D-97074 Würzburg, Germany

Background and Goals

As environmental issues and the cautious exploitation of natural resources become increasingly important, chemical processes must be developed to work more efficiently and selectively, thereby making possible a significant reduction in the amounts of undesired or even toxic waste products. One promising approach towards these goals is the use of metal catalysts. Below some of our recent achievements in the field of selective metal-catalyzed oxyfunctionalizations are highlighted.

Titanium-Catalyzed Oxidations

The Sharpless and Kagan oxidizing systems use the achiral *tert.*-butyl hydroperoxide or cumyl hydroperoxide as primary oxidants with tartrates as the source of chirality [1]. Recently, we have reported on the preparation of optically active secondary hydroperoxides by kinetic resolution catalyzed by horseradish peroxidase. It was, therefore, of interest to assess whether such chiral oxidants may be used as the source of chirality in the $Ti(O^iPr)_4$-catalyzed oxidation of allylic alcohols and sulfides. Unfortunately, the degree of chiral induction is rather limited for allylic alcohols; nevertheless, the addition of the diethyl 2-hydroxy-2-hydroxymethylmalonate (DHHM) in catalytic amounts raised the enantiomeric excess (ee) from 20-30 to up to 50% in the epoxidation of the dialkyl-substituted 3-methyl-2-buten-1-ol and geraniol by the (-)-(S)-1-phenylethyl and (-)-(S)-1-phenylpropyl hydroperoxides [2]. For a variety of phenyl-substituted allylic alcohols, however, the additive DHHM showed no significant improvement in enantioselectivity.

$$R^1{\diagdown}S{\diagdown}R^2 \quad \xrightarrow[\text{CCl}_4,\ \text{Ar},\ -20\ ^{\circ}\text{C, MS 4Å}]{\substack{\text{OOH} \\ R^3{\diagup}{\diagdown}R^4 \quad \text{(0.05 equiv.)} \\ /\ Ti(O^iPr)_4}} \quad R^1{\diagdown}\overset{O}{\underset{}{S}}{\diagdown}R^2 \quad + \text{ sulfone} \qquad (1)$$

$$ee \leq 75\%$$

Significantly higher ee values (Eq. 1) were observed for the oxidation of sulfides. This is presumably due to enantioselective oxidation of the sulfide to the corresponding sulfoxide and subsequent kinetic resolution of the latter through oxidation to the sulfone. The ee values obtained with the optically active hydroperoxides are comparable to those of the literature, and in the case of *n*-butyl *p*-tolyl sulfide even exceed them by ca. 40%.

Rhenium-Catalyzed Oxidations

To date, a wide range of applications of the versatile catalyst methyltrioxorhenium (MTO) have been developed or improved in our group. These include arene oxidation [3], epoxidation with MTO and the urea/hydrogen peroxide adduct (UHP) [4], and sulfoxidation [5]. The range of oxyfunctionalizations has now been extended to the oxidation of silanes to silanols. For SiH insertions, the oxidant MTO/UHP is superior to other literature methods since i) the oxidation is catalytic, ii) the formation of disiloxanes is surpressed (Eq. 2), and iii) optically active silanes are oxidized with complete retention of configuration (Eq. 3).

$$PhMe_2SiH \quad \xrightarrow[CH_2Cl_2, \text{ r.t., 8 h}]{MTO \text{ (cat.), [O]}} \quad PhMe_2SiOH \quad + \quad (PhMe_2Si)_2O \quad (2)$$

[O] = 85% H_2O_2	20 :	80
UHP	98 :	2

$$\underset{CH_3}{\overset{\alpha\text{-Naph}}{Ph^{\cdots\cdots}Si\!-\!H}} \quad \xrightarrow[CH_2Cl_2, \text{ r.t., 24 h}]{UHP, MTO \text{ (cat.)}} \quad \underset{CH_3}{\overset{\alpha\text{-Naph}}{Ph^{\cdots\cdots}Si\!-\!OH}} \quad (3)$$

ee ≤ 99%   yield = 97%, ee 94%

The high degree of retention suggests a transition state for the oxygen-atom transfer analogous to that already postulated for the SiH insertions by dioxiranes [6].

Manganese-Catalyzed Oxidations

Chiral (salen)Mn(III) complexes have received much interest in catalytic asymmetric oxidation. A broad spectrum of different applications has been covered, which we have now extended not only to the transformation of silyl enol ethers and ketene acetals [7], but also to the quite unreactive isoflavones. The latter have now been enantioselectively epoxidized (ee 42-82%) for the first time. Furthermore, chromenes may be epoxidized in moderate to high enantioselectivity by dioxirane as primary oxidant and catalyzed by (salen)Mn(III) complexes [8].

The silyl enol ethers and ketene acetals were oxidized with moderate to high ee values (≤89%) to the corresponding α-hydroxy ketones or esters (Eq. 4). Substrates with *cis* substituents in the ß position work best, while bulky substituents α to the siloxy functionality dramatically lower the enantiofacial control.

Other work on metal-catalyzed oxidations includes the nickel-catalyzed hydroxylation of 1,3-dicarbonyl compounds by dioxirane [9]. Furthermore, a highly diastereoselective, titanium-catalyzed olefinic epoxidation of diols (which cannot be employed directly in the Sharpless epoxidation without protection of the non-allylic hydroxy group) with β-hydro-

$$R^1\text{—}C(OSiMe_2R^3)=CR^2 \xrightarrow[\substack{\text{(salen)Mn(III) (cat.)}\\ \text{NaOCl, pH 11.3, PPNO}\\ \text{CH}_2\text{Cl}_2,\ 0\,^\circ\text{C}\rightarrow\text{r.t.},\ 24\,\text{h}}]{} \xrightarrow[\substack{\text{HCl}\\ \text{MeOH, r.t., 3 h}}]{} R^2\text{—CO—}\overset{*}{C}\text{H(OH)}R^1 \qquad (4)$$

R^1	R^2	R^3	ee (%)	config.
Ph	Et	Me	87	S
Ph	OMe	^{t}Bu	57	S
Ph	Ph	Me	12	R

peroxy alcohols as tridentate oxygen donors [10] was achieved. In this short summary we have presented some of the highlights of our current work on selective metal-catalyzed oxidations. Our encouraging results stimulate the further search for efficient, selective, and versatile catalytic oxidation methods in organic synthesis.

Acknowledgement

Support from the *Deutsche Forschungsgemeinschaft* (SFB 347 'Selektive Reaktionen Metall-aktivierter Moleküle') and the *Fonds der Chemischen Industrie* is gratefully appreciated.

References

1. (a) Katsuki, T.; Sharpless, K. B. *J. Am. Chem. Soc.* **1980**, *102*, 5974-5976. (b) Pitchen, P.; Duñach, E.; Deshmukh, M. N.; Kagan, H. B. *J. Am. Chem. Soc.* **1984**, *106*, 8188-8193.
2. Adam, W.; Korb, M. N. *Tetrahedron: Asymmetry* **1997**, *8*, 1131-1142.
3. (a) Adam, W.; Herrmann, W. A.; Lin, J.; Saha-Möller, C. R.; Fischer, R. W.; Correia, J. D. G. *Angew. Chem. Int. Ed. Engl.* **1994**, *33*, 2475-2477. (b) Adam, W.; Herrmann, W. A.; Lin, J.; Saha-Möller, C. R. *J. Org. Chem.* **1994**, *59*, 8281-8283. (c) Adam, W.; Herrmann, W. A.; Saha-Möller, C. R.; Shimizu, M. *J. Mol. Cat. A: Chemical* **1995**, *97*, 15-20.
4. Adam, W.; Mitchell, C. M. *Angew. Chem. Int. Ed. Engl.* **1996**, *35*, 533-535.
5. Adam, W.; Mitchell, C. M.; Saha-Möller, C. R. *Tetrahedron* **1994**, *50*, 13121-13124.
6. Adam, W.; Mello, R.; Curci, R. *Angew. Chem. Int. Ed. Engl.* **1990**, *29*, 890-891.
7. a) Adam, W.; Fell, R. T.; Mock-Knoblauch, C.; Saha-Möller, C. R. *Tetrahedron Lett.* **1996**, *37*, 6531-6534. b) Adam, W.; Fell, R. T.; Stegmann, V. R.; Saha-Möller, C. R. *J. Am. Chem. Soc.*, submitted.
8. a) Adam, W.; Jekö, J.; Lévai, A.; Nemes, C.; Patonay, T.; Sebök, P. *Tetrahedron Lett.* **1995**, *36*, 3669-3672s. b) Adam, W.; Jekö, J.; Lévai, A.; Majer, Z.; Nemes, C.; Patonay, T.; Párkányi, L.; Sebök, P. *Tetrahedron: Asymmetry* **1996**, *7*, 2437-2446.
9. Adam, W.; Smerz, A. K. *Tetrahedron* **1996**, *52*, 5799-5804.
10. Adam, W.; Peters, K.; Renz, M. *J. Org. Chem.* **1997**, *62*, 3183-3189.

Chemo- and Diastereoselective Epoxidations Catalyzed by Titanium-Containing Zeolites: Evidence for a Hydrogen-Bonded, Peroxy-Type Loaded Complex as Oxidizing Species

Waldemar Adam[a], Avelino Corma[b], Michael Renz[*,a,c]

[a] University of Würzburg, Institute of Organic Chemistry, Am Hubland, D-97074 Würzburg, Germany; [b] Instituto Tecnología Química, UPV-CSIC, Avd. de los Naranjos s/n, E-46022 Valencia, Spain; [c] Laboratoire de Chimie de Coordination du CNRS, 205 route de Narbonne, F-31077 Toulouse cedex 4, France

The last decades of the twentieth century will be historically declared as the age of "environmental consciousness". The imposition has been placed on the chemical industry to develop more effective and environmentally friendly processes. Since oxidations are important industrial reactions as a source of oxyfunctionalized molecules from a wide variety of cheap hydrocarbons, new selective oxidation catalysts play an increasingly important role in chemical manufacture. One group of the most promising materials are the metal-containing zeolites. The first titanium-containing zeolite (TS-1), which is an active oxidation catalyst, was synthesized in the early 1980′s by Enichem research workers. Since that time, many applications of such heterogeneous oxidation systems have been reported, which include the hydroxylation of arenes, epoxidation, CH insertion and heteroatom oxidation [1]. In continuation of our studies on selective, catalytic oxyfunctionalizations of organic substrates [2], we have investigated the chemoselectivity of the epoxidation of α-methylstyrene and of chiral allylic alcohols. The latter are ideal substrates for diastereoselective epoxidations. By comparison with established oxidants, useful mechanistic information on the much speculated oxidizing species in the TS-1 oxidation has now been obtained.

1 Chemoselectivity

In the epoxidation catalyzed by metal-doped zeolites, double-bond cleavage has been observed for styrene derivates [3]. In a careful study of the epoxidation of α-methylstyrene (1), it was shown that this is due to the labile nature of the epoxide, the primary oxidation product. The product of the double-bond cleavage, acetophenone (2d), results from the Grob fragmentation of the hydroperoxy alcohol 2a (Eq. 1) [4], which was formed by perhydrolysis of the epoxide. The cleavage reaction is suppressed by using methanol as solvent. In this case, methanolysis of the epoxide intermediate produces the monomethyl ether of the diol 2b, which persists the reaction conditions.

Ti-beta, H₂O₂ (0.29 equiv.) MeCN, 50 °C, 5 h

conv. 19%, yield > 90%

2a : **2b** : **2c** : **2d**
14 : 34 : 38 : 14

$$ \text{1} \xrightarrow[\text{MeCN, 50 °C, 5 h}]{\text{Ti-beta, H}_2\text{O}_2 \text{ (0.29 equiv.)}} \text{2a} + \text{2b} + \text{2c} + \text{2d} \tag{1}$$

Another side reaction is observed in the epoxidation of allylic alcohols catalyzed by titanium-containing zeolites. For substrates with methyl substituents, which generate stabilized allylic cations, migration and/or substitution of the hydroxy group is observed with the usually employed dilute, aqueous hydrogen peroxide solution (Eq. 2). These undesired reactions are avoided when concentrated 85% H₂O₂ is used for the epoxidations

TS-1 (100 wt%), H₂O₂ (30%), H₂O, 50 °C, 3 h

conversion > 95%, m.b. < 50%

56 : 30 : 14

$$ \text{(2)} $$

catalyzed by Ti-β and TS-1 [5]. For the latter hydrophobic zeolite, the 85% aqueous H₂O₂ may be replaced by the easier to handle and safer anhydrous crystalline adduct of urea and hydrogen peroxide (UHP). The hydrophilic Ti-β is deactivated by urea. Under these conditions, i.e. with TS-1/UHP and Ti-β/85% H₂O₂, the epoxy alcohols **4** are obtained in high yields and without any side products (Eq. 3).

$$ \text{3} \xrightarrow{\text{oxidant}} \textit{threo-}\textbf{4} + \textit{erythro-}\textbf{4} \tag{3}$$

2 Diastereoselectivity

The allylic alcohols **3** are epoxidized highly *threo*-selectively when 1,3-allylic (A^{1,3}) strain is present in the molecule, i.e. for the substrates **3c,d** (Table 1, entries 3 and 4). Allylic alcohols without allylic strain, i.e. **3b** (Table 1, entry 2), or with A^{1,2} strain such as substrate **3a** (Table 1, entry 1), give only a low diastereomeric excess, but here also the *threo* product is favored.

The *tert*-butyl derivatives **3e,f** are not epoxidized by TS-1 (Table 1, entries 5 and 6). Fortunately, the Ti-β zeolite with the larger pores was effective for the epoxidation of these sterically more demanding substrates. This confirms that the oxidation occurs inside of the zeolite and not on the outer surface [5b]. The observed diastereomeric ratios are similar to those obtained for the methyl-substituted ones.

Table 1 Diastereomeric ratios for the epoxidation of the allylic alcohols **3** with several oxidants

entry	R^1	R^2	R^3	R^4		TS-1 UHP	Ti-β H_2O_2 (85%)	m-CPBA	VO(acac)$_2$ TBHP	DMD
		substrate						diastereoselectivities		
1	Me	H	H	Me	**3a**	55 : 45	56 : 44	45 : 55	05 : 95	60 : 40
2	H	Me	H	Me	**3b**	65 : 35	64 : 36	64 : 36	29 : 71	53 : 47
3	H	H	Me	Me	**3c**	87 : 13	91 : 09	95 : 05	71 : 29	67 : 33
4	H	Me	Me	Me	**3d**	95 : 05	95 : 05	95 : 05	86 : 14	76 : 24
5	H	H	t-Bu	Me	**3e**	---	95 : 05	95 : 05	95 : 05	73 : 27
6	Me	H	H	t-Bu	**3f**	---	70 : 30	56 : 44	05 : 95	24 : 76

3 The Nature of the Oxidizing Species

Analogous to the peracids, for the Ti-containing zeolites and the allylic alcohols with $A^{1,3}$ strain, i.e. the derivatives **3c** and **3d**, the *threo*-epoxy alcohols **4c** and **4d** are obtained essentially exclusively (Table 1, entries 3 and 4). When $A^{1,3}$ strain is relaxed as in the (E)-derivative **3b**, diastereoselective control is essentially lost for both the zeolites and the peracid (Table 1, entry 2); neither does $A^{1,2}$ strain, as in substrate **3a**, manifest itself for both (Table 1, entry 1). Thus, the diastereomeric ratios for the four stereolabeled allylic alcohols **3a-d** (Table 1) show that the transition-state geometries for the oxygen transfer by titanium-containing zeolites and m-CPBA are structurally closely related (structure **A**, Figure 1). The peracid associates with the substrate by hydrogen bonding with an optimal dihedral angle (O–C–C=C) of 120° in the allylic alcohol; a similar arrangement applies to the titanium-catalyzed epoxidation (structure **A**, Figure 1). Consequently, through effective hydrogen bonding, the *threo*-epoxy alcohols **4** are obtained predominantly since $A^{1,3}$ strain dictates the preferred conformation of the allylic alcohol in the transition state, whereas $A^{1,2}$ strain is essentially negligible.

For the metal-catalyzed epoxidation by the homogeneous VO(acac)$_2$/TBHP oxidant, the allylic alcohol **3** is bound by a metal-alcoholate bond to the vanadium and the oxygen atom is transferred at an optimal dihedral angle (O–C–C=C) of 40 to 50°. In terms of diastereoselectivity this is expressed in a high preference of the *erythro* isomer for substrates **3a** and **3f** with $A^{1,2}$ strain (Table 1, entries 1 and 6). Comparison of these diastereoselectivities with the heterogeneous zeolite-catalyzed epoxidations brings out clear differences (Table 1). Thus, while for substrate **3a** (entry 1) no preference between *erythro* and *threo* isomers is noted, for derivative **3f** (entry 6) the *threo* isomer dominates. Clearly, although both allylic alcohols possess $A^{1,2}$ strain, different diastereoselectivities are displayed, in complete contrast to the homogeneous vanadium-catalyzed epoxidation, which is consistently *erythro*-selective. Unquestionably, metal-alcoholate bonding of the substrate in the loaded complex, as it operates for the homogeneous vanadium-catalyzed epoxidation, cannot apply to the heterogeneous catalytic process of the zeolites (structure **B**, Figure 1).

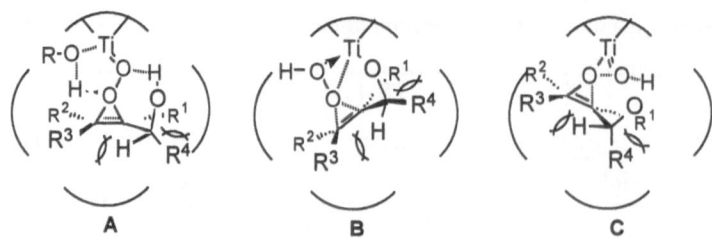

Figure 1 Proposed transition-state structures **A** – **C** for the heterogeneous epoxidation of allylic alcohols catalyzed by the titanium-containing zeolites TS-1 and Ti-β.

For DMD it has been estimated with conformationally fixed cyclic substrates that the optimal dihedral angle in the allylic alcohol lies between 130 and 140°. Since the hydroxy group is now further away from the oxygen-transferring site, hydrogen bonding is expectedly weaker and the diastereoselectivity for the $A^{1,3}$-strained substrates **3c** and **3d** drops substantially for DMD compared to Ti-beta and TS-1 (Table 1, entries 3 and 4). Also for the allylic alcohols with large $A^{1,2}$ strain, as the *tert*-butyl derivative **3f**, a significant discrepancy between the zeolites and DMD is observed. The *threo*-epoxy alcohol **4f** is obtained preferentially in the Ti-beta-catalyzed epoxidation (Table 1, entry 6), but the *erythro* isomer for DMD. From these differences in the diastereoselectivities we suggest that a peroxo-type transition state **C** (Figure 1) does not apply for the epoxidation by the titanium-containing zeolite.

On the basis of these comparative stereochemical data (Table 1), we conclude that the hydrogen-bonded, peroxy-type transition state geometry (structure **A**, Figure 1), as established for peracids, operates for the oxygen transfer catalyzed by Ti-containing zeolites.

Acknowledgement

The generous financial support of the Deutsche Forschungsgemeinschaft (SFB 347 "Selektive Reaktionen Metall-aktivierter Moleküle") and the Fonds der Chemischen Industrie is gratefully appreciated. M. R. thanks the DAAD for a 6-month doctoral fellowship to initiate this work in Valencia (Spain).

References

[1]   a) R. A. Sheldon, *Top. Curr. Chem.* **1993**, *164*, 21-43; b) E. Höft, H. Kosslick, R. Fricke, H.-J. Hamann, *J. Prakt. Chem.* **1996**, *338*, 1-15.

[2]   a) M. Prein, W. Adam, *Angew. Chem. Int. Ed. Engl.* **1996**, *35*, 477 - 494; b) W. Adam, M. Prein, *Acc. Chem. Res.* **1996**, *29*, 275-283; c) W. Adam, A. K. Smerz, *Bull. Chem. Soc. Belg.* **1996**, *105*, 581-599. d) W. Adam, C. M. Mitchell, *Angew. Chem. Int. Ed. Engl.* **1996**, *35*, 533-535.

[3]   a) S. B. Kumar, S. P. Mirajkar, G. C. G. Pais, P. Kumar, R. Kumar, *J. Catal.* **1995**, *156*, 163-166; b) J. S. Reddy, U. R. Khire, P. Ratnasamy, R. B. Mitra, *J. Chem. Soc., Chem. Commun.* **1992**, 1234-1235.

[4]   W. Adam, A. Corma, A. Martínez, M. Renz, *Chem. Ber.* **1996**, *129*, 1453-1455.

[5]   a) W. Adam, R. Kumar, T. I. Reddy, M. Renz, *Angew. Chem. Int. Ed. Engl.* **1996**, *35*, 880-882. b) W. Adam, A. Corma, T. I. Reddy, M. Renz, *J. Org. Chem.* **1997**, *62*, 3631-3637. c) W. Adam, A. Corma, A. Martínez, C. M. Mitchell, T. I. Reddy, M. Renz, A. K. Smerz, *J. Mol. Catal. A* **1997**, 117, 357-366.

Enantioselective Alkylperoxylation of Prochiral Allylic and Benzylic C–H Groups with *tert*-Butyl Hydroperoxide in the Presence of Chiral Copper Complexes

Manfred Schulz, Ralph Kluge, Feyissa Gadissa Gelalcha

Martin-Luther-Universität Halle-Wittenberg, Institut für Organische Chemie Merseburg, D-06099 Halle, Germany

Summary

First direct enantioselective *tert*-butylperoxylations of allylic and benzylic C–H compounds were achieved by oxidation with *t*BuOOH in the presence of optically active bisoxazoline copper complexes. Simple olefinic compounds, e.g. cyclohexene, α-angelica lactone and allylbenzene were converted to the corresponding peroxides in good yields and ee values between 9 and 20%. In the oxidation of 1-substituted cyclohexenes, mixtures of regioisomeric allylic peroxides with different ee values were obtained. The highest ee value (84 %, -50 °C) was observed in the case of the new peroxide, (*S*)-3-*tert*-butylperoxy-1-methyl-1-cyclohexene.

Introduction

The regio-, diastereo- and especially the enantioselective transformation of C–H groups to C–O functionalities is one of the most difficult problems in preparative organic chemistry. During the last few years significant progress was achieved in this area, especially in the enantioselective allylic acyloxylation reaction by employing optically active copper complexes (eq. 1). After the initially rather low ee values obtained by Denney (ee < 10%) [1] and Muzart (ee ≤ 30%) [2], the real breakthrough came in 1995 by the work of Pfaltz and

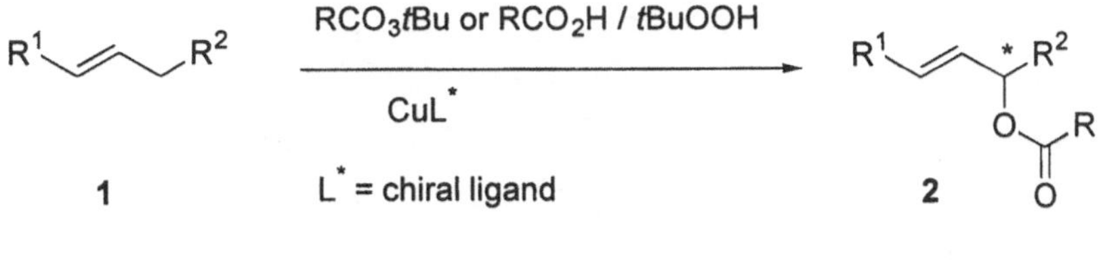

$$ \text{1} \xrightarrow[\substack{\text{CuL}^* \\ \text{L}^* = \text{chiral ligand}}]{\text{RCO}_3t\text{Bu or RCO}_2\text{H} / t\text{BuOOH}} \text{2} \qquad (1) $$

coworkers [3] (ee up to 84% of **2**) by the introduction of chiral bisoxazoline copper(I) complexes. The hitherto highest ee value (93%) was achieved very recently by Katsuki by employing a trisoxazoline copper(II) complex [4].

In contrast, hitherto no example was reported for the related enantioselective peroxylation reaction (eq. 2), although the corresponding racemic reaction is known for more than 40 years, discovered by Kharasch, Fono, Treibs and Pellmann [5-7] and later investigated by other authors (e.g. Kochi [8], Mimoun [9] and Minisci [10]).

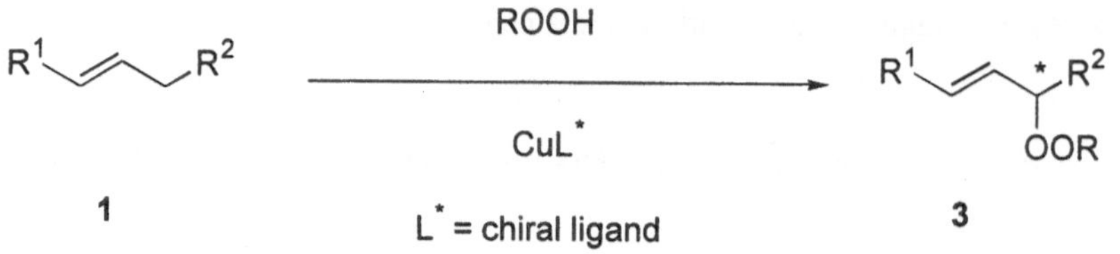

Moreover, no synthetic (nonenzymatic) method for the enantioselective conversion of pro-chiral C–H compounds to optically active peroxides (or hydroperoxides) is known as far.

Results

Recently we could show that the allylic peroxides **3** are useful substrates for the diastereoselective functionalization of the double bond [11]. For example, by starting from cyclohexene (**4**), three asymmetric centers are introduced in only two or three steps with high diastereoselectivity. Therefore, the enantioselective oxidation of allylic C–H compounds to optically active ally-lic peroxides should be of interest.

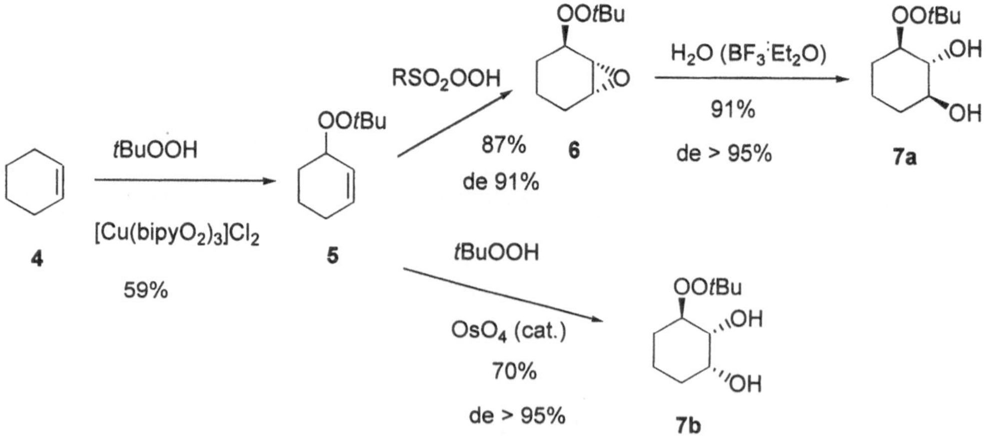

Initial oxidations of cyclohexene (**4**) with *t*BuOOH in the presence of (*S*)-amino acid copper complexes led to rather disappointing results, the peroxide **5** had < 4% ee. Higher ee values were expected by using the chiral bisoxazoline ligands **8** and **9**, which have been effective in the acyloxylation reaction [3]. Indeed, the oxidation of cyclohexene (**4**), α-angelica lactone (**10**) and allylbenzene (**12**) with *t*BuOOH/ Cu(I)OTf/ **8** led to the optically active peroxides **5**, **11** and **13a** with ee values of 9 to 20%.

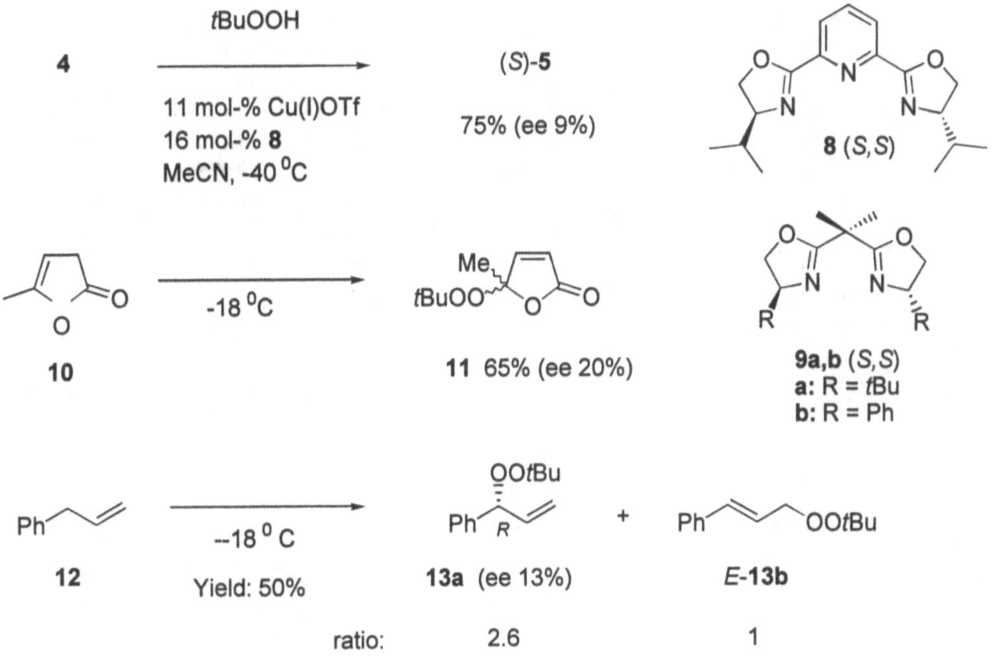

Oxidations of 1-substituted cyclohexenes **14a-c** with *t*BuOOH/ Cu(I)OTf/ **8** led to mixtures of regioisomeric peroxides (**15-17a-c**) which were separated by flash chromatography and had considerably different ee values (Table 1). It is evident that the ee values and the regioisomeric ratios are influenced by the R substituent in **14a-c**. Thus, the more bulky *tert*-butyl group led to the preferred formation of the isomer **16** and to decreased enantioselec-tivity. Unexpected high ee values (up to 84%) were obtained for the peroxide **16a** and, surprisingly, **15a** and **16a** had opposite configurations. It should be pointed out that the use of different bidentate bisoxazoline ligands (e.g. **9a,b**), which gave comparable or higher ee values in the acyloxylation (relative to ligand **8**) [3], led to drastically lower enantioselec-tivities (ee < 10%) in the peroxylation of **14a**.

The differences between the peroxylation and the acyloxylation reaction are more readily understood, when the results of the benzoyloxylation of **14a** are compared with the results in Table 1 (entries 1 and 2). Different regio- and enantioselectivities and, in the case of **15a** and **18**, opposite configurations were obtained.

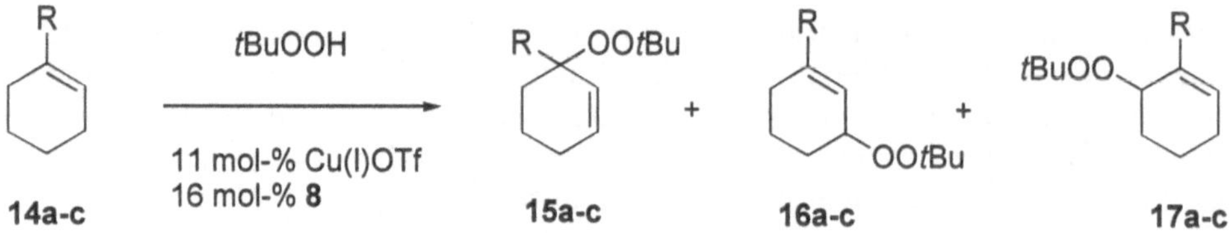

14a-c → **15a-c** + **16a-c** + **17a-c**

*t*BuOOH, 11 mol-% Cu(I)OTf, 16 mol-% **8**

Table 1 Enantioselective Peroxylation of 1-Substituted Cyclohexenes **14a-c**.

Substrate (R)	Solvent	Temp. [°C]	Time [d]	Conversion[a] (Yield)[b] [%]	Product Composition (ee, Config.)[c] [%]		
					15	**16**	**17**
14a (CH$_3$)	MeCN	0	14	82 (72)	35 (20, *R*)	33 (32, *S*)	32 (0)
14a (CH$_3$)	MeCN	-36	23	44 (52)	51 (23, *R*)	23 (66, *S*)	26 (0)
14a (CH$_3$)	Acetone	-40	13	19 (50)	49 (27, *R*)	19 (75, *S*)	32 (0)
14a (CH$_3$)	Acetone	-50	20	25 (54)	56 (32, *R*)	18 (84, *S*)	26 (10)
14b (Ph)	MeCN	-20	23	61 (55)	28 (n.d.[d])	55 (5, *S*)	17 (< 5)
14c (*t*Bu)	MeCN	-20	27	76 (67)	5 (n.d.[d])	85 (20, *S*)	10 (n.d.[d])

a) Conversion of *t*BuOOH. b) Sum of isolated products (rel. to conversion). c) Determined by chiral GC analysis and by reduction (DIBAH) to the allylic alcohols. d) Not determined.

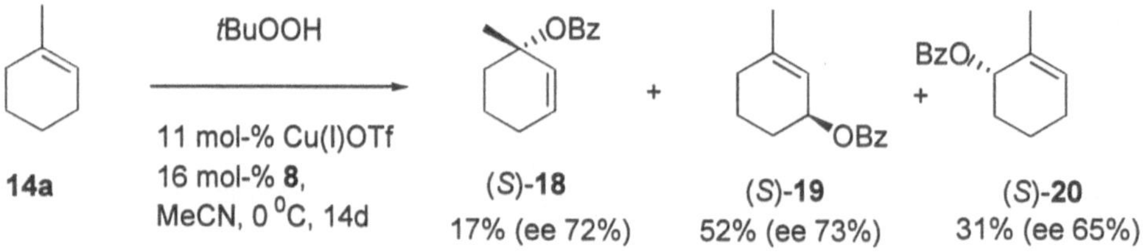

14a → **(S)-18** 17% (ee 72%) + **(S)-19** 52% (ee 73%) + **(S)-20** 31% (ee 65%)

*t*BuOOH, 11 mol-% Cu(I)OTf, 16 mol-% **8**, MeCN, 0 °C, 14d

(no reaction at -40 °C)

The peroxylation and the acyloxylation are competetive reactions, when mixtures of *t*BuOOH and carboxylic acids are used. It was shown in the oxidation of cyclohexene (**4**) that the product ratio (peroxide **5** / allylic acetate **21**) is influenced by the ligand as well as by the *t*BuOOH/ AcOH ratio (Table 2). The peroxylation reaction is favored at lower temperatures.

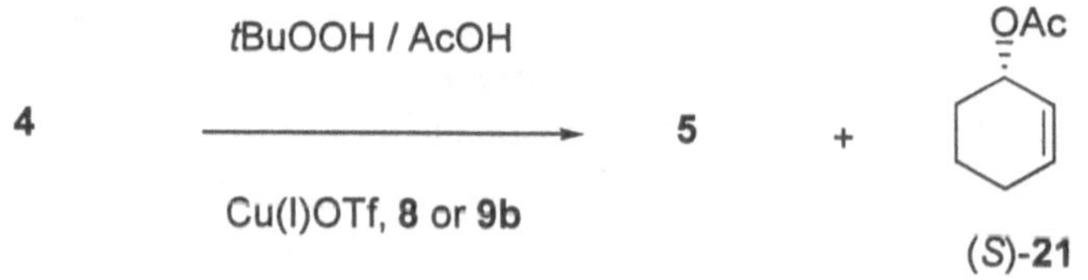

4 → **5** + **(S)-21**

*t*BuOOH / AcOH, Cu(I)OTf, **8** or **9b**

Table 2 Oxidation of Cyclohexene with *t*BuOOH / AcOH and Chiral Copper Complexes.

Ligand	Temp. [°C]	Ratio		Product Ratio	
		*t*BuOOH /	AcOH	5	(S)-21 (ee, %)[a)]
8	65	1	2.5	1	2.4 (66)
8	65	1	4	1	2.5 (41)
8	65	1.2	1	2.16	1 (15)
8	0	1	2.5	1	0
9b	65	1	2.5	1	1 (47)

a) Ee of **21** (chiral GC); the ee of **4** was < 4% in all cases.

The diastereoselectivity of the peroxylation reaction with the system *t*BuOOH/ Cu(I)OTf/ **8** was tested in the oxidation of (-)-α-pinene (**22**). The introduction of the peroxy group occured mainly *anti* to the dimethylsubstituted bridge (*exo* attack).

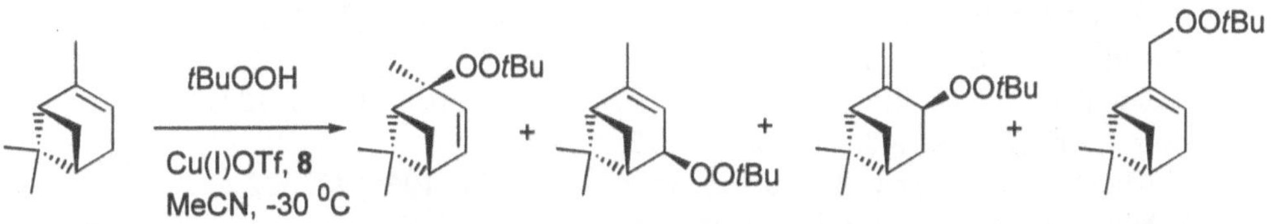

Our results represent the first examples for an asymmetric induction in the alkylperoxylation of allylic C–H bonds. It is evident that the chiral copper complex must be involved in the peroxylation step, the details are still unknown. It is well established that *t*BuO⁻, *t*BuOO⁻ and allylic radicals are intermediates in this oxidation [8, 10]. The formation of a Cu(III) allyl complex by a very fast reaction of Cu(II) and the allylic radical (k ≈ 10⁷-10⁹ l mol⁻¹ s⁻¹, com-pare ref. [12]) was postulated. The different results of the peroxylation and acyloxylation and the diastereoselectivity of the pinene oxidation are inconsistent with an intramolecular $S_{N2'}$-like peroxy-group transfer (path A; analogous to the "Beckwith-acyloxylation-mecha-nism" [13]) or with the external nucleophilic attack on a Cu(III) allyl complex (path B; ana-logous to the reaction of Pd(II) allyl complexes [14]).

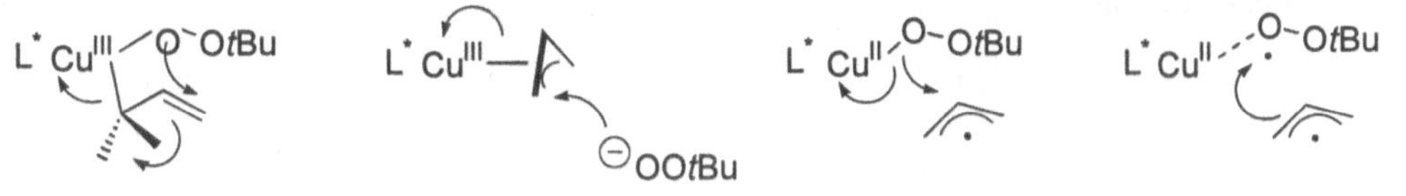

path A   path B   path C   path D

As alternative mechanisms, which avoid the formation of a labile Cu(III) intermediate, the radical attack of the allylic radical on a (peroxy)copper(II) complex (substitution reaction, path C; Cu(II)OOR complexes, stable below -20 ^{0}C, have been isolated [15]) or on a com-plexed peroxy radical (path D; metal-coordinated peroxy radicals were detected by ESR spectroscopy [16] and the important role of oxy radicals coordinated to Cu(II) in biochemistry, e.g. of *galactose oxidase*, is well established [17]) may apply.

Acknowledgement

This work was supported by the Deutsche Forschungsgemeinschaft (SFB 347 "Selektive Reaktionen Metall-aktivierter Moleküle") and the Fonds der Chemischen Industrie.

References

1. D.B. Denney, R. Napier, A. Cammarata *J. Org. Chem.* **1965**, *30*, 3151.
2. J. Muzart *J. Mol. Catal.* **1991**, *64*, 381.
3. A.S. Gokhale, A.B.E. Minidis, A. Pfaltz *Tetrahedron Lett.* **1995**, *36*, 1831.
4. T. Katsuki, K. Kawasaki *Tetrahedron* **1997**, *53*, 6337.
5. M.S. Kharasch, P. Pausen, W. Nudenberg *J. Org. Chem.* **1953**, *18*, 322.
6. M.S. Kharasch, A. Fono *J. Org. Chem.* **1958**, *23*, 324.
7. W. Treibs, G. Pellmann *Chem. Ber.* **1954**, *87*, 1201.
8. J.K. Kochi *Organometallic Mechan. and Catal.*, Academic Press, New York **1978**.
9. L. Saussine, E. Bruzi, H. Mimoun *J. Am. Chem. Soc.* **1985**, *107*, 3534.
10. F. Minisci, F. Fontana, S. Araneo, F. Rucepero, S. Banfi, S. Quici *J. Am. Chem. Soc.* **1995**, *117*, 226.
11. M. Schulz, R. Kluge, S. Liebsch, J. Lessig, M. Halik, F. Gadissa *Tetrahedron* **1996**, *52*, 13151.
12. H. Cohen, D. Meyerstein *Inorg. Chem.* **1987**, *26*, 2342.
13. A.L.J. Beckwith, A.A. Zavitsas *J. Am. Chem. Soc.* **1986**, *108*, 8230.
14. A. Pfaltz in *Stereoselective Synthesis*, E. Ottow, K. Schöllkopf, B.-G. Schulz (Eds.), Springer Verlag, Berlin Heidelberg **1993**, 15.
15. N. Katajima, T. Katayama, K. Fujisawa, Y. Iwata, Y. Moro-oka *J. Am. Chem. Soc.* **1993**, *115*, 7872.
16. R. Prikryl, A. Tkac, L. Malik, L. Omelka, K. Vesely *Coll. Czech. Chem. Commun.* **1976**, *40*, 104.
17. R.M. Wachter, M.P. Montague-Smith, B.P. Branchaud *J. Am. Chem. Soc.* **1997**, *119*, 7743 and refs. cit.

Catalytic Activation of Hydrogen Peroxide and Bistrimethylsilyl Peroxide for the Oxidation of Olefins and Aromatic Hydrocarbons

D. Kleinhenz, C. Jost, G. Wahl, J. Sundermeyer

Fachbereich Chemie der Philipps-Universität Marburg, Hans-Meerwein-Straße, D-35032 Marburg, Germany.

1 Introduction

The catalytic oxidation of various organic compounds is of outstanding importance in synthetic organic chemistry [1]. In particular the epoxidation of olefins and the oxidation of aromatic hydrocarbons attract considerable interest as the resulting epoxides and p-benzoquinones serve as important oxyfunctionalized building blocks in the synthesis of fine chemicals and biologically important natural products, e.g. 2-methyl-naphthoquinone (vitamin K_3).

We are concerned with the catalytic activation of hydrogen peroxide and its much less examined silyl analogue bistrimethylsilyl peroxide (BTSP) for the oxyfunctionalisation of organic substrates. The advantages of hydrogen peroxide are low price and the formation of ecologically harmless water as the reduction product. BTSP, which can be produced [2] and stored in pure form, exhibits an extraordinary thermal stability and a good solubility in organic solvents. Its inertness towards olefins and aromatics prevented a synthetic brake through in the use of this aprotic and save substitute for 100% hydrogen peroxide.

2 Results and discussion

As catalysts we tested d^0 oxo and peroxo complexes of group 6 and 7 metals - in particular of molybdenum, tungsten and rhenium. Our interest focussed on the influence of complex type, ligand nucleophilicity and lipophilicity and substitution pattern on the oxidation catalysis.

Besides Mimoun-type complexes $[MO(O_2)_2L_n]$ (M = Mo, W, n = 1,2) [3] we employed compounds of the type $[MO_2Cl_2L_n]$ (M = Mo, W, n = 1,2) as oxidation catalysts. Furthermore, in the activation of BTSP we tested the rhenium complexes $[Me_3SiOReO_3]$, $[CH_3ReO_3]$ and $[Re_2O_7L_2]$ [4].

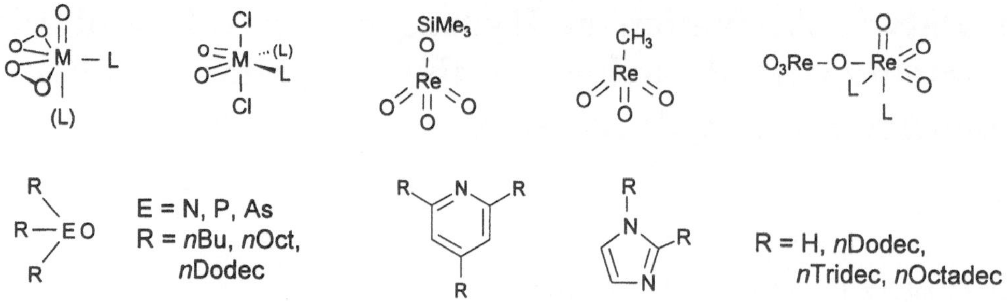

SiMe$_3$

CH$_3$

E = N, P, As
R = nBu, nOct, nDodec

R = H, nDodec, nTridec, nOctadec

Complex types and ligands employed in the activation of H$_2$O$_2$ and BTSP (M = Mo, W).

These metal complexes were stabilized by different neutral N- and O-donor ligands L. Among those were trialkylamine oxides, trialkylphosphine oxides, trialkylarsine oxides and a series of different alkylsubstituted pyridines, imidazoles and pyrazoles.

In order to overcome the low catalytic performance of Mimoune-type complexes in the activation of hydrogen peroxide we developed a coupled oxidation / phase transfer catalysis [5]. The organic phase serves as "oxidation zone" and contains the olefine, about 95% of the catalyst and the epoxide, which is easily separated and protected from hydrolysis. The aqueous phase contains the protic oxidant and serves as "loading zone". Depending on the lipophilicity of ligands the perhydrolysis of the metal oxo function is expected to take place at the phase boundary or in the aqueous phase. By applying tenside-like ligands for this extraction process into the non-aqueous phase, the activity of molybdenum and tungsten peroxides approaches that of the homogeneous methylrhenium trioxide / H$_2$O$_2$ / tBuOH oxidation system developed by Herrmann [6].

Such a surface active and fully characterized catalytically active species is
shown as follows:

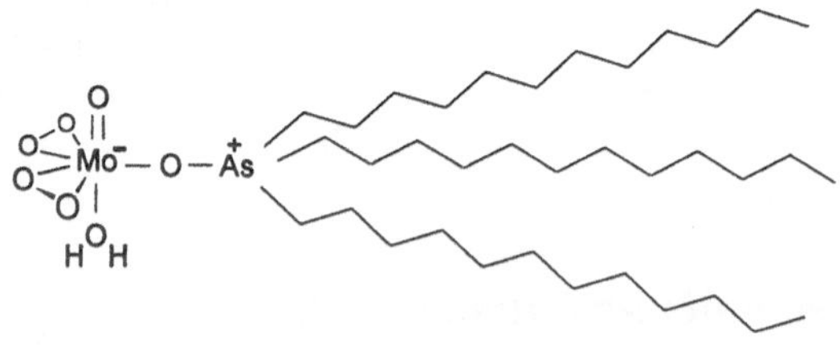

In contrast the catalytic activation of bistrimethylsilyl peroxide (BTSP) takes advantage of the perfect solubility of the oxidant in organic media. Prior to this study, there were no reports on the d^0 metal catalyzed activation of BTSP in the oxidation of olefins and aromatics [7]. A good catalyst is able to convert its oxo function into a peroxo function by elimination of siloxane, which is a bad ligand and does not inhibit the catalysis as water does. As a very strong oxenoid a silylperoxy intermediate is assumed to be involved:

$$O=MX_n \rightleftharpoons \ \text{(silylperoxide intermediate)} \ \rightleftharpoons \ \text{(peroxo complex)}$$

A series of different parameters significantly influences the oxidation catalysis. In particular the nature of the donor ligand, that stabilizes the oxidation catalyst, shows a remarkable effect on the catalyst activity. In this respect we could determine that in the two-phase oxidations [MO(O$_2$)$_2$L] complexes with trialkylarsine oxides - preferably with *n*-dodecyl substituents - were superior to those with corresponding trialkylamine oxides. The lowest activity was obtained with trialkylphosphine oxide complexes. This order was completely

Table 1 Activation of ROOR (R = H, SiMe$_3$)[a] for the oxidation of olefins and aromatic hydrocarbons

exp.	Substrate	product	catalyst	time (h)	cv. (%)	yi. (%)	m.b. (%)
1	(cyclooctene)	(cyclooctene oxide)	[WO$_2$Cl$_2$L(H$_2$O)][b]	26	78	76	98
2	(1-octene, C$_6$H$_{13}$)	(epoxide, C$_6$H$_{13}$)	[MoO(O$_2$)$_2$L][b]	24	67	63	96
3	(1-octene, C$_6$H$_{13}$)	(epoxide, C$_6$H$_{13}$)	[MoO(O$_2$)$_2$L][b]	24	55	52	97
4	(cyclopentene)	(cyclopentene oxide)	[WO$_2$Cl$_2$L$_2$][b]	24	100	95	95
5	(cyclooctene)	(cyclooctene oxide)	[MoO(O$_2$)$_2$L][b]	23	100	95	95
6	(1-octene, C$_6$H$_{13}$)	(epoxide, C$_6$H$_{13}$)	[Re$_2$O$_7$L$_2$][b]	3	76	71	95
7	(durene)	(quinone)	[Re$_2$O$_7$L$_2$][b]	2	100	58	58
8	(methylnaphthalene)	(naphthoquinone isomers)	[MeReO$_3$][b]	2	100	78	78
9	(anthracene)	(anthraquinone)	[Re$_2$O$_7$L$_2$][b]	2	100	54	54
10	(trimethoxytoluene, MeO)	(methoxyquinone, MeO)	[MoO$_2$Cl$_2$(dme)L][b]	18	84	33	49

[a] R = H (w = 30%; exp. 1-3), R = SiMe$_3$ (exp. 4-10). [b] L = ON(*n*Dodec)$_3$ (exp. 1), 3,5-Bis(*n*Tridec)pyridine (exp. 2), 2-(*n*Dodec)imidazole (exp. 3), OP(*n*Bu)$_3$ (exp. 4-10). [c] molecular ratio = ROOR : organic substrate : [catalyst] = 4 : 1 : 0.002 (exp. 1); 4 : 1 : 0.04 (exp. 2-3); 1 : 1 : 0.02 (exp. 4); 1 : 1 : 0.04 (exp. 5); 1 : 1 : 0.001 (exp. 6); 5 : 1 : 0.02 (exp. 7-9) and 3 : 1 : 0.02 (exp. 10); solvent = H$_2$O / CHCl$_3$ (exp. 1-3); CHCl$_3$ (exp. 4-10); temperature = 60°C; argon atmosphere (exp. 4-10). [d] conversion (cv.) and yield (yi.) were determined by product isolation (exp. 9) or by GC (Carlo Erba HRGC 5300, 25 m DB-5, 0.32 mm / 0.52 µm, internal standards); mass balance (m.b.).

reversed in the one-phase BTSP oxidations. In this case the trialkylphosphine oxide complexes reveal by far the best catalytic activity.

By the depicted methods cyclic and terminal olefins can be epoxidized in almost quantitative yield and very good selectivity with hydrogen peroxide or BTSP. Moreover it was found that some of the depicted d^0 oxo and peroxo catalysts are able to catalyze the oxidation of aromatic hydrocarbons to the corresponding p-benzoquinones with BTSP as oxidant. The yields and mass balances are similar compared with the well-known $MeReO_3$ / H_2O_2 system [8]. Some of the results of the oxidations are shown in table 1.

In summary we have found that appropriate ligand design can in fact make complexes of molybdenum and tungsten catalytically active in hydrogen peroxide activation by applying a phase transfer process. On the other hand bistrimethylsilyl peroxide is the oxidant of choice for the molybdenum, tungsten and rhenium catalyzed selective oxidations of olefins and aromatic hydrocarbons in homogeneous phase. Readily available and cheap dirhenium heptoxide which, due to hydrolysis, is catalytically inhibited in protic medium becomes highly active in aprotic medium in the presence of BTSP and phosphine oxides.

Acknowledgements

Financial support by the Deutsche Forschungsgemeinschaft (SFB 347 and Schwerpunktprogramm "Peroxidchemie") as well as fruitful collaborations with project C-2 (Prof. Kiefer), B-1 (Prof. Bringmann) and A-1 (Prof. Adam) are gratefully acknowledged.

4 References

[1] M. Hudlicky, *Oxidations in Organic Chemistry (ACS Monogr.)* **1990**.

[2] W. P. Jackson, *Synlett* **1990**, *9*, 536.

[3] H. Mimoun, I. S. de Roch, L. Sajus, *Bull. Soc. Chim. Fr.* **1969**, *5*, 1481.

[4] W. A. Herrmann, *J. Organomet. Chem.* **1995**, *500*, 149.

[5] M. Schulz, J. H. Teles, J. Sundermeyer, DE 195.33.331.4, **1995**; M. Schulz, J. H. Teles, J. Sundermeyer, G. Wahl, WO 10054, **1995**.

[6] W. A. Herrmann, R. W. Fischer, D. W. Marz, *Angew. Chem.* **1991**, *103*, 1706; *Angew. Chem. Int. Ed. Engl.* **1991**, *30*, 1638; b) W. A. Herrmann, R. W. Fischer, W. Scherer, M. U. Rauch, *Angew. Chem.* **1993**, *105*, 1209; *Angew. Chem. Int. Ed. Engl.* **1993**, *32*, 1157.

[7] The use of BTSP in metal catalyzed S-oxidation has been reported: W. Adam, D. Golsch, J. Sundermeyer, G. Wahl, *Chem. Ber.* **1996**, *129*, 1177.

[8] W. Adam, W. A. Herrmann, J. Lin, C. R. Saha-Möller, W. Fischer, J. Correia, *Angew. Chem.* **1994**, *106*, 2545; *Angew. Chem. Int. Ed. Engl.* **1994**, *33*, 2475.

Ti-TADDOLates and Related Complexes in Asymmetric Synthesis: Ring-Opening Reactions and Cu-Catalyzed Conjugate Additions of *Grignard* Reagents

Dieter Seebach*, Konstanze Gottwald, Georg Jaeschke, Laurent Audergon, Arkadius Pichota

Laboratorium für Organische Chemie, ETH Zürich, Universitätstrasse 16, CH - 8092 Zürich, Switzerland

1 Ti-TADDOLate Mediated Stereoselective Alkoxide Transfer

Ti-TADDOLates **1** have been used successfully as chiral *Lewis*-acids in many different types of reaction, for example in additions of diethylzinc to aldehydes, [2+2] cycloadditions, Diels-Alder and ene reactions [1,2].

Moreover, the combination of a *Lewis*-acidic and a nucleophilic functionality within the same molecule makes TADDOLates also suitable auxiliaries for an alkoxide transfer from the chiral ligand sphere of **1** to a *Lewis*-acid activated carbonyl group. This alkoxide transfer can be used for the preparation of enantiomerically pure carboxylic esters by "desymmetrization" of *meso*-compounds [3,4] as well as by kinetic resolutions [5,6].

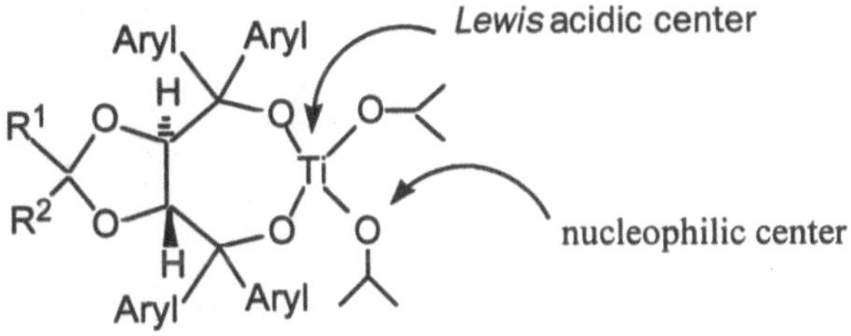

1 (**1a**: $R^1 = R^2 = CH_3$, Aryl = Ph)

1.1 Kinetic Resolution of Azlactones

Enantioenriched α-amino acids can be prepared *via* Ti-TADDOLate mediated kinetic resolution of azlactones with *in situ* racemization of the substrate. The resolution of different azlactones of type **2** leads to highly enantioenriched *N*-benzoylated phenylalanine derivatives **3**, which can be further enriched by one recrystallization (er > 98:2). Seven different azlactones have thus been enantioselectively converted to the corresponding *N*-benzoyl-phenylalanine analogs **3**. A substoichiometric version of the reaction is achieved by recycling the catalyst with isopropanol.

Enantioselective ring-opening of *meso*-anhydrides and -sulfonylimides by Ti-TADDOLates

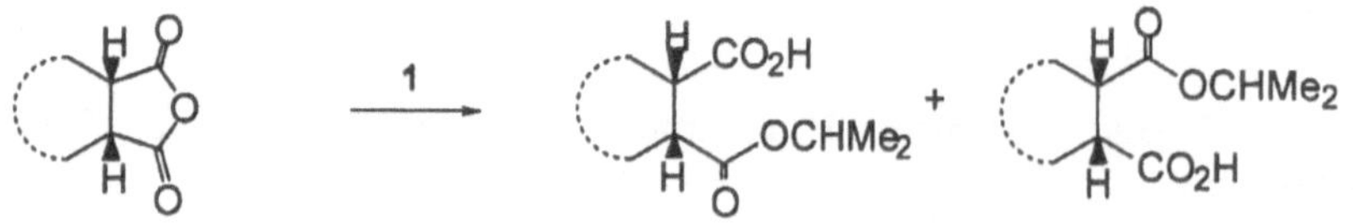

yield 73 to 92 %
er 95:5 to 99:1 [3]

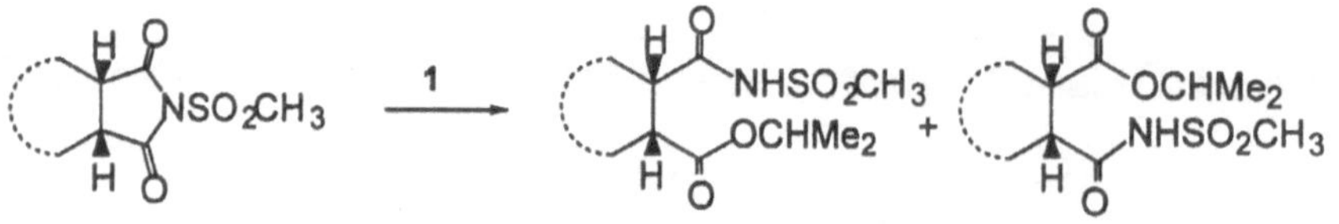

yield 63 to 92 %
er 87:13 to 97:3
after recrystallization : er > 99.5:0.5 [4]

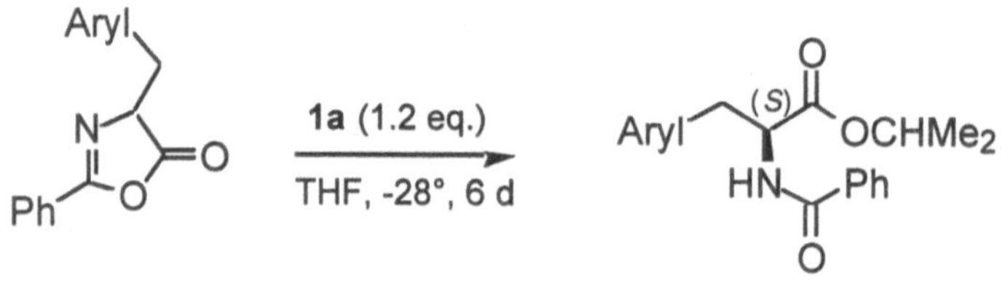

for Aryl = Ph
yield 75%, er 85:15
after recrystallization:
yield 25 %, er 98: 2

1.2 Enantiomer Differentiating Reaction of Ti-TADDOLate 1a with a Chiral Anhydride

Enantiomer differentiating reactions, in which different constitutional isomers are formed selectively from enantiomers, are rare in organic synthesis [7]. In previous publications, we have reported on the highly enantioselective ring-opening of cyclic *meso*-anhydrides [3] and the analogous enantiomer differentiating reaction of a chiral bicyclic anhydride [6]. A further example is the ring-opening of anhydride **4**. In a combination of kinetic resolution and regioselective nucleophilic attack it leads to enantioenriched half-esters **5** and **6** which are transformed to the corresponding amido-esters with (*S*)-1-(2-naphthyl)-ethylamine.

Nucleophilic attack occurs at the *Si*-carbonyl group of both enantiomers, giving rise to the same stereochemical course as with *meso*-anhydrides and -sulfonylimides.

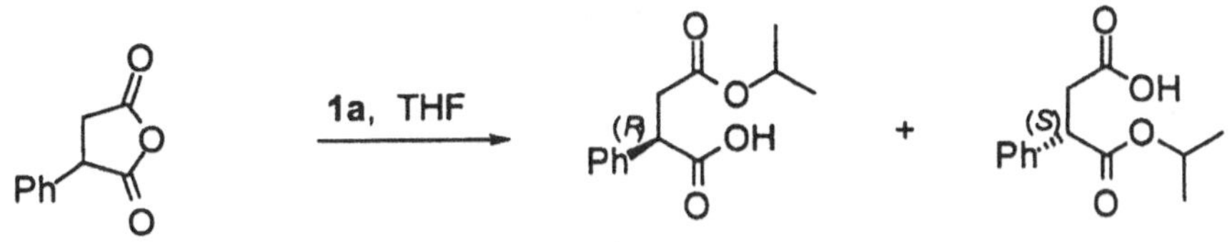

er and rs were determined after conversion to the amido-esters with	Temp.	5 er	6 er	5:6 rs (from HPLC)	Conversion
	rt	64:36	87:13	78:22	> 90%
	-28°	75:25	95:5	86:14	73%

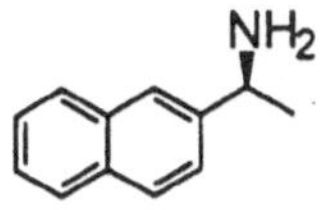

2 Ti-TADDOLates on Their Way to Late Transition Metals

As part of our research program directed towards TADDOL-derived ligands containing "soft" heteroatoms as the chelating units, we synthesized the thio-TADDOL-analogs **7** and **8** [8].

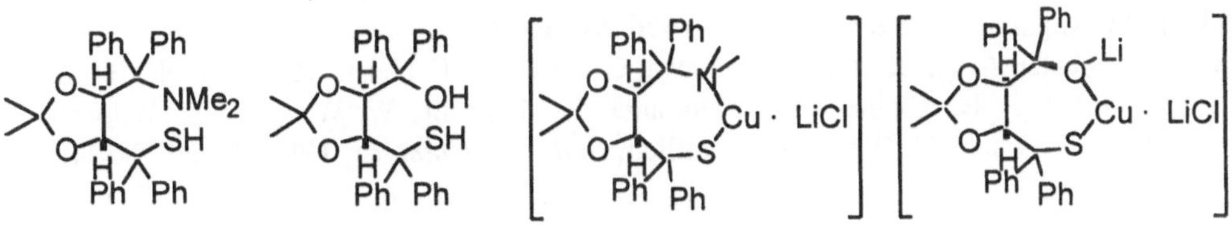

Following the concept of hard and soft bases and acids by *Pearson*, these ligands should prefer complexation with soft transition metals such as copper. In order to test the potential of **7** and **8** in asymmetric synthesis, we investigated the corresponding copper complexes **A** and **B** as catalysts in the enantioselective addition of *Grignard* reagents to cyclic enones.

A and B proved to be very effective catalysts for this reaction: the 1,4-additon products were formed exclusively and were isolated in high yields. The enantiomer ratios of the products were determined either by GC analysis or by ^{13}C-NMR analysis after conversion to the corresponding aminals with (1*R*,2*R*)- or (1*S*,2*S*)-diphenylethylendiamine.

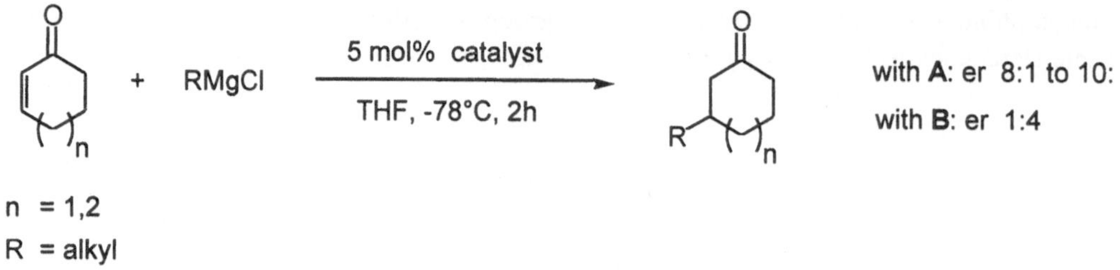

n = 1,2
R = alkyl

The copper complex **A** turned out to give higher enantiomer ratios than **B**. Surprisingly, the absolute configuration of the products was reversed on going from **A** to **B**, although both complexes have the same (*R,R*)-configuration.

3 References

[1] Synthesis of TADDOLs: A. K. Beck, B. Bastani, D. A. Plattner, W. Petter, D. Seebach, H. Braunschweiger, P. Gysi, L. LaVecchia, *Chimia* **1991**, *45*, 238-244.
[2] Applications of TADDOLates in asymmetric synthesis: a) R. Dahinden, A. K. Beck, D. Seebach, in: *Encyclopedia of Reagents for Organic Synthesis*; L. A. Paquette, Ed.-in-Chief; Wiley: Chichester, GB, 1995, 2167-2170 b) D. Seebach, R. Dahinden, R. E. Marti, A. K. Beck, D. A. Plattner, F. N. M. Kühnle, *J. Org. Chem.* **1995**, *60*, 1788-1799, and references cited herein.
[3] D. Seebach, G. Jaeschke, Y. M. Wang, *Angew. Chem.* **1995**, *107*, 2605-2606.
[4] D. Ramon, G. Guillena, D. Seebach, *Helv. Chim. Acta* **1996**, *79*, 875-894.
[5] K. Narasaka, F. Kanai, M. Okudo, N. Miyoshi, *Chem. Lett.* **1989**, 1187-1190.
[6] D. Seebach, G. Jaeschke, K. Gottwald, K. Matsuda, R. Formisano, D. A. Chaplin, M. Breuning, G. Bringmann, *Tetrahedron* **1997**, *53*, 7539-7556.
[7] H. B. Kagan, *Croatica Chem. Acta* **1996**, *69*, 669-680.
[8] D. Seebach, A. K. Beck, M. Hayakawa, G. Jaeschke, F. N. M. Kühnle, I. Nägeli, A. B. Pinkerton, P. B. Rheiner, R. O. Duthaler, P. M. Rothe, W. Weigand, R. Wünsch, S. Dick, R. Nesper, M. Wörle, V. Gramlich, *Bull. Soc. Chim. Fr.* **1997**, *134*, 315-331.

Transition Metal Peroxo Complexes as Versatile Tools in the Oxidative Ring Closure of Substituted 4-Pentenols - The Synthesis of Tetrahydrofurans and Tetrahydropyrans

J. Hartung*, P. Schmidt

Institut für Organische Chemie, Am Hubland, D-97074 Würzburg, Germany

1 Introduction

Multiply substituted tetrahydrofurans and tetrahydropyrans occur widely in nature. Several of these oxolane or oxixane derivatives bear heteroatom substituents in β-position to the ether oxygen. Prominent candidates which belong to this family of compounds range from relatively simple structures such as (+)-muscarine (**1**), one of the physiologically active constituents of the fly agaric [1], to rather complex polyether toxins, for instance thyrsiferol (**2**), which has been isolated from the marine alga *Laurencia thyrsiferia* [2]. Although the biosynthetic origins of these heterocycles are as diverse as their structures, many secondary metabolites of marine organisms are likely to originate from respective alkenols which are converted presumably in the presence of monooxygenase enzymes into the desired β-functionalized ethers [3].

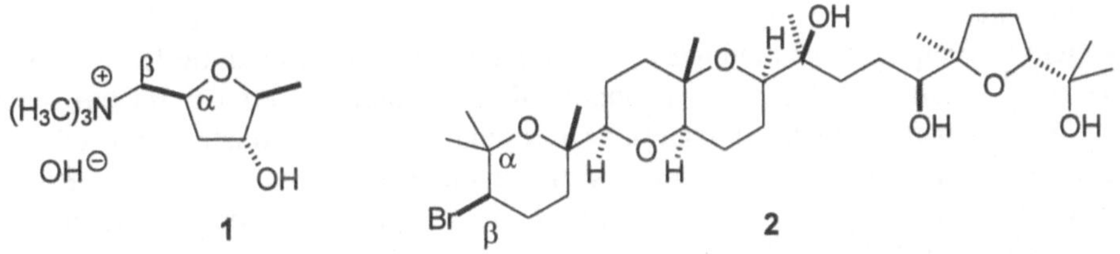

Figure 1 Naturally occurring β-functionalized oxolanes and oxixanes

The achievements of nature concerning the formation of cyclic ethers are still a challenge to synthetic organic chemistry. It is surprising to note that the seemingly simple task of complete stereo- and regiocontrol in oxidative ring closures of 4-pentenols to the respective *cis-* or *trans*-disubstituted tetrahydrofurans (**4** or **5**; Figure 2), or even to tetrahydropyrans (**6** or **7**) has not been satisfactorily solved yet. In order to address this question the present study on oxidative ring closures of substituted 4-pentenols was initiated. As principal methodology transition metal mediated oxidations were chosen in order to elaborate functional mimics for peroxidase enzymes as tools in synthetic organic chemistry.

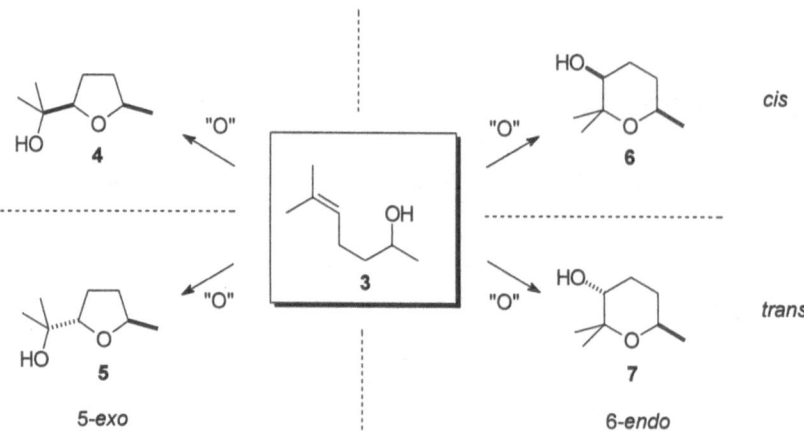

Figure 2 Modes of ring closures for oxidative transformations of 6-methyl-5-hepten-2-ol (**3**)

2 Theoretical Considerations

In order to gain insight into the thermochemistry of the desired synthetic transformations a detailed combined experimental and theoretical conformational analysis on tetrahydro-furans in general and the β-hydroxylated ethers **4-7** in particular was performed [4]. The assessment of computational methods available indicated that the ab initio HF/6-31G* level of theory is satisfactory for the description of the geometry *and* the relative heats of formation of oxolanes and of oxixanes. Our results show that oxidative ring closures of pentenol **3** to tetrahydrofurans **4,5** are slightly favoured by thermochemistry whereas the formation of tetrahydropyrans **6,7** should rely in the gas phase on processes which proceed under kinetic control (Table 1).

Table 1. Calculated relative heats of formation of heterocycles **4-7** (ab initio HF/6-31G*)

Compound	E_{tot} [Hartree][a,b]	ZPVE [Hartree][b,c]	E_{rel} [kJ/mol]
4 (*cis-5-exo*)	-461.73237	0.25144	1.84
5 (*trans-5-exo*)	-461.73307	0.25157	≡ 0.00
6 (*cis-6-endo*)	-461.72996	0.25169	8.17
7 (*trans-6-endo*)	-461.72709	0.25203	15.72

[a] Total energies (without temperature corrections) based on HF/6-31G* optimized structures including ZPVE corrections (unscaled) - [b] 1 Hartree = 2627.27 kJ/mol - [c] unscaled values

3 Conversion of Substituted Pentenols into Heterocyclic Compounds

Based on our computational results *trans*-2,5-disubstituted tetrahydrofurans should be accessible via oxidative ring closures under thermodynamic control. Thus, our model pentenol 1-phenyl-5-methyl-4-hexen-1-ol (**8**) was reacted in a two phase system ($H_2O/CHCl_3$) with a mixture of ammonium vanadate, hydrogen peroxide and potassium bromide. These experimental parameters were chosen for two reasons. First, Bartlett [5] had demonstrated that iodine cyclizations are reversible under comparable conditions. Next, we intended to mimic the synthesis of β-bromoethers comparable to biohalogenations by vanadium dependent bromoperoxidases [6]. Much to our surprise this reaction afforded the tetrahydropyran *trans*-**10** as the main cyclic ether along with minor amounts of tetrahyfurans **9** in a *cis-trans* ratio of 72 : 28.

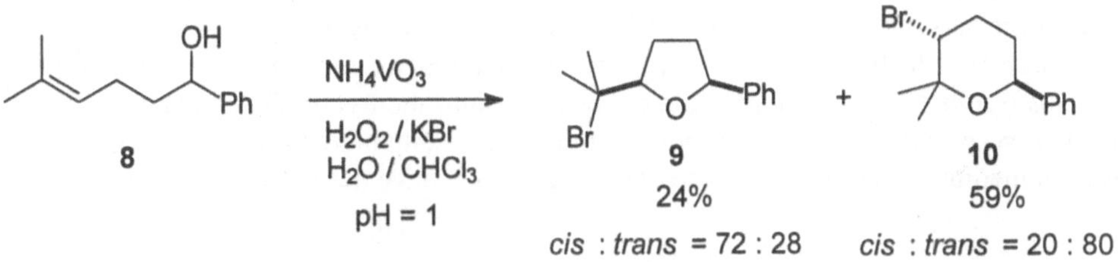

A more selective synthesis of *cis*-2,5-disubstituted tetrahydrofurans was readily achieved if vanadium(V) coordination compounds such as **11** [7] were employed in anhydrous media. The remarkable stereo- and regioselectivity of this procedure can be explained by coordination of the alkenol **8** to a vacant site of the transition metal which was followed by oxygen transfer from the η^2 bonded hydroperoxide ligand (epoxidation) and successive rearrangement via backside attack of the hydroxyl group.

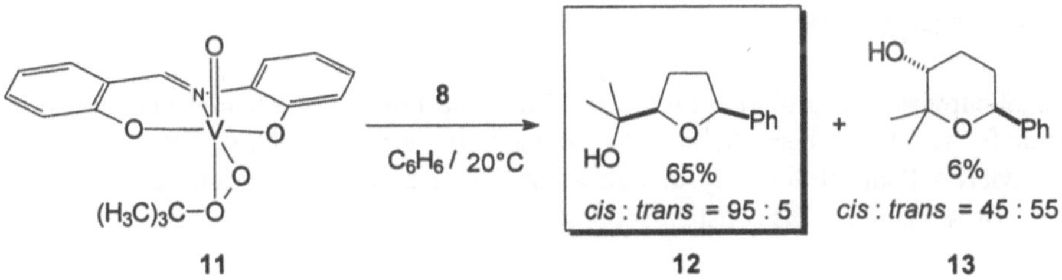

Molybdenum coordination compounds, such as $MoO_2(acac)_2$ (Hacac = pentane-2,4-dione) are able to convert the pentenol **8** in the presence of *tert*-butyl hydroperoxide (TBHP) into the 6-*endo* product **13** in 63% yield. The origins for this reversal in regiocontrol are not fully understood yet and are subject to further investigation.

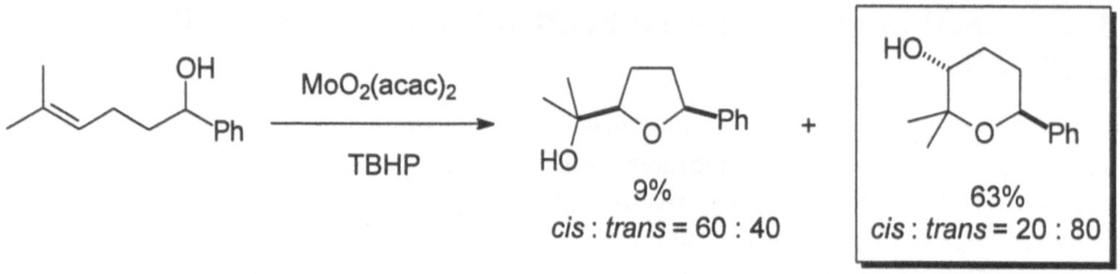

4 Enzymatic Transformations of 4-Pentenols using Chloroperoxidase

In order to extend our methodology towards an enantioselective synthesis of tetrahydrofurans or tetrahydropyrans, pentenol **8** was subjected to enzymatic reactions involving monooxygenases. Chloroperoxidase (CPO) from *Caldariomyces fumago* uses porphyrin bound iron as cofactor and is known to act as monooxygenase towards xenobiotic substrates [8]. This enzyme was efficiently able to convert **8** into the hydroxylated ethers **12,13** using TBHP as secondary oxidant and a potassium bromide containing citrate buffer (pH = 4.8) as reaction medium. Enantiomeric excesses of **12** and **13** have not been determined yet.

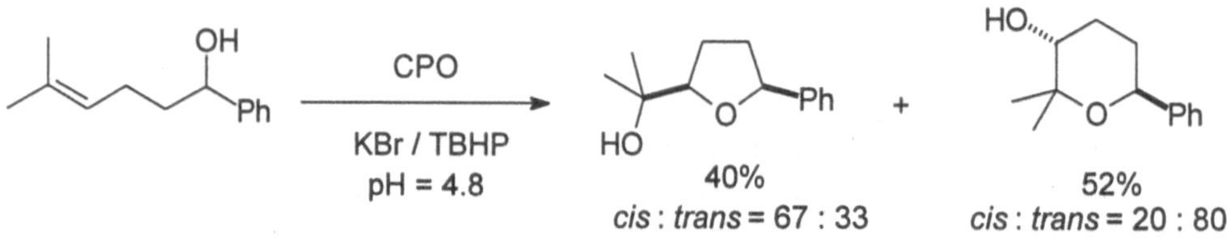

Acknowledgement

Financial support was provided by the DFG and the Fonds der Chemischen Industrie. We are grateful to Dipl.-Chem. Miriam Münchbach for recording GLC/MS-spectra, to LM-Chem. Marko Schmitt for helpful discussions on enzymatic reactions and to Prof. Dr. Gerhard Bringmann for his constant support.

References

[1] P. C. Wang, M. M. Joullié in *The Alkaloids* (Ed. A. Brossi), Academic Press, New York, **1984**, *23*, pp 327-380.
[2] J. W. Blunt, M.P. Hartshorn, T. J. McLennan, M. H. G. Munro, W. T. Robinson und S. C. Yorke, *Tetrahedron Lett.* **1978**, 69-72.

[3] M. J. Garson, *Chem. Rev.* **1993**, *93*, 1699-1733.

[4] H. Fueß, J. Hartung, P. Schmidt, I. Svoboda, *Eur. J. Org. Chem.*, in preparation.

[5] P. A. Bartlett in *Asymmetric Synthesis* (Ed. J. D. Morrison), Academic Press, New York, **1984**, pp 411-453.

[6] V. R. Hedge, G.C.G. Pais, R. Kumar, P. Kumar und B. Pandey, *J. Chem. Res. (S)* **1996**, 62-63.

[7] H. Mimoun, M. Mignard, P. Brechot und L. Saussine, *J. Am. Chem. Soc.* **1986**, *108*, 3711-3718.

[8] H. L. Holland, *Organic Synthesis with Enzymes,* VCH, Weinheim, **1992**, pp 153-231.

Kinetic Resolution of Dihydronaphthalenes by Jacobsen Epoxidation: Evidence for Polar Intermediates and Structure Elucidation by Calculation of CD Spectra

Torsten Linker*, Frank Rebien, Jürgen Kraus, and Gerhard Bringmann*

Institute of Organic Chemistry, University of Würzburg, Am Hubland, D-97074 Würzburg, Germany

The epoxidation of double bonds with optically active (salen)manganese(III) complexes (Jacobsen epoxidation) has become one of the most attractive methods in asymmetric synthesis [1]. Although many applications for various unfunctionalized olefins exist in literature, the influence of polar substituents at the allyl position remains unexplored till now. During the course of our studies on the synthesis of podophyllotoxin derivatives [2], we examined the kinetic resolution of the dihydronaphthalenes rac-2. Indeed, high selectivities (k_{rel} = 8.6) were obtained with the catalyst (S,S)-1a (Table 1), and the desired epoxide (+)-3 was isolated in enantiomerically pure form after one recrystallization [3]. The absolute configurations of all products were established by comparison of experimental and calculated CD spectra, a method which was successfully applied for other optically active compounds [4].

Interestingly, the reaction with m-chloroperbenzoic acid (MCPBA) affords one epoxide 3 exclusively (entry 1), whereas product mixtures were obtained in the presence of the catalyst 1a (entries 2-4).

Table 1 Jacobsen epoxidation of racemic dihydronaphthalenes 2.

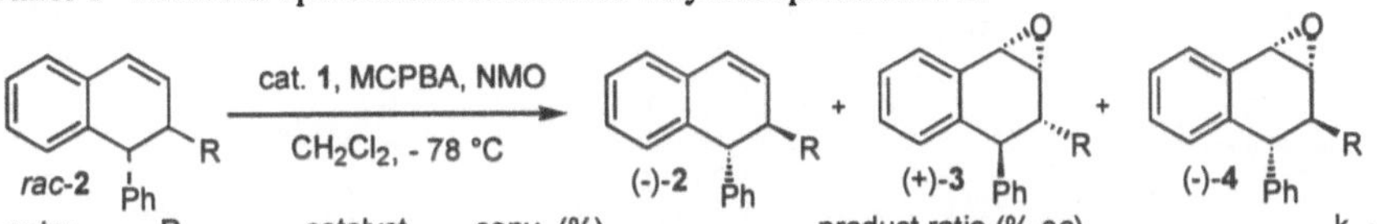

entry	R	catalyst	conv. (%)	product ratio (% ee)			k_{rel}
1	CO$_2$Me	-	>98	<2 : >96 (rac)	:	<2	-
2	CO$_2$Me	1a	26	74 (20) : 20 (93)	:	6	8.6
3	CO$_2$Me	1a	81	19 (99) : 44 (75)	:	37	6.4
4	CO$_2$H	1a	24	76 (8) : 3	:	21	1.8
5	CO$_2$H	1b	>98	<2 : <2	:	>96 (rac)	-

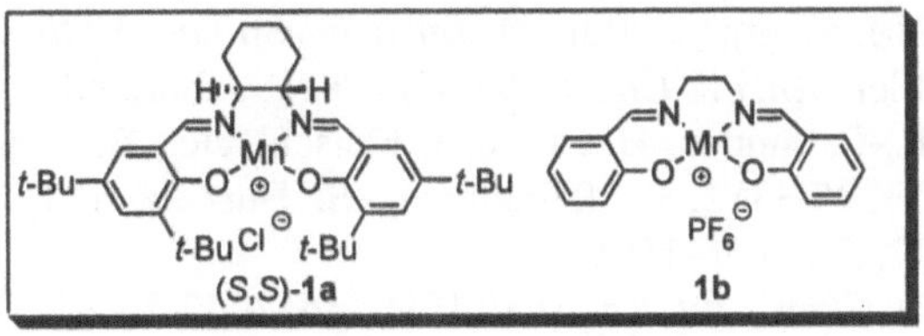

This result can best be rationalized by manganaoxetane intermediates during the Jacobsen epoxidation (Scheme 1). Thus, only steric interactions are operative in intermediate **A**, whereas electrostatic repulsions disfavor manganaoxetane **B**. Indeed, the reaction of the carboxylic acid (R = CO_2H) affords the epoxide **4** as the major product, since polar interactions become predominant, which demonstrates the importance of the allylic substituent in the Jacobsen epoxidation. Finally, the sterically less demanding achiral catalyst **1b** yields the epoxide **4** as the sole product (entry 5).

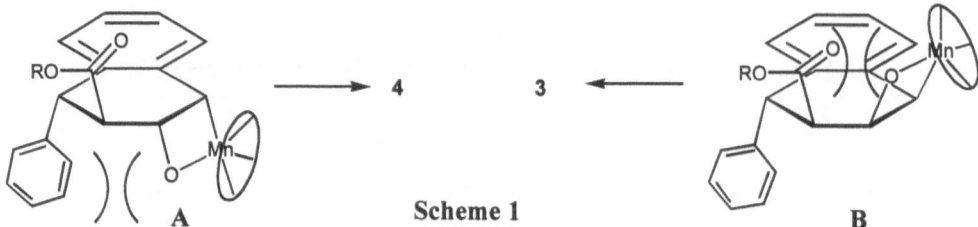

Scheme 1

In conclusion, the kinetic resolution of dihydronaphthalenes affords enantiomerically pure epoxides, which can serve as precursors for podophyllotoxin analogues. A strong influence of allylic substituents on the stereoselectivities was observed, which provides evidence for polar manganaoxetane intermediates. These results are not only important for the mechanistic understanding of the Jacobsen epoxidation [5], but allow the controlled diastereoselective synthesis of both epoxides **3** and **4** by simply adding an achiral catalyst **1b** (entry 1 vs. 5).

Acknowledgement

This work was generously supported by the Deutsche Forschungsgemeinschaft (SFB 347 "Selektive Reaktionen Metall-aktivierter Moleküle", Li 556/2-2, and a Heisenberg fellowship for T. L.).

References

[1] W. Zhang, J. L. Loebach, S. R. Wilson, E. N. Jacobsen, *J. Am. Chem. Soc.* **1990**, *112*, 2801-2803; E. N. Jacobsen in *Catalytic Asymmetric Synthesis* (Ed. I. Ojiama), VCH, Weinheim, **1993**, pp 159-202; T. Katsuki, *J. Mol. Catal.* **1996**, 87-107.
[2] T. Linker, K. Peters, E.-M. Peters, F. Rebien, *Angew. Chem., Int. Ed. Engl.* **1996**, *35*, 2487-2489; T. Linker, F. Rebien, G. Tóth, *J. Chem. Soc., Chem. Commun.* **1996**, 2585-2586; T. Linker, M. Maurer, F. Rebien, *Tetrahedron Lett.* **1996**, *37*, 8363-8366.
[3] T. Linker, F. Rebien, *Angew. Chem., Int. Ed. Engl.*, submitted.
[4] G. Bringmann, K.-P. Gulden, H. Busse, J. Fleischhauer, B. Kramer, E. Zobel, *Tetrahedron* **1993**, *49*, 3305-3312; G. Bringmann, S. Busemann, K. Krohn, K. Beckmann, *Tetrahedron* **1997**, *53*, 1655-1664.
[5] T. Linker, *Angew. Chem., Int. Ed. Engl.* **1997**, *36*, 2060-2062.

One-Electron Oxidation of Metal Enolates

Armin Burghart, Andreas Haeuseler, Anja Langels, Rolf Söllner, Michael Schmittel*

Institut für Organische Chemie der Universität Würzburg, Am Hubland,
D-97074 Würzburg, Germany

During the last decades metal enolates of the general formula M-O-C=C have adopted an important standing as versatile synthetic carbon nucleophiles in stereoselective aldol and Michael reactions.[1] In contrast, the umpolung of metal enolates through one-electron oxidation is completely unknown for many metals such as M = Ti, Zr, etc. although the resulting radical cations should offer interesting electrophilic features that could be exploited in carbon-carbon bond formation processes, quite in analogy to the recent chemistry of silyl and stannyl enol ether radical cations. Herein, we report on the first characterization of titanium and zirconium enolate radical cations in solution, the clean formation of metallocene cations through mesolytic cleavage and a new strategy for stereoselective carbon-carbon bond formation through redox umpolung in titanium and silicon bisenolates.

1   Formation of metallocene cations

1.1   Preparation of isolable titanium and zirconium enolates

Hitherto, titanium and zirconium enolates have been mostly prepared *in situ* and only isolated in a few cases as a result of their high sensitivity against hydrolysis.[2] In order to gain some information about the radical cations of such enolates, we prepared a series of isolable titanium and zirconium enolates with exceptional stability towards hydrolysis. We reasoned that metal enolates derived from stable β,β-diarylenols[3] should exhibit an increased stability due to the electronic stabilization and steric shielding by the two aryl units in the β-position, as compared to those generated from simple carbonyl compounds.

   The reaction of various sterically congested sodium enolates, generated by quantitative deprotonation of stable enols,[3] with titanocene dichloride, zirconocene dichloride or dimethyl zirconocene afforded a series of isolable titanium and zirconium enolates 1-10.[4,5]

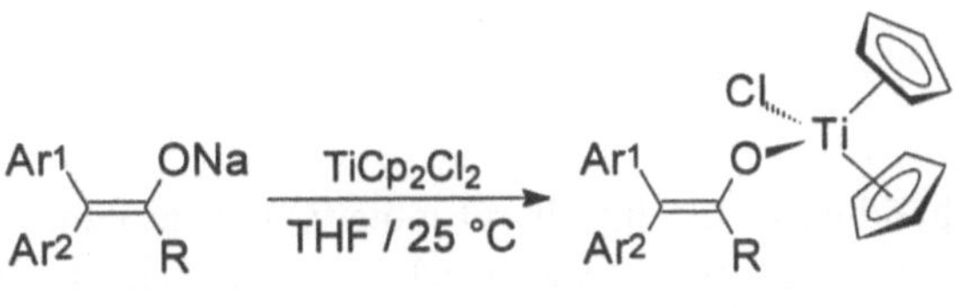

Ar1	Ar2	R		yield
Mes	Mes	H	1	81%
Mes	Mes	'Bu	2	62%
Mes	Mes	Ph	3	62%
Mes	Ph	H	4	32%

Scheme 1 Synthesis of Titanium Enolates

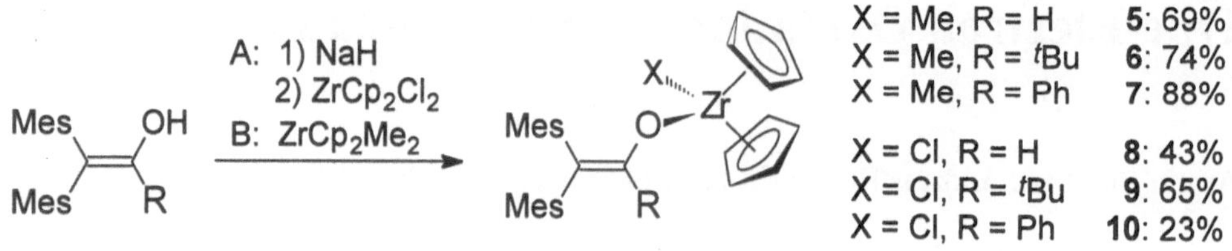

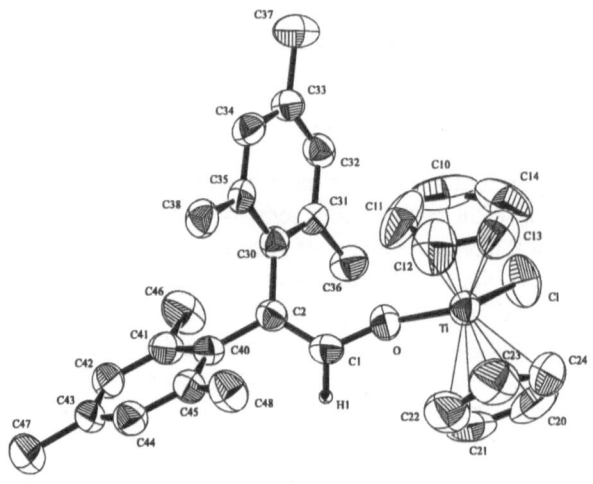

Scheme 2: Synthesis of Zirconocene Enolates

The crystal structure of **1** could be determined.[6] Notably, the mesityl rings of **1** are not coplanar with the central C-C double bond, but form dihedral angles of 57.7(3)° and 59.8(4)°, resulting in an almost perpendicular [83.1(2)°] arrangement of the two mesityl groups. The bond lengths C1-C2 [1.336(5) Å] and C1-O [1.328(5) Å] are typical for the enolato moiety. The geometry about the titanocene unit is undisturbed with even the O-Ti-Cl angle [96.5(1)°] being within the usual range observed for TiCp$_2$X$_2$ structures. Thus, the high steric congestion about the enolato moiety does not convey a compression and deformation of the TiCp$_2$Cl subunit. The large Ti-O-C1 angle in combination with the Ti-O bond length suggests a comparatively high Ti-O multiple bond order.

Figure 1 ORTEP Plot of **1**

Due to the considerable steric shielding of the β,β-diaryl moiety, all the titanium enolates exhibit an outstanding stability towards hydrolysis, which increases with the higher steric demand of the substituents at the C-C double bond. The kinetics of the hydrolysis, which is pseudo-first-order in THF/water (1 : 1) and acetonitrile/water (1 : 1), was investigated by UV spectroscopy. The pseudo-first-order rate constants measured in these solvent mixtures are in the range of $6.4 \cdot 10^{-4}$ s^{-1} < k_1 < $1.1 \cdot 10^{-3}$ s^{-1}. For comparison, the hydrolysis of sterically unshielded titanium enolates towards hydrolysis is about 1000 times faster.[6]

1.2 First characterization of titanium and zirconium enolate radical cations and their mesolytic O-M bond cleavage to metallocene cations

With pure titanium and zirconium enolates at hand we were able to characterize for the first time their radical cations and investigate their follow-up reactions.[4,5] Depending on the kinetic stability of the enolate radical cations it was possible to monitor reversible waves using standard or fast scan cyclic voltammetry. In table 1 the reversible half-wave potentials are depicted along with some data on the kinetic stability of the intermediate radical cations (k_f). In addition, to unequivocally establish the nature of these intermediates we have recorded the EPR spectra of some of the enolate radical cations.[7]

Table 1 Data and One-Electron Oxidation Results of Some Selected Metal Enolates

	2	3	6	9	10
$E_{\frac{1}{2}}$ [V$_{Fc}$]	+ 0.26	+ 0.31	+ 0.34	+ 0.43	+ 0.44
k_f [s^{-1}] (in CH$_2$Cl$_2$)	$1.1 \cdot 10^{-1}$	$3.6 \cdot 10^{-2}$	$5.0 \cdot 10^{2}$	$3.3 \cdot 10^{2}$	$3.1 \cdot 10^{2}$
Yield of Benzofuran	95% of **12**	86% of **13**	60% of **12**	90% of **12**	88% of **13**

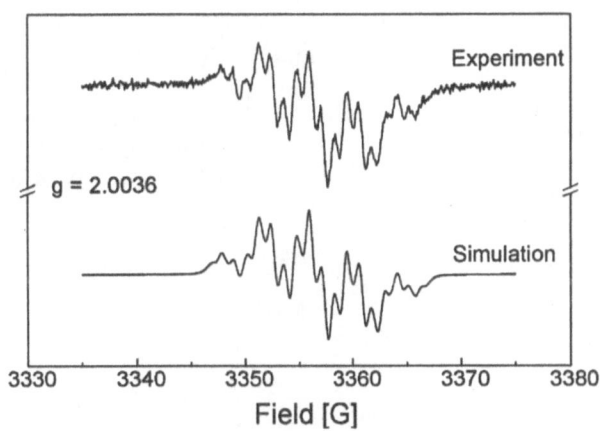

Figure 1. EPR spectrum of titanium enolate radical cation **4**$^{\bullet+}$ recorded in dichloromethane at -100 °C. Digital simulation is in good agreement with the spectrum obtained from the EPR experiment. Since no spin density resides on the titanocene fragment titanium enolate radical cations are electrophilic radicals which react with nucleophiles most likely at the β-carbon. Although the mesityl is more electron-rich than the phenyl group, spin density is higher on the phenyl moiety.

When the metal enolates were reacted with 200 mol% of a defined one-electron oxidant ([Fe(phen)$_3$](PF$_6$)$_3$ or NOSbF$_6$) the corresponding benzofurans were obtained in good yield (see table 1).[5,7]

11: R = H
12: R = ^{t}Bu
13: R = Ph

1-3, 5-10   11-13

Scheme 3 One-Electron Oxidation of Metal Enolates.

According to mechanistic investigations[8] formation of the benzofuran is initiated by a mesolytic O-M bond cleavage process on the stage of the metal enolate radical cation. The latter can either cleave to the α-carbonyl radical and a metallocene cation (mechanism A, scheme 4) or α-carbonyl cation and a metallocene radical (mechanism B). Since the oxidation potentials of α-carbonyl radicals are higher than those of the metallocene radicals,[9] the cleavage can be predicted to proceed according to mechanism A.

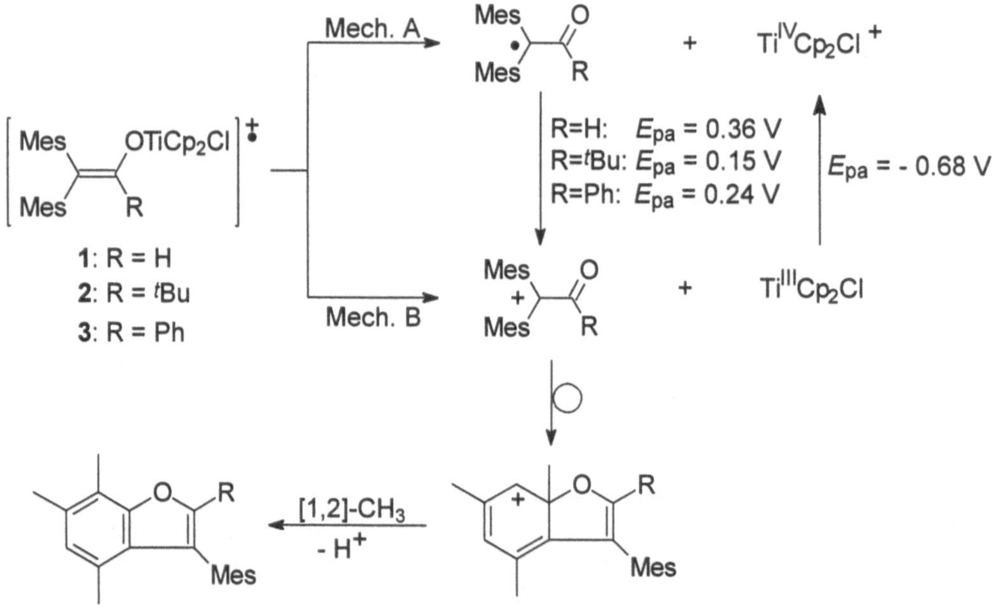

1: R = H
2: R = ^{t}Bu
3: R = Ph

R=H: E_{pa} = 0.36 V
R=^{t}Bu: E_{pa} = 0.15 V
R=Ph: E_{pa} = 0.24 V

E_{pa} = - 0.68 V

[1,2]-CH$_3$
- H$^+$

Scheme 4 Mechanism of Benzofuran Formation and Mesolytic Bond Cleavage Modes.

76

The above mesolytic bond cleavage reaction of titanium and zirconium enolate radical cations is currently probed as alternative approach to titanocene and zirconocene cations, which serve as active catalysts in the Ziegler-Natta polymerization of olefins.

2   Stereoselective carbon-carbon bond formation

Although the oxidative coupling of enolates and other enol derivatives has been exploited over many years as a valuable route to 1,4-dicarbonyl compounds, three severe limitations have prevented its wider use in organic synthesis up to now: i) diastereoselectivity is low, ii) enantioselective procedures have not been developed yet and iii) cross coupling between two different enol derivatives cannot be achieved in high yield.[10]

While these aspects have been partly solved in singular cases, we now wish to report on the elaboration of a novel *useful concept* (scheme 5) that may prove to be applicable for a wider variety of substrates since diastereoselectivity arises from intramolecularizing the oxidative carbon-carbon bond formation: The two enolate units are covalently linked to the same tether M and after umpolung of one of them through single electron transfer oxidation (SET) the above reaction sequence to 1,4-diketones takes place.[11]

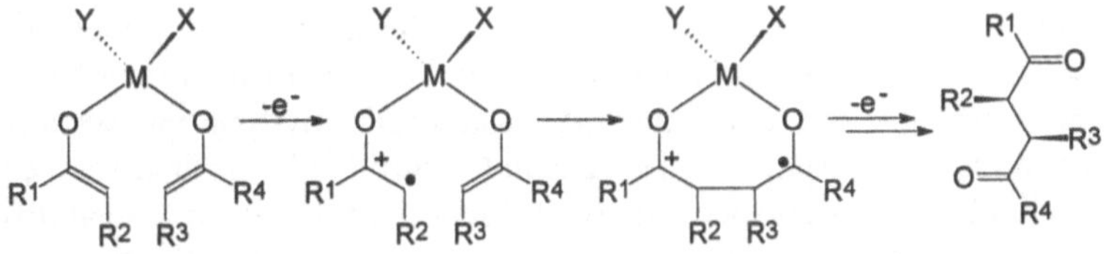

Scheme 5 The Concept.

We have decided to probe this conception with M = Si, Ti and hence have synthesized a series of various silicon and titanium bisenolates using standard procedures. Chemical oxidation proceeded smoothly at room temperature within 1 min in case of the titanium bisenolates **14,15** and over 12 h for the silicon bisenolates **18,19** to afford the diketones **16,17** in acceptable to good yields (table 2). The slower reaction with **18,19** is readily understandable in light of their much higher oxidation potentials. Rewardingly, diketone d,l-**17** was formed in high diastereomeric excess (de = 82%) from **19** supporting the utility of our strategy.

First results with **14,15,18,19** proposed to continue with other silicon bisenolates and to increase the steric bulk about the silicon tether as was realized in **20** and **21**. As with **15,19** all the enolate units in **20-22** exhibited a Z-configuration as derived from their ^{1}H NMR spectra and as further corroborated from a X-ray structure analysis on **20**.[11]

Interestingly, with the sterically bulkier substrates **20,21** the simple diastereoselectivity in the oxidative diketone formation could be improved up to de = 97% (table 2). The mediocre yield of **17** from **21** (given on a molar basis) indicates, however, that not all 4 enolate groups in **21** can be used for the diketone formation. Hence, it will be of importance to further probe the concept of steric bulk to increase diastereoselectivity in bisenolate cyclizations using two unoxidizable groups R at the silicon.

Table 2 Oxidation of titanium and silicon bisenolates using various SET oxidants.[11]

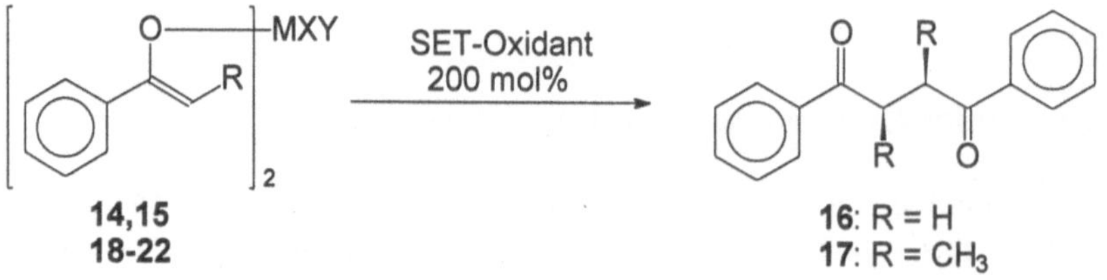

bisenolate	R	MXY	E_{pa} [V_{Fc}]	SET oxidant	product (yield)	de
14	H	TiCp$_2$	n.d.	[Fe(phen)$_3$](PF$_6$)$_3$	**16** (17%)	--
15	CH$_3$	TiCp$_2$	0.35	[Fe(phen)$_3$](PF$_6$)$_3$	**17** (39%)	0%
18	H	SiMe$_2$	1.29	[Fe(phen)$_3$](PF$_6$)$_3$	**16** (48%)	--
19	CH$_3$	SiMe$_2$	1.11	[Fe(phen)$_3$](PF$_6$)$_3$	**17** (57%)	82%
20	CH$_3$	SiMe[O-Prop]	1.13	Ce(NH$_4$)$_2$(NO$_3$)$_6$	**17** (60%)	97%
21	CH$_3$	Si[O-Prop]$_2$	1.16	Cu(OTf)$_2$	**17** (44%)	95%
22	CH$_3$	Ti[(+)-DET]	n.d.	Cu(OTf)$_2$	**17** (n.d.)	69%

[(+)-DET] = (+)-diethyl tartrate, [O-Prop] = [(Z)-(Me)HC=C(Ph)O].

From our investigations a rather clear picture of the mechanism has emerged. Accordingly, carbon-carbon bond formation takes place at the stage of the bisenolate radical cation and it seems reasonable to propose that the diastereoselectivity in the oxidative coupling of the silicon and titanium bisenolates **19-22** is not controlled by ground-state conformational preferences but by transition state energy differences of the two diafacial differentiating approaches. Indeed, simple transition state model considerations favor the *lk* approach of two enolate moieties in the bisenolate radical cation over the *ul* approach (scheme 6) since steric repulsion of the eclipsed methyl and phenyl substituents can be avoided in the former.

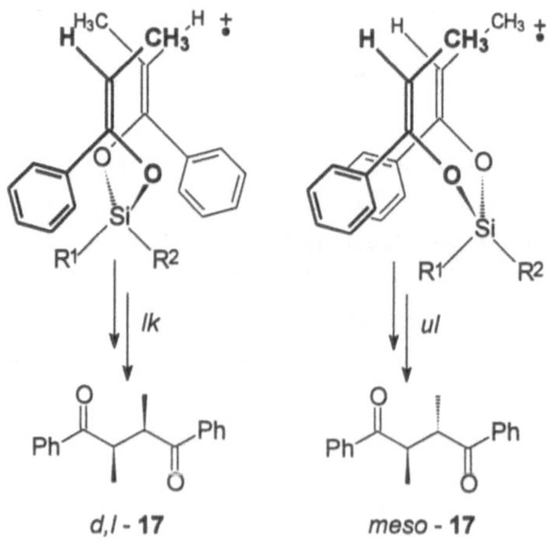

Scheme 6 Cyclization of the bisenolate radical cation.

78

In conclusion, we have been able to demonstrate that metal enolate radical cations offer interesting follow-up reactions, either mesolytic bond cleavage to metallocene cations or stereoselective carbon-carbon bond formation.

References

1. M. B. Smith, *Organic Synthesis*, Chapter 9, McGraw Hill, Singapore **1994**; I. Paterson, in *Comprehensive Organic Synthesis*; C. H. Heathcock, Ed. , Vol. *2*, 301, Oxford **1991**.
2. M. T. Reetz, *Organotitanium Reagents in Organic Synthesis*, Springer, Berlin **1986**.
3. M. Schmittel, A. Langels, *Liebigs Ann.* **1996**, 999; Z. Rappoport, S. E. Biali, *Acc. Chem. Res.* **1988**, *21*, 442.
4. M. Schmittel, R. Söllner, *Angew. Chem. Int. Ed. Engl.* **1996**, *35*, 2107.
5. M. Schmittel, R. Söllner, unpublished results.
6. M. Schmittel, H. Werner, O. Gevert, R. Söllner, *Chem. Ber.* **1997**, *130*, 195.
7. M. Schmittel, R. Söllner, *Chem. Ber.* **1997**, *130*, 771.
8. M. Schmittel, H. Trenkle, *Chem. Lett.* **1997**, 299; M. Schmittel, J.-P. Steffen, A. Burghart, *Chem. Commun.* **1996,** 2349; M. Schmittel, A. Burghart, M. Keller, *J. Chem. Soc., Perkin Trans.* 2, **1995**, 2327.
9. M. Röck, M. Schmittel, *J. Chem. Soc., Chem. Commun.* **1993**, 1739.
10. C. Alvarez-Ibarra, A. G. Csákÿ, B. Colmenero, M. L. Quiroga, *J. Org. Chem.* **1997**, *62*, 2478; T. Langer, M. Illich, G. Helmchen, *Synlett* **1996**, 1137 and cited references.
11. M. Schmittel, A. Burghart, W. Malisch, J. Reising, R. Söllner, *J. Org. Chem.*, submitted for publication.

C-C and C-H Bonds

New Explorations in Metal-Catalyzed Reactions

Peter M. Maitlis, Michael L. Turner, Ruhksana Quyoum, Helen C. Long and
Anthony Haynes

The Department of Chemistry, The University of Sheffield, Sheffield S3 7HF, U.K.

Abstract

The carbonylation of methanol to acetic acid homogeneously catalyzed in solution by
rhodium or iridium iodides is used to illustrate the methods we have used to quantify a
catalytic process in terms of a simple cyclic series of stoichiometric organometallic
reactions. Similar principles are now being applied to the study of the hydrogenation of
carbon monoxide to linear alkenes and alkanes (Fischer-Tropsch reaction) under *heterogeneous* conditions. To understand the polymerisation process, we first developed model
dirhodium organometallic systems to define the pathway for propene production from the
combination of two methylenes and a methyl. Labelling studies led to the postulate of a
Rh-σ-vinyl intermediate and reasons for this were proposed. Vinyl-metal complexes play
important roles in many chemical processes and we have applied these ideas to the
hydrogenation of carbon monoxide over rhodium. Addition of C_2 labelled vinylic probe
molecules to CO hydrogenation showed that vinylic species initiate the polymerisation of
surface methylenes in the Fischer-Tropsch reaction and offer a new and quantitative
understanding of the chemistry underlying the process.

The development of catalytic cycles in homogeneous catalysis

The fundamental rules governing the stoichiometric reactions of organo-transition metal
chemistry have for some time also been used to understand and quantify reactions of
organic substrates catalyzed by metal complexes in *homogeneous* solution.

Thus, for example, a typical catalytic process can be analysed into a cyclic series of
stoichiometric reactions, each of which can in principle be measured and quantified. The
reaction which is the slowest determines the rate of the overall process. This method of
working was used by Halpern to define hydrogenation reactions catalyzed by Wilkinson's
catalyst [1].

The Sheffield group applied this methodology to the industrially important carbonylation of methanol to acetic acid which is catalyzed by Group 8 iodides. The rhodium
catalyzed reaction was first developed by the Monsanto company and the cycle (Figure 1)

proposed by Forster and his coworkers [2]. It is composed of four organometallic and two organic reactions. The organometallic processes are:

i) The **oxidative addition** of methyl iodide to a low valent rhodium centre (**A** to **B**) is relatively slow (and hence can be rate-determining),

$$[Rh(CO)_2I_2]^- + MeI \rightarrow [MeRh(CO)_2I_3]^-$$

ii) In the **migratory insertion** (**B** to **C**) the methyl migrates onto the carbonyl ligand. This is a fast step for rhodium,

$$[MeRh(CO)_2I_3]^- \rightarrow [MeCORh(CO)I_3]^-$$

iii) In the **carbonylation** step (**C** to **D**) the carbon monoxide is introduced into the cycle,

$$[MeCORh(CO)I_3]^- + CO \rightarrow [MeCORh(CO)_2I_3]^-$$

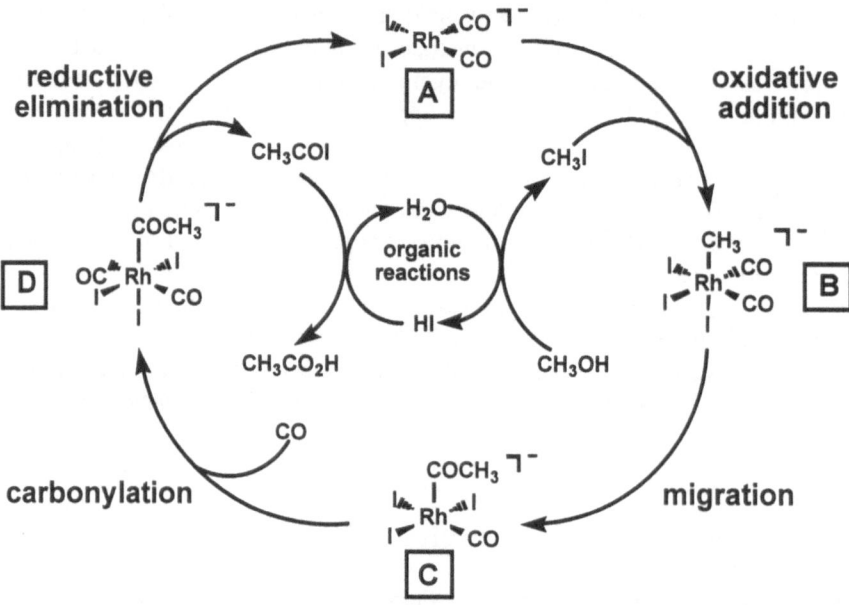

Figure 1 The cataylytic cycle for the Rh/I⁻ catalyzed carbonylation of methanol to acetic acid.

lastly, iv) the *reductive elimination* step (**D** to **A**) ends the cycle, gives the product and regenerates the starting metal complex, **A**, which then recycles,

$$[MeCORh(CO)_2I_3]^- \rightarrow [Rh(CO)_2I_2]^- + MeCOI$$

Thus in this series of reactions methyl iodide is converted into acetyl iodide. Two organic reactions are needed to complete the cycle, one is the hydrolysis of acetyl iodide to acetic acid, the desired product, and HI. The other is the conversion of methanol with HI into methyl iodide,

$$MeCOI + H_2O \rightarrow MeCO_2H + HI$$

$$HI + MeOH \rightarrow MeI + H_2O.$$

The reactions add up to give overall

$$MeOH + CO \rightarrow MeCO_2H.$$

Using special techniques (including FTIR spectroscopy and ^{13}C NMR at low temperatures, combined with extensive labelling, both ^{13}C and D) we have detected and structurally characterised the key organometallic intermediate $[MeRh(CO)_2I_3]^-$, (B), and have measured both its rate of formation (from MeI and $[Rh(CO)_2I_2]^-$(A)) and its rate of disappearance [3]. We showed that migratory insertion in $[MeRh(CO)_2I_3]^-$ is an order of magnitude faster than the reductive elimination of MeI, and that the low steady state concentration of $[MeRh(CO)_2I_3]^-$ during oxidative addition is primarily due to the rapid migratory insertion (**B** to **C**), and not to a fast back-reaction (**B** to **A**).

More recently we have applied our techniques to the new iridium catalyzed methanol carbonylation (the BP Chemicals' *Cativa* process). Although rather similar cycles can be written for Ir to those for rhodium, the details of the two processes are quite different and lead to considerable consequences. We have measured the initial oxidative addition step for both metals,

$$[M(CO)_2I_2]^- + MeI \rightarrow [MeM(CO)_2I_3]^-$$

which is *ca.* 120 times as fast for M = Ir as for M = Rh. In complete contrast, however, the migration is a factor of *ca.* 10^5 faster for Rh(III) than for Ir(III) at 100 °C in an aprotic solvent. Thus the "rate determining" step which determines the overall speed of the cycle is the oxidative addition for rhodium and the migration for iridium. As a consequence there is a build-up of the methyliridium species $[MeIr(CO)_2I_3]^-$ which can thus give rise to undesirable by-products (for example, methane [4]).

Neutral as well as anionic species play important roles for iridium; by contrast, only anionic species seem significant for rhodium. The overall rate can be improved and by-product formation decreased by accelerating the rate of the migration step in the iridium catalyzed process; this requires the formation of a neutral iridium intermediate where the migration is enhanced. Work that we and BP Chemicals in Hull have carried out has pinpointed several additives which have this effect and systems have been developed which accelerate both the model system and the actual catalytic process.

Both in this work and in the rhodium reactions we find good quantitative correlations in the activation parameters between the rate of the (rate determining) step that we have investigated and the overall catalytic cycle. This gives us considerable confidence that the model is a good one and the approach is useful.

1 Heterogeneously catalyzed reactions: the Fischer-Tropsch reaction

1.1 Introduction

Reactions catalyzed heterogeneously on metals are very much more complex to study and understand, partly because (even today) metal surfaces are not well understood, and because of the multiplicity of sites available on the metal surface. Interaction of the metal with the support also plays a vital part and indeed supports have been likened to the ligands employed in homogeneous systems for that reason. The analogy between a catalyst metal particle and a molecular cluster has been extensively commented on and structurally the similarities are certainly considerable. However, most transition metal clusters are rather inert due to the presence of strongly bound ligands (*e.g.* CO, Cp, *etc.*) on the periphery; by contrast metal surfaces are highly reactive. Thus the analogy falls down when the all important reactivities are considered.

We have taken the unusual step of focussing our attention on the possibility that even on a surface the metal atoms react largely at single sites. The reaction that we have chosen to study is a very complex one, the hydrogenation of carbon monoxide to linear alkenes and alkanes, the Fischer-Tropsch (FT) process. This is carried out at elevated temperatures, $\geq$ 200 °C, sometimes under pressure, but always over a catalyst. Industrially the most used catalysts are cobalt and iron; ruthenium and rhodium are also very active. The process was discovered in 1926 [5] and was extensively used in Germany during the war and in South Africa in the 1960's and 1970's to make gasoline from poor quality coal via gasification. In the last few years there has been a remarkable revival of interest in the process since it offers a route to very pure and hence non-polluting hydrocarbons. In each case the feedstock is now natural gas, methane. Shell have set up a plant in Malaysia to make "middle distillates" [6] and over the last year Exxon and a consortium led by Syntroleum in the United States have announced new and very economic FT processes [7].

Although quite complex, the FT process can be analysed into several steps, each of which can be understood in terms of simple organometallic model processes occurring on single atoms or, occasionally, on groups of two or three metal atoms.

We started our work some years ago, expanding the work of Roly Pettit. He and his collaborators first carried out some model studies on organometallic complexes and later did interesting "probe" experiments in which they showed the intermediacy of surface methylenes [8] in the formation of the linear alkenes and alkanes by adding (unlabelled) diazomethane to a labelled ^{13}CO gas stream over cobalt metal. They interpreted the data in terms of a mechanism in which methylene (CH_2) was formed from CO on a metal surface and was then polymerised; the initiator was a surface H or methyl (Figure 2). The model for propene formation, one of the most prevalent FT products, thus consisted of a fragment of surface bearing one methyl and two methylenes.

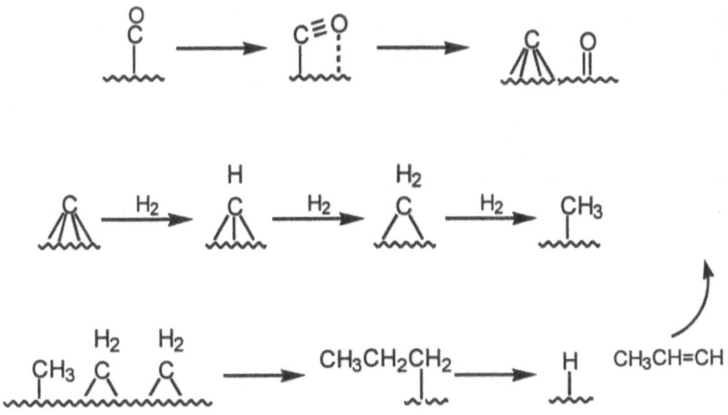

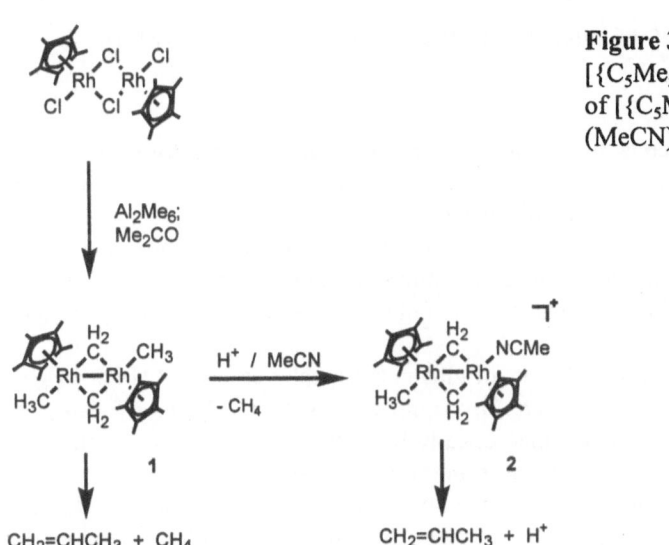

Figure 2 Schematic to illustrate the for-mation of surface methylene from CO and the suggested *alkyl* chain process for the polymerisation of the methylene (after ref. 8)

1.2 Model studies in Sheffield

At that time, we had synthesised a dirhodium complex bearing pentamethylcyclo-pentadienyl ligands in which the two metal atoms each bore a methyl and were bridged by two methylenes, complex **1**, [9] Figure 3. This was a perfect model to test the Pettit ideas on the mechanism of the Fischer-Tropsch polymerisation reaction.   The thermal decomposition of **1** required quite high temperatures, but gave largely methane and propene. Further investigations were made much easier by our discovery that the decomposition proceeded much more easily on the cation **2**; this allowed detailed labelling studies which showed that the propene (and the methane) were derived from the methyl and methylene groups and not from the solvent or the C_5Me_5 ligands [10].

Figure 3 The synthesis of $[\{C_5Me_5Rh\}_2(\mu\text{-}CH_2)_2(CH_3)]$ **1**, and of $[\{C_5Me_2Rh\}]_2(\mu\text{-}CH_2)_2(CH_3)\text{-}(MeCN)]^+$ **2**, (ref. 9)

87

However, the D-labelling found in the propene derived from $[\{C_5Me_5Rh\}_2(\mu\text{-}CH_2)_2(CD_3)(MeCN)]^+$ was $CDH_2CH=CD_2$ (ca. 90%); the product expected on a simple $Me+CH_2+CH_2$ coupling model is $CD_3CH=CH_2$, which was only formed to a small extent. The formation of 90% $CDH_2CH=CD_2$ was inconsistent with a simple process in which methyl coupled with two methylenes, an *alkyl* model, such as predicted on the basis of the Pettit proposals. Indeed the only rational mechanism leading to the distribution of label found in the product involved the intermediacy of a Rh-σ-vinyl, which couples with a μ-methylene, for example as shown in Figure 4.

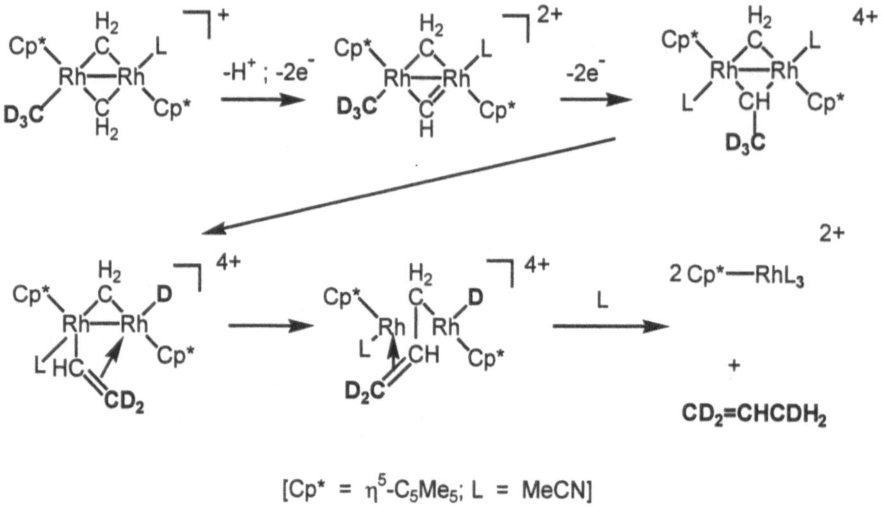

$[Cp^* = \eta^5\text{-}C_5Me_5;\ L = MeCN]$

Figure 4 Proposed path for the formation of $CD_2HCH=CD_2$ from $[\{C_5Me_5Rh\}_2(\mu\text{-}CH_2)_2(CD_3)(MeCN)]^+$

That very surprising result led us to rethink the mechanism of the polymerisation step in the Fischer-Tropsch reaction itself. Was it possible that this might also involve vinylic intermediates? The idea of vinylic intermediates while unexpected at first, has been slowly gaining ground, assisted by the very elegant work of Werner and his colleagues [11], and further work showed that such vinylic couplings are important in other reactions too. These results suggested the possible involvement of vinyl/alkenyl intermediates in the Fischer-Tropsch reactions themselves.

We believe that such a reaction path has its origins in the greater ease with which an sp^3 carbon couples to an sp^2 (vinyl, carbonyl, *etc.*) by comparison with a sp^3 plus sp^3 (alkyl + alkyl) coupling. This has been explained theoretically by Calhorda, Brown and Cooley for coupling on Ni(II) or Pd(II) [12] and developed further by ourselves [13].

In view of this result for the model reaction we investigated the effect of adding vinylic probes to Fischer-Tropsch reactions, initially over rhodium. We chose this kinetic approach, adding doubly labelled probe molecules (*e.g.* $^{13}CH_2=^{13}CHX$) to the Fischer-Tropsch reaction, since heterogeneous catalytic reactions are so difficult to investigate directly. If the vinylic probes were closely related to the intermediate species involved in the surface reactions, they should readily be incorporated.

We used the doubly ^{13}C-labelled vinyls, $^{13}CH_2=^{13}CHBr$, $^{13}CH_2=^{13}CH_2$, and $(^{13}CH_2=^{13}CH)_4Si$, as probes since they would give maximum unambiguous mechanistic information. Analysis of the mass-spectra of the organic products would immediately show, i) whether the $^{13}C_2$-probe had been incorporated into the products, and to what degree, ii) whether C-C cleavage of the probe to C_1 was significant, and whether any hydrocarbons were formed by such paths, iii) whether the C_2-probe initiated or propagated methylene oligomerisation or both, and iv) what other organic reactions were occurring on the surface.

The results from these experiments were completely clear: $^{12}CH_2=^{12}CH$- and $^{13}CH_2=^{13}CH$- from the vinyl probes were very obviously incorporated, while ethyl from ethyl bromide or tetraethylsilane, was not. While there was substantial incorporation of $^{13}C_2$ from initiation of the polymerisation, there was no evidence either for the intermediacy of vinyl in propagation, or of significant amounts of cleavage of the vinyl initiator to C_1 fragments.

The alkenyl mechanism for the Fischer-Tropsch polymerisation on rhodium

From the $^{13}C_2$-incorporation data we developed a new *alkenyl mechanism* for Fischer-Tropsch polymerisation of surface methylenes [13]. There is general agreement that the adsorption, activation, and cleavage of CO to a surface carbide species, and the hydrogenation of the carbide to surface methyne and methylene, proceed as shown in Figure 2. However to accommodate the new data showing the incorporation of C_2 vinyl (from vinyl bromide) but not of ethyl (from ethyl bromide), it is therefore proposed that the participation of vinyl is an integral part of the surface polymerisation mechanism. This is shown in Figure 5 in the form of a catalytic *alkenyl* cycle, where the first step (top left) is the formation of a surface vinyl. Each succeeding propagation step is the (irreversible) linear homologation of a surface alkenyl species with surface methylene. Although it is inappropriate to make direct comparisons of this polymerisation reaction with the carbonylation cycle in Figure 1, this step is another type of *migration* reaction. It is followed by an isomerisation of the resulting allylic species to another alkenyl. Growth is terminated by reaction with surface hydrogen, giving the 1-alkene directly, another form of *reductive elimination*. For simplicity the diagram is restricted to the linear propagation; however, the formation of some branched chain isomers, as well as coupled products, can easily be accommodated. It may also be noted that as industrial Fischer-Tropsch reactions are carried out under more severe conditions (*e.g.* pressure), further reactions occur, for example the reabsorption of product alkenes into the cycles.

A simple mathematical treatment of the steady state kinetics for the system, based upon the propagation and termination steps of the catalytic cycle in Figure 5, has been developed and accounts well for the observed effect of $^{13}C_2$ vinyl initiation [14]. Good agreement has been found between the theoretical and the experimental, and the degree of initiation of various coverages of labelled surface vinyl is related to the amounts of ^{13}C label found in the product 1-alkenes. For $^{13}C_2H_3Br$ there is some 30% incorporation of $^{13}C_2$ into the C_3 and *ca.* 11% incorporation of $^{13}C_2$ into the C_4 hydrocarbons.

We also suggest that in the absence of a vinylic initiator, the initial formation of a surface vinyl (-CH=CH$_2$) occurs by the reaction of a surface methyne ($\equiv$CH) and a surface methylene (>CH$_2$). Although this reaction does not yet seem to have been demonstrated on a surface, it has been modelled in organometallic complexes [15].

We have also extended our studies to ruthenium, a more conventional Fischer-Tropsch catalyst. The results are strikingly similar to those obtained for rhodium [16]. Ruthenium is *ca.* 2-3 times more active for Fischer-Tropsch reactions under comparable conditions, and high levels of incorporation of $^{13}C_2$, consistent with our vinyl initiation model, are again seen when $^{13}C_2H_3Br$ is used as probe; $^{13}C_2H_4$ is as effective a probe molecule as $^{13}C_2H_3Br$, but again ethyl from C_2H_5Br is not incorporated. The other major difference is that over Ru there is some incorporation of $^{13}C_1$, by C-C cleavage of the probe at reaction temperatures >180 °C.

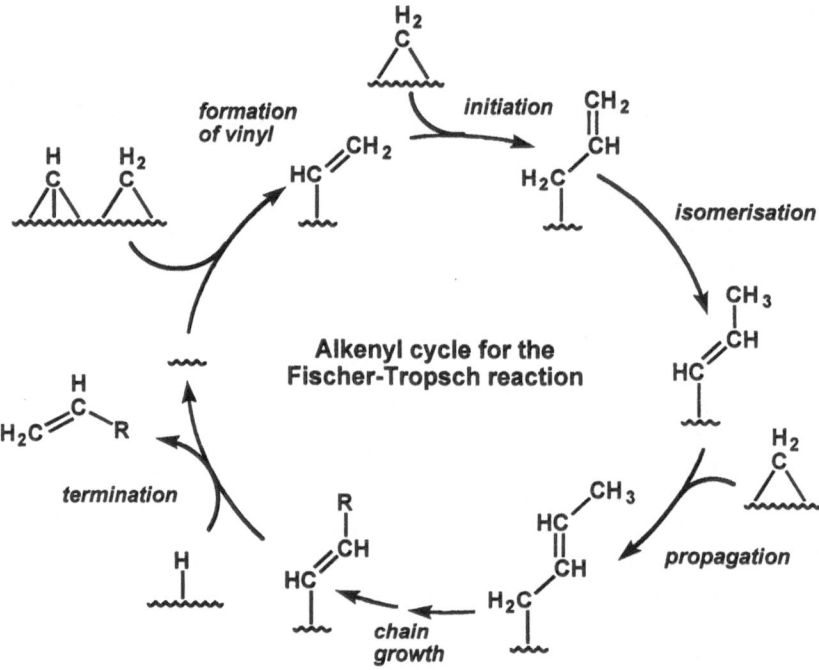

Figure 5 The alkenyl cycle proposed for methylene oligomerisation in the Fischer-Tropsch reaction

$^{13}C_2$ distribution in pentane

Although we were able to say with great certainty that two ^{13}C atoms are incorporated into the hydrocarbon products when $^{13}C_2$-vinylic probes are added, defining their precise positions had been beyond our ability. Very recently however we have been able to meaningfully analyse the fragmentation patterns of the mass-spectra of the n-alkanes produced [17]. This has not been easy: the FT reactions give a large range of products. Analyses of the lighter hydrocarbons ($C_{<4}$) do not yield useful results and the heavier hydrocarbons are formed in very small amounts under our reaction conditions, and quantitative MS are difficult to obtain. Further, the analysis is predicated on the hypothesis that the fragmentation occurs equally in all C-C single bonds, but not necessarily elsewhere. Thus analysis of olefins is ruled out, at least at this stage. We are thus left with the need to analyse an n-alkane produced in sufficient amount to be easily investigated and yet of sufficient length to give a meaningful result.

For this reason we have chosen to analyse the n-pentane formed in probe initiated FT reactions. Three arrangements of two ^{13}C and three ^{12}C nuclei can be defined: i) they are arranged completely randomly in the molecule, ii) they are in the 1,2-positions only, and iii) they are in the 3,4-positions only.

The MS fragmentation we have considered is that leading to C_2 and C_3 ions, namely,

$$C_5H_{12} - e^- \rightarrow C_5H_{12}^{\bullet+}$$

$$C_5H_{12}^{\bullet+} \rightarrow C_2H_5^{\bullet} + C_3H_7^+$$

$$C_5H_{12}^{\bullet+} \rightarrow C_2H_5^+ + C_3H_7^{\bullet}$$

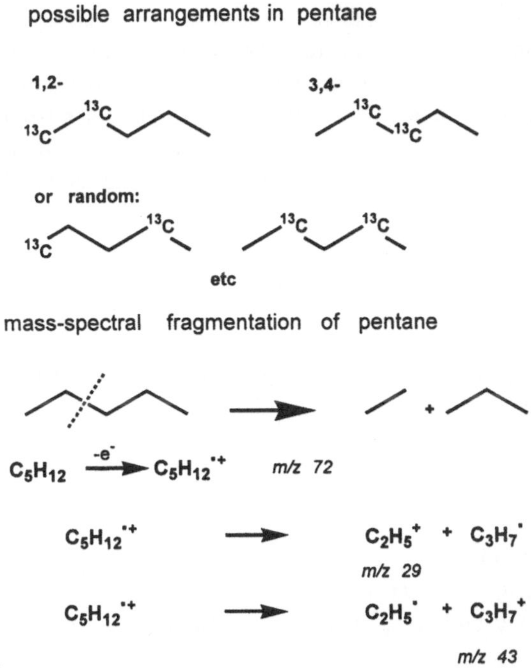

possible arrangements in pentane

1,2-

3,4-

or random:

etc

Figure 6 Arrangement of two ^{13}C nuclei in pentane and the mass-spectrometric fragmentation of $^{12}C_5H_{12}$ to C_2 and C_3 fragments

mass-spectral fragmentation of pentane

$$C_5H_{12} \xrightarrow{-e^-} C_5H_{12}^{\bullet+} \qquad m/z\ 72$$

$$C_5H_{12}^{\bullet+} \longrightarrow C_2H_5^+ + C_3H_7^{\bullet}$$
$$m/z\ 29$$

$$C_5H_{12}^{\bullet+} \longrightarrow C_2H_5^{\bullet} + C_3H_7^+$$
$$m/z\ 43$$

Each of the possible arrangements noted above leads to a different predicted abundance of $^{13}C_1$ and $^{13}C_2$ in the C_2 and C_3 ions. For example, a 1,2- arrangement should give substantial amounts of $^{13}C_2$ in both fragments, while a 3,4- arrangement should give $^{13}C_2$

91

in the C_3 but no significant amount of $^{13}C_2$ in the C_2 fragment (Figure 6). However each cation should then contain significant $^{13}C_1$, which should be absent if there is a 1,2-arrangement. A similar but more complex analysis can be carried out for the random arrangement. Again this would be expected to show some $^{13}C_1$ in the C_2 and C_3 fragments, which are absent in a 1,2-arrangement. Thus a very simple diagnostic test that can easily be applied to distinguish the 1,2-arrangement from the other possibilties is the amount of $^{13}C_1$ present in $C_2H_5^+$ and $C_3H_7^+$. Our data show that the pentane derived from rhodium catalyzed FT reaction, with added $^{13}C_2$-vinyl bromide or ethene as probes, contains largely a 1,2-arrangement of the ^{13}C labels. By contrast, the data obtained from similar reactions over the Ru catalysts indicate that a 1,2-arrangement is not the only one present and agree better with a mixture of random and 3,4-.

This result confirms very nicely i) that the vinyl probe does indeed initiate the methylene polymerisation and does not take part in the propagation over Rh catalysts, and ii) that there is cleavage and rearrangement when the FT reactions are carried out over Ru catalysts.

Summary

Our studies have shown that the model catalytic cycles that can be drawn for reactions homogeneously catalyzed by metals in solution can lead to important *quantitative* results for the overall process. They also offer a useful guide to the sort of catalytic cycles that can be set up for the reaction *heterogeneously* catalyzed on surfaces. These are also based on models derived from organo-transition metal chemistry and treat the metal surface as if it were an assembly of (single) reactive sites. In the case of the Fischer-Tropsch hydrogenation of carbon monoxide, a new mechanism, embodied in the *alkenyl* cycle, has been proposed and demonstrated. Not only does it account for the broad features of the overall reaction but intimate details such as the amount and location of incorporated labelled molecules can be quantitatively established.

Acknowledgements

We thank the EPRSC, the Royal Society and BP Chemicals for generous support of this work and Dr. Andrew Cutler for thought-provoking ideas.

References

1. J. Halpern, *Inorg. Chim. Acta*, **1981**, *50*, 11.
2. T. W. Dekleva and D Forster, *Adv. Catal.*, **1986**, *34*, 81.
3. P. M. Maitlis, A. Haynes, G. J. Sunley and M. J. Howard, *J. Chem. Soc., Dalton Trans.*, **1996**, 2187.
4. T. Ghaffar, J. Charmant, A. Haynes and P. M. Maitlis, *unpublished.*
5. F. Fischer and H. Tropsch, *Brennstoff-Chem.*, 1926, *7*, 97; *Chem. Ber.*, **1926**, *59*, 830.
6. E. W. Kuipers, C. Scheper, J. H. Wilson, I. H. Vinkenburg and H. Oosterbeek, *J. Catal.*, **1996**, Vol. 158, 288.

7. *Business Week,* May 19, **1997**.
8. R. C. Brady III and R. Pettit, *J. Am. Chem. Soc.*, **1980**, *102*, 6181; **1981**, *103*, 1287.
9. P. M. Maitlis, *J. Organomet. Chem.*, **1995**, *500*, 239; K. Isobe, A. Vázquez de Miguel, P. M. Bailey, S. Okeya and P. M. Maitlis, *J. Chem. Soc., Dalton Trans.*, **1983**, 1441; S. Okeya, N. J. Meanwell, B. F. Taylor, K. Isobe, A. Vázquez de Miguel and P. M. Maitlis, *J. Chem. Soc., Dalton Trans.*, **1984**, 1453.
10. I. M. Saez, N. J. Meanwell, A. Nutton, K. Isobe, A. Vázquez de Miguel, D. W. Bruce, S. Okeya, D. G. Andrews, P. R. Ashton, I. R. Johnstone and P. M. Maitlis, *J. Chem. Soc., Dalton Trans.*, **1986**, 1565; I. M. Saez, D. G. Andrews and P. M. Maitlis, *Polyhedron*, **1988**, *7*, 827.
11. T. Braun, O. Gevert and H. Werner, *J. Am. Chem. Soc.*, **1995**, *117*, 7291.
12. M. J. Calhorda, J. M. Brown and N. A. Cooley, *Organometallics*, **1991**, *10*, 1431.
13. P. M. Maitlis, H. C. Long, R. Quyoum, M. L. Turner and Z-Q. Wang, *J. Chem. Soc., Chem. Commun.*, **1996**, 1.
14. M. L. Turner, H. C. Long, A. Shenton, P. K. Byers and P. M. Maitlis, *Chemistry - A European Journal*, **1995**, *1*, 549; M. L. Turner, P. K. Byers, H. C. Long and P. M. Maitlis, *J. Am. Chem. Soc.*, **1993**, *115*, 4417.
15. D. L. Davies, M. J. Parrott, P. Sherwood and F. G. A. Stone, *J. Chem. Soc., Dalton Trans.*, **1987**, 1201.
16. H. C. Long, M. L. Turner, P. Fornasiero, J. Kašpar, M. Graziani and P. M. Maitlis, *J. Catal.*, **1997**, *167*, 172.
17. M. L. Turner, R. Quyoum and P. M. Maitlis, unpublished results.

The Chemical Reactivity of some Alkenyl Complexes of Iridium

E. Carmona, F. M. Alías, M. L. Poveda and M. Sellin

Departamento de Química Inorgánica-Instituto de Investigaciones Químicas, Universidad de Sevilla-Consejo Superior de Investigaciones Científicas C/ Américo Vespuccio s/n, Isla de la Cartuja, 41092 Sevilla, Spain

The activation of olefinic substrates by transition metal compounds is a very important process in organometallic chemistry, both from the fundamental and from the applied viewpoints [1]. A conceptually simple reaction an alkene can undergo in the presence of a suitable metal fragment is the activation of a vinylic C-H bond to give an alkenyl complex, [M]-CR=CR'R'' [2]. Compounds of this type have been known for many years and have been extensively studied. They have found many applications in organic and organometallic synthesis and are proposed as active intermediates in some catalytic processes, relevant in Fischer-Tropsch synthesis [3]. The interest that this class of compounds arises in part from the possibility of modulating the reactivity of the C=C double bond by changing the nature of the metal fragment. For late transition metal alkenyl complexes nucleophilic behaviour at the β carbon is often found, an observation which suggests an important contribution of the carbenic resonance form [M]$^+$= CR—CR'R''R'R'', to their overall electronic structure.

For some years, our group has been involved [4-10] in the study of the chemical reactivity of alkenyl complexes of Rh and Ir that contain an ancillary hydrotris(pyrazolyl)-borate-type ligand, Tp'. Here we wish to summarize our work in this area concentrating specifically on the following topics: (i) the formation of hydrido-alkenyl complexes of Ir(III) that contain in addition an olefin ligand and the study of their C-C coupling reactions; (ii) the activation of the C-H bonds of some organic molecules (benzene and cyclic ethers); (iii) the coupling of the alkenyl fragment with the -C≡N multiple bond of nitriles and (iv) the protonation of the coordinated alkenyl (or alkenyl type) ligand to generate cationic Ir(III)-alkylidene complexes.

Formation of Alkene-Alkenyl Complexes of Iridium and their C-C coupling reactions

We have reported recently [7] that the bis(ethene) complex, Tp*Ir(C$_2$H$_4$)$_2$ **1**, (Tp* = Tp^{Me}2 = tris(3,5-dimethylpyrazol-1-yl)hydroborate ligand) can be prepared readily in an one-pot reaction by treatment of [Ir(μ-Cl)(coe)$_2$]$_2$ with C$_2$H$_4$ at low temperatures, followed by the addition of KTp*. Chemical and spectroscopic studies indicate that the Tp* ligand has tridentate coordination, both in solution and in the solid state. This was also confirmed by a single-crystal X-ray analysis, which shows that the five-coordinate iridium centre has a distorted trigonal-bipyramidal environment, with the two molecules of C_2H_4 occupying one of the equatorial and one of the axial positions.

Compound **1** can be activated thermally, or photochemically, under mild conditions, to give the hydrido-vinyl complex **2**. As discussed below, the thermal reaction yields also the crotyl derivative Tp*IrH(η^3-C_3H_4Me) **3** as a mixture of *endo* or *exo* allyl isomers, that contain a *syn* or *anti* Me group [7]. In turn, the photochemical activation, when effected in C_6H_6, also provides the phenyl complex Tp*IrH(C_6H_5)(C_2H_4), by involvement of the solvent, but **2** can be obtained in almost quantitative yield by the low-temperature (-60 °C) irradiation of frozen, dilute cyclohexane suspensions of **1**.

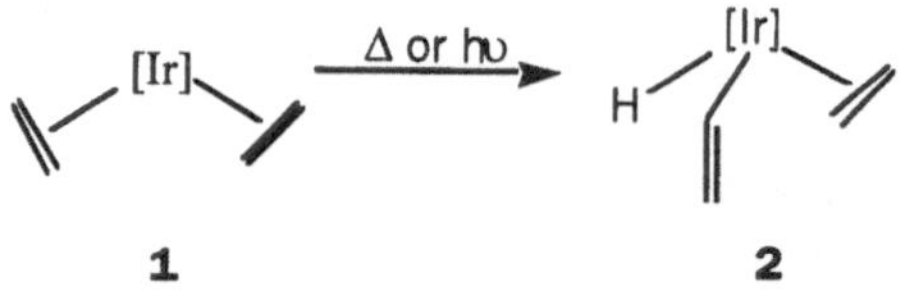

Interestingly, the analogous reactions of [Ir(μ-Cl)(coe)$_2$]$_2$ with propene and 1-butene, in the presence of KTp* yield directly the C-H activation products (olefin = propene, **4**; 1-butene, **5**) as a mixture of the *cis* and *trans*-alkenyl isomers, as shown below for the reaction with C_3H_6. In this case the two trans stereomers are formed in a 20:1 ratio, but for the less abundant cis isomer only one stereomer could be detected [7].

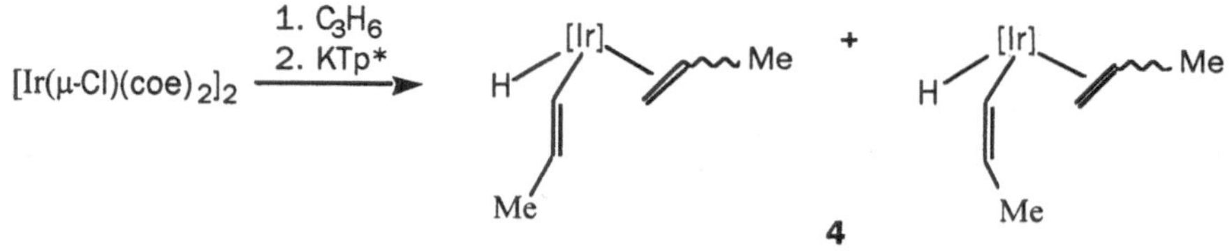

For the above systems and for other related ones containing the unsubstituted Tp ligand and/or different alkenes, the Ir(III) hydrido-alkenyl complexes are thermodynamically more stable than the corresponding Ir(I)-alkene isomers. At least for ethylene, this order of stability contrasts with that found in the analogous cyclopentadienyl complexes where the opposite order of stability is found. Since this olefinic activation occurs without a change in the coordination mode of the Tp' ligand, we propose that the harder nature of the Tp' ligands (compared to the softer cyclopentadienyls) and their propensity to enforce six-coordination at the metal, a situation that is highly favourable for Ir(III), are key features in the stabilization of the Ir(III) hydride-alkenyl complexes.

The formation of **2** by thermal activation of **1** is only the first step of this rearrangement. Heating cyclohexane solutions of the latter compound at 60 °C leads to the gradual and irreversible formation of the hydrido-crotyl **3** (as already indicated, as a mixture of stereomers). Careful NMR monitoring of this transformation reveals that **2** is the active species in this rearrangement, an observation which was further demonstrated by the independent conversion of pure samples of **2** into the same mixture of hydrido-crotyls, under identical reaction conditions.

$$\text{Tp*Ir(C}_2\text{H}_4)_2 \xrightarrow[\text{C}_6\text{H}_{12}]{60\,^\circ\text{C}} \text{Tp*IrH(C}_2\text{H}_3\text{)(C}_2\text{H}_4) \xrightarrow[\text{C}_6\text{H}_{12}]{60\,^\circ\text{C}} \text{Tp*IrH(}\eta^3\text{-C}_3\text{H}_4\text{Me)}$$

$$\quad\quad\quad\;\, \textbf{1} \quad\quad\quad\quad\quad\quad\quad\quad\quad\quad \textbf{2} \quad\quad\quad\quad\quad\quad\quad\quad\quad \textbf{3}$$

$$\quad\quad\quad\quad\quad\quad\quad\quad\quad\quad\quad\quad\quad\quad\quad\quad\quad\quad\quad\quad\quad \text{(exo-anti > 80\%)}$$

An analogous coupling of the alkene and alkenyl ligands occurs when the alkenyl complexes **4** and **5** are heated under similar conditions. Compounds containing six- and eight-carbon allylic chains, respectively, are formed as mixtures of the corresponding *endo* or *exo*, *syn* or *anti* stereomers. Mechanistic studies on these reactions suggest the involvement of only Ir(III) intermediates [7]. The key step responsible for the C-C coupling reaction is proposed to be the migratory insertion of a vinylidene ligand into the Ir-alkyl bond resulting from the formal insertion of the olefin into the Ir-H bond.

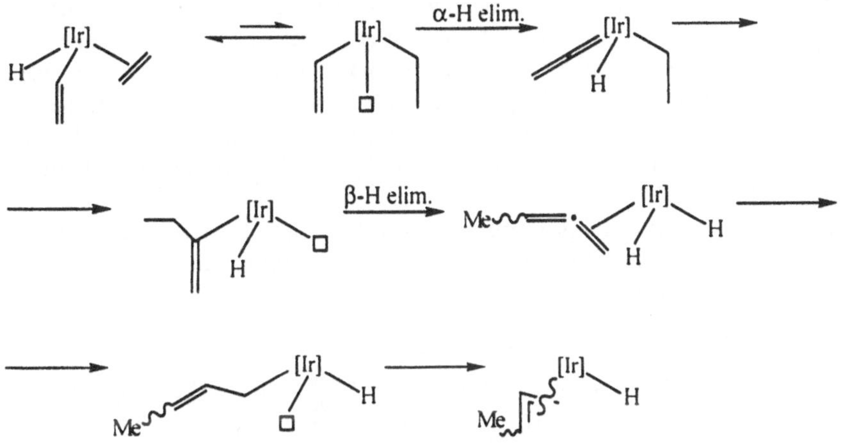

The activation of the C-H bonds by the alkenyl complex 2

Although the hydrido-vinyl compound **2** is able to activate the C-H bonds of different organic substrates, for the sake of simplicity only those reactions involving tetrahydrofuran (THF) and benzene, as examples of cyclic ethers and aromatic molecules, repectively, will be briefly discussed.

When a pure sample of **2** is heated in THF, at 60 °C, for 8h, a 1:1 mixture of the hydrido-crotyl Tp*IrH(η^3-C$_3$H$_4$Me) **3**, and a new compound **6**, is obtained [8]. The same mixture forms starting from the bis(ethene) compound **1**, albeit longer reaction times are needed. Salient spectroscopic features of relevance for the characterization of **6** are an IR stretching at ca. 2130 cm^{-1}, concurrent with a high-field, ^{1}H NMR resonance (δ -17.90), indicative of an Ir-H funcionality, as well as a low field ^{13}C NMR signal at 258.8 ppm, characteristic of a carbene ligand. These observations coupled with the results of 1D and 2D ^{1}H-^{1}H and ^{1}H-^{13}C NMR experiments suggest formulation of **6** as a hydride-oxycarbene species. Its formation constitutes an unprecedented double dehydrogenation of one of the α-methylene groups of tetrahydrofuran. In a formal sense, one of the abstracted H atoms ends

up as a hydride ligand while the other becomes incorporated into the n-butyl fragment that results from the coupling of the two molecules of ethylene. The structure suggested for this complex was later demonstrated by X-ray studies, which showed Ir-H and Ir=C distances of 1.8(1) and 1.881(5) Å, respectively [8]. Mechanistic studies suggest the formation of an Tp*Ir(C$_2$H$_5$)(C$_2$H$_3$) intermediate (perhaps stabilized by an agostic β-ethyl interaction), which in view of the above results is trapped by tetrahydrofuran only with moderate efficiency.

A very similar experimental procedure, but using C$_6$H$_6$ as the reaction solvent, provides different, unexpected results. High yields of a white crystalline compound **7**, along with the sparingly soluble, binuclear species **8**, are obtained [9].

The formation of these compounds involves the activation of two molecules of C$_6$H$_6$ by a single metal site, a transformation for which there is little precedent in the analogous cyclopentadienyl systems. An additional remarkable feature of this reaction is the generation of thermally robust Ir(III)-N$_2$ complexes. The presence of the coordinated N$_2$ molecule in **7** and **8** is indicated by a strong absorption at 2190 cm^{-1} in the IR spectrum of the former and at 2130 cm^{-1} in the Raman spectrum of the latter. The binuclear derivative **8** was additionally characterized by X-ray crystallography [9].

The N_2 ligand in **7** can be substituted by PMe_3 or CO, although heating at 60 °C for 3h, and 2-3 atm of CO are needed. Interestingly, compound **7** appears to be also a suitable species for the activation of cyclic ethers and provides the corresponding hydride-oxycarbene when heated in THF. Mechanistic studies aimed at clarifying the formation of

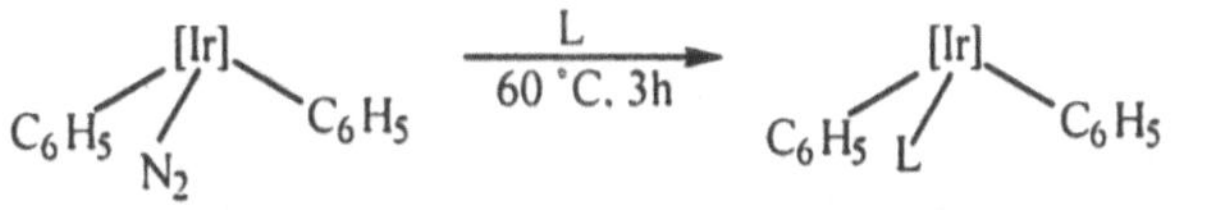

compound **7** point to the participation of only Ir(III) intermediates, and moreover highlight the high efficiency of this system for the activation of the C-H bonds of C_6H_6.

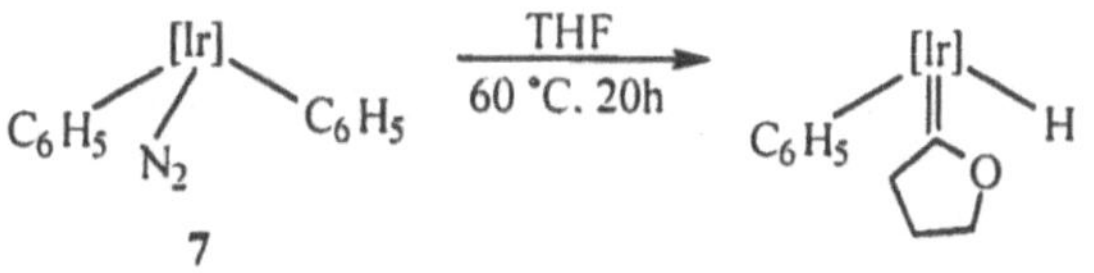

Coupling of coordinated alkenyl and nitrile fragments

When either compound **1** or **2** are heated at 60 °C in MeCN, C-H activation, with concomitant migratory C_2H_4 insertion into the resulting Ir-H bond and nitrile coordination occur. The new alkyl-alkenyl compound **9** can be isolated in high yields [7], but heating its MeCN solutions at higher temperatures, in the presence of small amounts of water produces

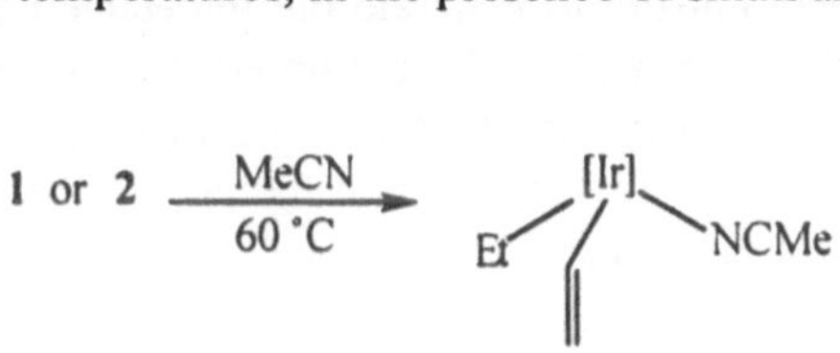

amounts of water produces a red crystalline material **10**, for which spectroscopic data suggest a delocalized iridapyrrol structure [10]. Thus, the iridium-bonded methine unit is responsible for resonances at δ 10.71 and 191.3 in the ^1H and $^{13}C\{^1H\}$ NMR spectra, respectively. These signals are clearly in a range intermediate between those of metal-carbene and metal-vinyl resonances (see canonic forms **I** and **II**). X-ray studies are also in accord with this proposal, as suggested for example by the planarity of the five-membered metallacycle or by the short Ir-C distance within the cycle of 1.86(2) [11].

99

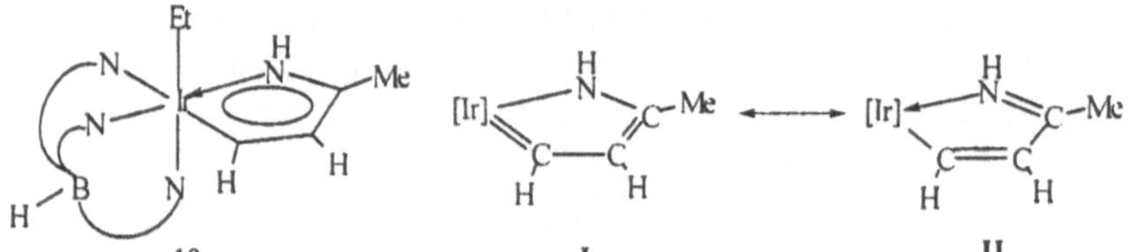

This 3+2 cycloaddition reaction appears to be quite general. The analogous transformation of the propenyl complex Tp*IrH(CH=CHMe)(C$_3$H$_6$) shows remarkable selectivity and gives compound **11**, as the only detectable product. Similar Ir-alkenyl derivatives of the unsubstituted Tp ligand, as well as [Ir]-allyl complexes (for example the crotyl species **3**) also undergo this coupling reaction. Not unexpectedly organic nitriles other than acetonitrile behave similarly.

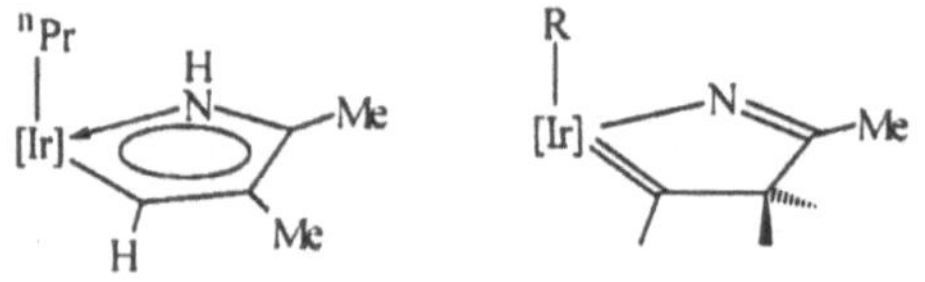

Kinetic studies show that the process is catalyzed by water, a graphical representation of k_{obs} vs. [D$_2$O] shows saturation kinetics [10]. In addition, the coupling is intramolecular, less than 15% of CD$_3$CN incorporation being observed when compound **9** (with natural isotopic composition) is heated in neat CD$_3$CN under saturation conditions. These and other observations suggest that the reaction occurs by direct coupling of the [Ir]-CH=CH$_2$ fragment (recall the importance of the dipolar canonic structure [Ir] = CH-CH$_2$) with the coordinated molecule of acetonitrile, to generate the undetected iridacycle **III**, the role of water being the catalysis of the **III** → **10** tautomerism. Eventhough the formation of metallapyrrole derivatives by different synthetic methodologies finds precedent in the literature, this coupling of the β-vinyl carbon and the acetonitrile ligand constitutes a new kind of reactivity of metal-vinyl fragments.

Generation of cationic ethylidene complexes of Ir(III)

The coupling reaction discussed in the preceding section may be interpreted in terms of nucleophilic reactivity at the alkenyl β-C. Recalling, once again, the importance of the carbenic resonance form, [Ir]$^+$ = CH—C$^-$H$_2$ in this species, their protonation is expected to give cationic ethylidene species, [Ir]$^+$ = CH—C$^-$H$_3$. This approach has proved successful and has provided unique electrophilic Ir(III) ethylidene-hydride and ethylidene-ethyl complexes. We have studied their α-migratory insertion reactions and the relative migratory aptitudes of hydride and ethyl ligands in this transformation.

The compound Tp*Ir(CH=CH$_2$)H(PMe$_3$) **12** can be obtained by the thermal activation of the Ir(I)-olefin adduct Tp*Ir(C$_2$H$_4$)(PMe$_3$) [12] while the analogous ethyl derivative Tp*Ir(CH=CH$_2$)(C$_2$H$_5$)(PMe$_3$) **13** is best prepared by the direct reaction of **2** with PMe$_3$ [7]. Protonation of **12** with [H(OEt$_2$)$_2$][BAr$_4$'] (Ar' = 3,5-C$_6$H$_3$(CF$_3$)$_2$) [13] occurs rapidly and reversibly at low-temperatures and gives the hydride-alkylidene compound Tp*Ir(=CHMe)H(PMe$_3$)$^+$ **14** in the form of two rotamers that interconvert very rapidly at -60 °C. The barrier to rotation is $\Delta G^{\ddagger} \approx$ 10 kcal/mol. Salient spectroscopic features of **14** are ^{1}H NMR resonances at δ ca. 20 and -16 for the C*H*Me and Ir-bound hydrogen atoms, respectively, as well as a low-field ^{13}C{^{1}H} NMR resonance at δ 325.0.

The protonation of the analogous ethyl-vinyl compound **13**, under similar conditions, yields Tp*Ir(=CHMe)(C$_2$H$_5$)(PMe$_3$)$^+$ **15**, once again in the form of two rotamers which undergo exchange at a somewhat slower-rate (-35 °C, $\Delta G^{\ddagger} \approx$ 12 kcal/mol). Both alkylidene species **14** and **15** undergo migratory insertion reactions to give the cationic hydrido-olefin complexes Tp*IrH(olefin)(PMe$_3$)$^+$ (olefin = C$_2$H$_4$, 2-butene, respectively for **14** and **15**). A comparative analysis of the rates of these transformations, carried out at -47 °C by ^{1}H NMR monitoring of solutions of the ethylidene derivatives in CD$_2$Cl$_2$ shows that both rearrangements follow first order kinetics and are characterized by k_H and k_{Et} values of 2.0 × 10^{-4} and 2.5 × 10^{-5} s^{-1}, respectively ($\Delta G_H^{\ddagger} \approx \Delta G_{Et}^{\ddagger} \approx$ 17 kcal/mol) [14]. Both on theoretical grounds and on the basis of previous studies [15], the =CH(Me) migratory insertion into Ir-H was expected to proceed faster than into Ir-C$_2$H$_5$. Further experimental work and theoretical studies, which are now in progress, are clearly needed to explain this striking observation.

The partial vinylic character of the iridapyrrol structures described in the previous section (see complex **10**, and resonance form **II**) suggests they might also undergo protonation at the β-vinylic carbon. Thus, upon treatment of **10** with [H(OEt$_2$)$_2$][BAr$_4$'] or with triflic acid, HTfO, a smooth reaction ensues that proceeds almost quantitatively to give **16b** as the final reaction product [16].

The overall transformation may be explained by assuming protonation at the β–carbon of the iridapyrrol ring, followed by migration of the Et group onto the resulting alkylidene carbon. After that, β-H elimination and isomerization of the Ir-olefin linkage would account for the formation of **16b**. The structure proposed for this species on spectroscopic grounds has been confirmed by X-ray studies [16]. It is interesting to mention that in this case, the proposed alkylidene intermediate can not be detected, even if the protonation is effected at -90 °C. This may be due, at least in part, to the fact that the alkylidene ligand has the correct orientation to undergo attack by the migrating Et group and to the availability of an adequately oriented (toward the vacant coordination site) β-H atoms in the resulting alkyl.

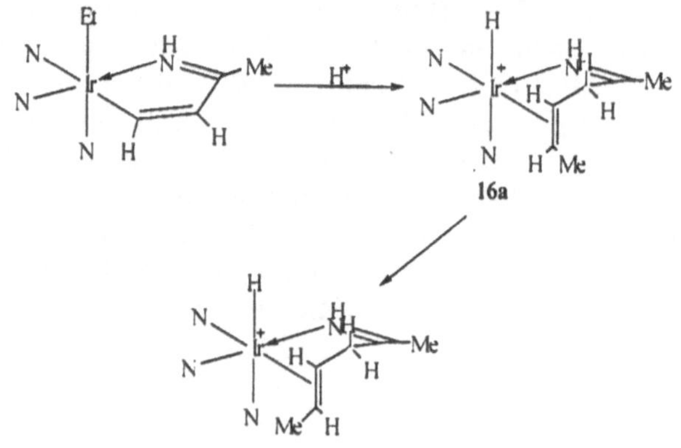

16a

16b

At variance with the above results, the analogous protonation of the hydride-iridapyrrol complex Tp*(H)Ir(C(Me)-C(Me)-C(Me)-NH) yields the cationic hydride-alkylidene compound **17**. The ^{1}H NMR spectrum of **17** shows the expected high-field ^{1}H NMR signal (δ - 15.68), characteristic of the Ir-H whilst in the ^{13}C{^{1}H} NMR spectrum a singlet at 324.5 ppm is clearly indicative of the presence of an alkylidene functionality. X-ray studies [16] reveal an Ir=C< distance of 1.80(3) Å.

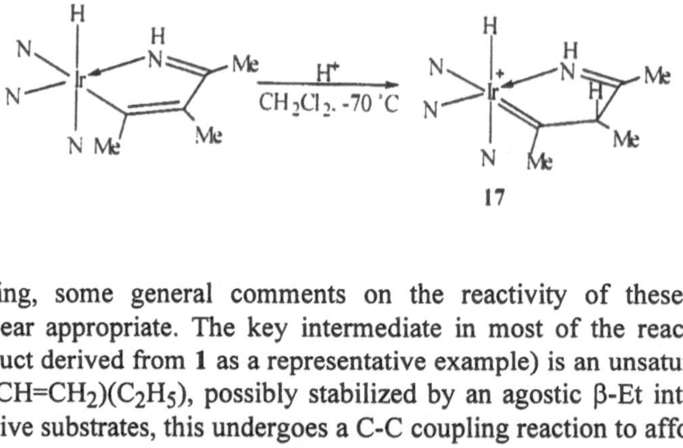

17

Before closing, some general comments on the reactivity of these Ir(III)-alkenyl compounds appear appropriate. The key intermediate in most of the reactions described (taking the product derived from **1** as a representative example) is an unsaturated species of the type Tp*Ir(CH=CH$_2$)(C$_2$H$_5$), possibly stabilized by an agostic β-Et interaction. In the absence of reactive substrates, this undergoes a C-C coupling reaction to afford the hydride-crotyl **3**, but C$_6$H$_6$ traps it in a very efficient manner and undergoes C-H activation to produce the bis(phenyl) compound. Tetrahydrofuran undergoes a most unusual regioselective double C-H activation to the oxicarbene, but partial conversion to the hydride-crotyl does also take place. Trapping by nitriles is very efficient, the polarity of the coordinated N≡C- allowing C-C coupling with formation of iridapyrrol structures. Finally, the nucleophilicity of the alkenyl β-carbon atom allows the generation of cationic alkylidene complexes of Ir(III). In this respect, it should be noted that, to our knowledge, this system provides the first example of a comparative study of the migratory aptitudes of H and C$_2$H$_5$ groups onto the coordinated =CR$_2$ moiety of electrophilic alkylidenes. The observation of rather similar rates for both transformations is unprecedented and not yet fully understood.

Acknowledgements

We thank the Ministerio de Educación y Ciencia and the Junta de Andalucía for financial support. We are very grateful to our co-workers, whose names appear in the references, for their important contributions to this project.

References

[1]   Collman, J. P.; Hegedus, L. S; Norton, J. R.; Finke, R. G. *Principles and Applications of Organotransition Metal Chemistry*; University Science Books: Mill Valley, CA, **1987**.

[2]   Arndtsen, B. A.; Bergman, R. G.; Mobley, T. A.; Peterson, T. H. *Acc. Chem. Res.* **1995**, *28*, 154.

[3]   Maitlis, P. M.; Long, H. C.; Quyoum, R.; Turner, M. L.; Wang, Z. *J. Chem. Soc., Chem. Commun.* **1996**, 1.

[4]   Pérez, P. J.; Poveda, M. L.; Carmona, E. *Angew. Chem., Int. Ed. Engl.* **1995**, *34*, 231.

[5]   Paneque, M.; Taboada, S.; Carmona, E. *Organometallics* 1996, *15*, 2678.

[6]   Paneque, M.; Poveda, M. L.; Rey, L.; Taboada, S.; Carmona, E.; Ruiz, C. *J. Organomet. Chem.* **1995**, *504*, 147.

[7]   Alvarado, Y.; Boutry, O.; Gutierrez, E.; Monge, A.; Nicasio, M. C.; Poveda, M. L.; Pérez, P. J.; Ruíz, C.; Bianchini, C.; Carmona, E. *Chem. Eur. J.*, **1997**, *6*, 860.

[8]   Boutry, O.; Gutiérrez, E.; Monge, A.; Nicasio, M. C.; Pérez, P. J.; Carmona, E. *J. Am. Chem. Soc.* **1992**, *114*, 7288.

[9]   Gutiérrez, E.; Monge, A.; Nicasio, M. C.; Poveda, M. L.; Carmona, E. *J. Am. Chem. Soc.* **1994**, *116*, 791.

[10]   Alvarado, Y.; Daff, P. J.; Pérez, P. J.; Poveda, M. L.; Sánchez-Delgado, R.; Carmona, E. *Organometallics* **1996**, *15*, 2192.

[11]   Alvarado, Y.; Atencio, R.; Daff, P. J.; Pérez, P. J.; Poveda, M. L.; Sánchez-Delgado, R.; Carmona, E. unpublished results.

[12]   Pérez, P. J.; Poveda, M. L.; Rey, L.; Carmona, E. unpublished results.

[13]   Brookhart, M.; Grant, B.; Volpe, A. F. Jr *Organometallics* **1992**, *11*, 3920.

[14]   Alías, F. M.; Poveda, M. L.; Sellin, M.; Carmona, E. to be submitted.

[15]   Parkin, G.; Bunel, E.; Burger, B. J.; Trimmer, M. S.; Van Asselt, A.; Bercaw, J. E. *J. Mol. Catal.* **1987**, *41*, 21.

[16]   (a) Francisco M. Alías. PhD Thesis, University of Sevilla, **1997**. (b) Alías, F. M.; Gutiérrez, E.; Monge, A.; Poveda, M. L.; Sellin, M.; Carmona, E. to be submitted.

Asymmetric Allylic Substitutions with Pd Complexes of Phosphinooxazolines as Ligands - Preparative and Mechanistic Aspects

Günter Helmchen*, Henning Steinhagen, Michael Reggelin[‡] and Steffen Kudis

Organisch-Chemisches Institut, Universität Heidelberg, D-69120 Heidelberg, Germany
[‡]Organisch-Chemisches Institut, Universität Frankfurt D-60439 Frankfurt/Main, Germany

Abstract: Phosphinooxazolines were found to be highly effective ligands in Pd catalyzed asymmetric substitutions of allylic compounds. Malonates, amines and nitronates were employed as nucleophiles. Mechanistic aspects were studied by NMR and x-ray crystal structure analysis. Particular emphasis was placed on the identification of Pd-bound intermediates in order to gain a clear understanding of the enantioselective step in the catalytic cycle.

1 Introduction

Pd catalyzed asymmetric C-C and C-N bond forming substitutions at allylic compounds are being employed by many research groups (ref. 1). These reactions essentially involve oxidative addition of a Pd° fragment to a chiral, racemic allylic derivative to yield a π-allyl complex (Scheme 1) that with a nucleophile furnishes a chiral product. It was logical to try to achieve enantioselectivity in this preparatively useful reaction with the help of chiral ligands L*. Remarkably, the C_2-symmetric chelate diphosphines giving excellent results in hydrogenations, i.e., CHIRAPHOS, BINAP etc., gave disappointing results in allylic substitutions, particularly so with cyclic allylic substrates. Only in the early 1990s, it was demonstrated with bisoxazolines (ref. 2) and new types of diphosphines (ref. 3) that high enantioselectivity is possible with proper combinations of a substrate and a C_2-symmetric ligand.

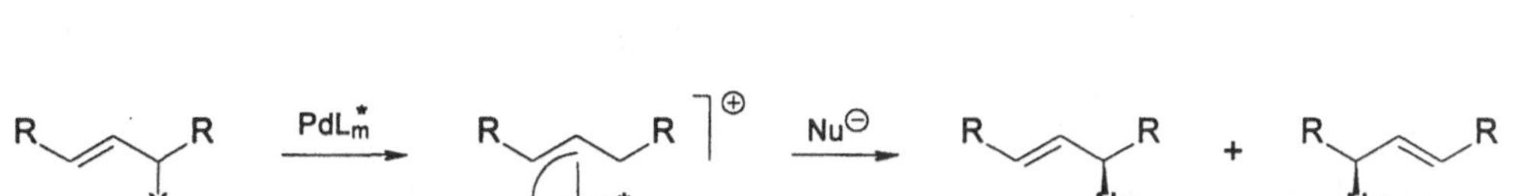

Scheme 1

In our own work we launched an attempt to develop Pd complexes of non-C$_2$-symmetric ligands for allylic substitutions. Our approach was inspired by work of Faller, who had realized remarkable high enantioselectivity in stoichiometric substitutions at cyclopentadienyl-molybdenum complexes. Fallers success is due to the use of ligands (CO, NO) with different electronic rather than steric properties (ref. 4). In catalysis this approach would involve the use of a chiral chelate ligand with two electronically distinct donor centers. This was apparently first probed by Caesarotti with the ligand PROLOPHOS with two slightly, by bonding to O or N, differentiated P atoms (ref. 5). A fairly low level of enantioselectivity was obtained. We felt that a more pronounced difference in electronic as well as steric properties was required and, therefore, chose the combination of a hard, N, and a soft, P, S or Se, donor. Realization of this proposal relied on the proven usefulness of the oxazoline moiety. Aryl groups were preferred as substituents at P because triarylphosphines are normally stable to air. These considerations led to the development of phosphinooxazoline (PHOX) ligands **1** (ref. 6). The same concept was independently pursued by the groups of Pfaltz (ref. 7), Williams (ref. 8) and, with a different type of P-N chelate ligands (QUINAP), J.M. Brown (ref. 9a).

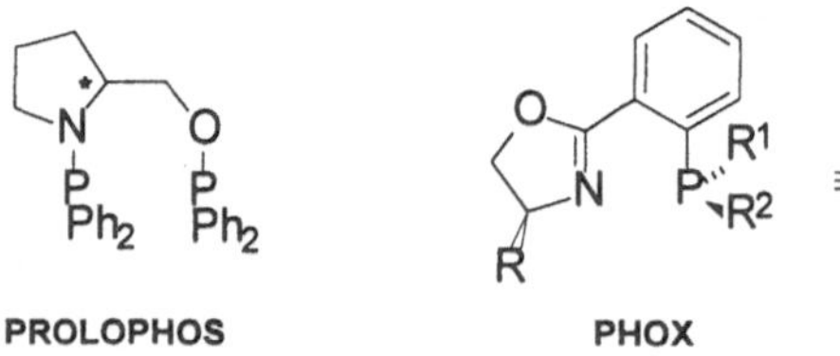

PROLOPHOS   PHOX   **1**

2 Preparation of Phosphinooxazolines

Oxazolines are available from amino alcohols which can be prepared from the chiral pool of natural amino acids (Scheme 2). There are many established routes from amino alcohols to oxazolines (ref. 10). Usually, one step procedures are employed (ref. 11, 12, 13). However, better yields are often achieved with a three step procedure involving formation of an N-acyl amino alcohol, activation of the OH group and ring closure with base (ref. 14).

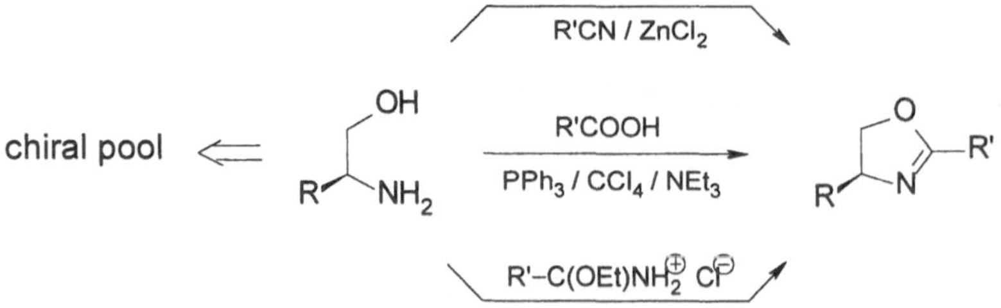

Scheme 2

Introduction of phosphorus is described in Scheme 3 (ref. 14). *Nucleophilic* substitution of fluorine with a diarylphosphide proceeds with 70-90% yield. In the case of stereogenic phosphorus, with, *e.g.*, Ph and 1-naphthyl or 2-biphenyl substituents, ca. 7:3 mixtures of diastereomers are formed which can be easily separated by flash chromatography or crystallization. *Electrophilic* phosphorus and also sulfur and selenium compounds can be reacted with the Grignard compound obtained from the corresponding bromo derivative and activated magnesium. Yields with halophosphines are only 30-50%, but the P-diastereomers are formed with selectivity of $\geq 85:15$. Configurations at phosphorus were determined by crystal structure analysis.

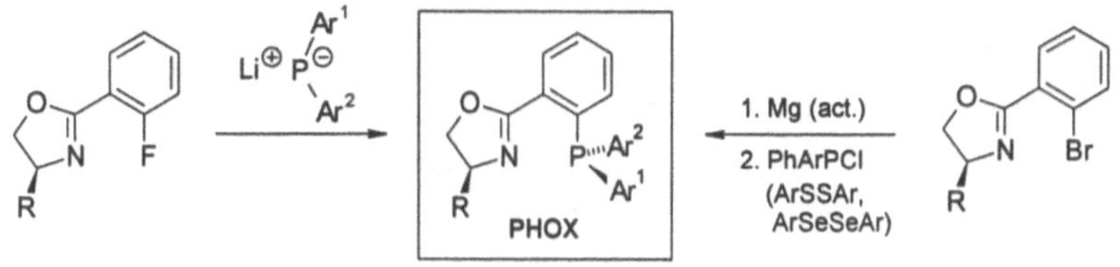

	R	Ar^1	Ar^2
A	*i*Pr	2-biphenylyl	phenyl
epi-A	*i*Pr	phenyl	2-biphenylyl
B	*i*Pr	2-biphenylyl	3,5-(CF$_3$)$_2$-C$_6$H$_3$
epi-B	*i*Pr	3,5-(CF$_3$)$_2$-C$_6$H$_3$	2-biphenylyl
C	phenyl	2-biphenylyl	phenyl
epi-C	phenyl	phenyl	2-biphenylyl

Scheme 3

3 Mechanistic Aspects

Malonates are the nucleophiles most often used in allylic substitutions (ref. 6, 7, 8). In addition, a variety of other nucleophiles were investigated: amines (ref. 15), N-acylamides (ref. 15), nitro compounds (ref. 16), and sulfinates (ref. 17). These nucleophiles are less reactive than malonates; however, enantioselectivities are very similar and the steric course is *grosso modo* independent of the nucleophile.

Rationalizing the steric course of the nucleophilic substitution is difficult because there are two diastereomeric π-allyl complexes, designated *exo*- and *endo*-isomer here (**4x, 4n** in Scheme 4). The products can be formed via four pathways and the preferred product can arise by reaction at the allylic C *trans* to P of the *exo* or *cis* to P at the *endo* isomer. For the decision between these possibilities, a postulate of Bosnich (ref. 18) was helpful. Based on the assumption of an early transition state, this postulate states that the more abundant isomer is the more reactive one (ref. 19). The more abundant is generally the *exo* isomer (see below). In conjunction with the known configuration of the products of allylic substitutions it is deduced that the nucleophile preferentially attacks the carbon *trans* to P (ref. 20).

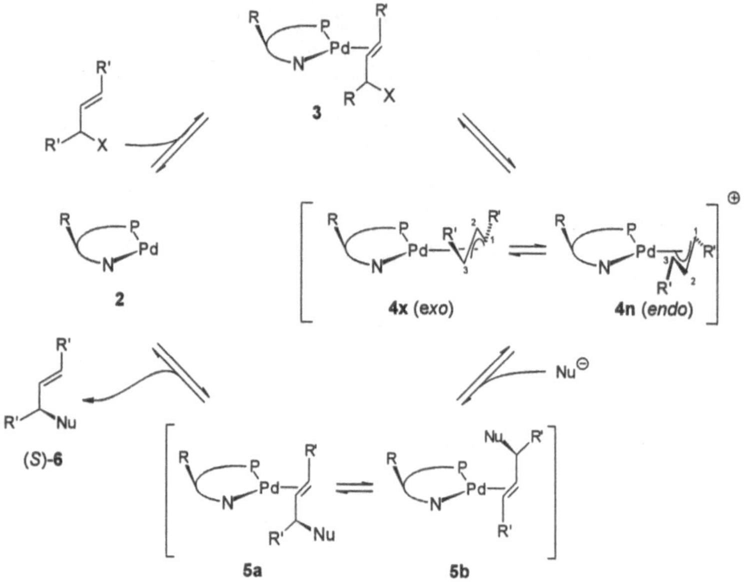

Scheme 4

There is so far no direct proof for this mechanistic proposal. There is, however, support by circumstantial evidence from ^{13}C NMR shifts, NMR studies on the interconversion of *exo* and *endo* diastereomers, and x-ray crystal structures (ref. 20). The crystal structures, for examples see Figure 1, allowed us to understand why *exo* isomers are more stable than *endo* isomers. There are several important general observations: (*i*) The "inner" chelate cycle PdNCCCP is non-planar. (*ii*) A consequence of non-planarity is conformational nonequivalence of the substituents at P, one is axially, the other equatorially arranged. Aryl ring planes are nearly perpendicular to each other, with the axial group pointing its edge, the equatorial its face to the metal. (*iii*) The substituent of the oxazoline ring occupies an axial position in a way that only the equatorial H can interact with the allylic moiety. The dominating interaction is the one with the equatorial aryl group at P. Minimization of this interaction is the reason for preference of the *exo* over the *endo* diastereomer.

The analysis of x-ray structures of π-allyl complexes was helpful for an understanding of equilibria of π-allyl complexes and structural aspects important for selectivity. In addition, the reaction course was studied by modern 2D NMR spectroscopic methods in order to gain a more detailed insight into the mechanism of enantiodiscrimination in the catalytic reaction (ref. 21).

Initially the stoichiometric reaction between the π-allyl complexes **4** (R = *i*Pr, R' = Ph; 10:1 mixture of **4x** and **4n**) and sodium dimethyl malonate as a nucleophile was examined. A sample containing the π-allyl complexes **4** and the nucleophile was prepared at low temperature and warmed up inside the NMR probehead to rt. The progress of the reaction was then monitored by ^{31}P NMR spectroscopy (Figure 2). During the course of the reaction **4x** and **4n** are always in rapid equilibrium. As first new species a compound with a singlet at δ = 11.18 ppm appears, whose concentration reaches a maximum already after 90 s before it

108

is consumed whithin a few minutes. This compound is the Pd^0 alkene complex **5a**. Consumption of **5a** is accompanied by the appearance of a new species with a characteristic AB spin system ($\delta_A^{korr} = 9.41$, $\delta_B^{korr} = 12.91$, $^2J_{P,P} = 128$ Hz) which is the main component after a reaction time of ca. 1 h. We assign structure **7** to this long-lived intermediate. The precise geometry of this complex could not yet be determined.

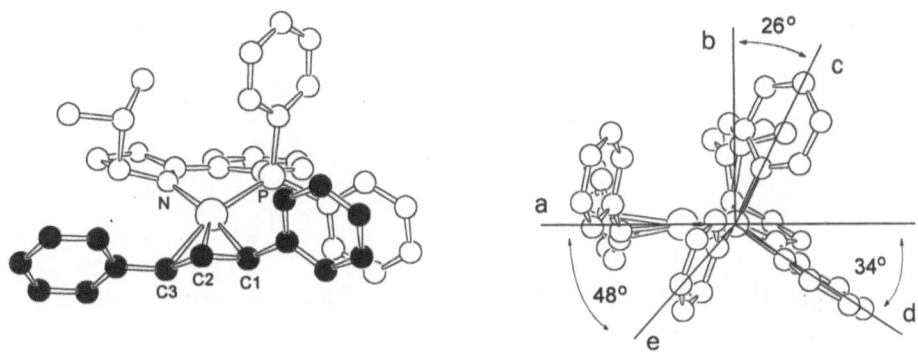

Only traces of the metal-free product (*S*)-**6**, which is formed from **7**, were detected in the reaction mixture when the reaction was stopped at a conversion of approximately 50%. The concentration of (*S*)-**6** only increased when conversion exceeded 50%. Therefore, complex **7** is a stable by-product of the stoichiometric reaction. For the determination of the constitution of the mechanistically meaningful transient species **5a**, the reaction was carried out with $^{13}C_3$-labeled NaDMM at -20 to -30 °C and stopped after 2 min by cooling to -78 °C. A sample prepared in this way contained ca. 75% of the phosphorus in the form of the Pd^0 alkene complex **5a** and was stable for several weeks at -78 °C in an inert atmosphere.

Assignment of the resonances of all NMR-active nuclei was possible by use of $^{13}C_3$-labeled malonate with a large set of 2D NMR experiments (^{1}H,^{1}H-COSY, -TOCSY, -NOESY, -ROESY, ^{13}C,^{1}H-HSQC, -HMBC, -HMQC-TOCSY and ^{31}P,^{1}H-HMBC). By quantitative analysis of NOE- and ROE data information on distances of H nuclei could be

Figure 1 Front view (left) and side view (right) of the x-ray crystal structure of complex **4x** (R = *i*Pr, R' = Ph)

obtained, which is only in accordance with conformer **5a** and not with **5b** (Scheme 5). For example, characteristic is the NOE between 2-H and 17-H which is altered during the transformation of **4x** to **5a**. The value of this NOE contact in **5a** corresponds to a reduction of the distance between 2-H and 17-H by 1.4 Å, as compared with the distance in **4x** determined by x-ray diffraction analysis.

Statements concerning the mechanism of the allylic substitution require the following plausible assumptions: (a) The attack of malonate at **4x** under formation of **5a** proceeds via a "least motion" reaction path, i.e., a rotation of 30° from **4x** to **5a** is the main process, (b) rotation of the alkene fragment relative to the N-Pd-P-plane in complex **5a** is sufficiently slow so that equilibration between **5a** and **5b** is slow compared with the rate of their formation.

Assumption (b) is supported by experiments in which the reaction was carried out in a temperature gradient (-78 °C to rt) and monitored by ^{1}H NMR spectroscopy. In this experiment broadening of the resonances of complex **5a**, which would be expected for a dynamic exchange process, was not found. Accepting assumption (b), the configuration of the more reactive allyl complex **4x** is conserved in the Pd^0 alkene complex **5a**. Considering the known absolute configuration of the product, an attack of the nucleophile *trans* to phosphorus at the *exo* π-allyl complex **4x** can be derived. This result is in accordance with earlier interpretations (ref. 9, 20a), but is here based on a precisely characterized Pd^0 alkene product complex in the Pd-complex catalyzed allylic substitution.

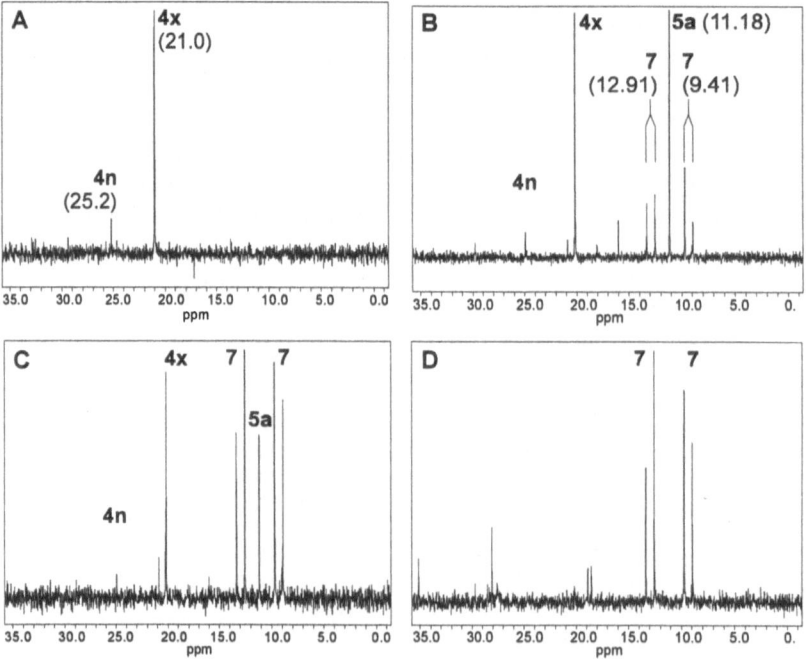

Figure 2 ^{31}P{^{1}H} NMR-Spectra recorded at various reaction times (-60 °C). **A**: equilibrating p-allyl complexes **4x** and **4n**; **B, C, D**: reaction mixtures 370 s, 500 s and 1 h after addition of the sodium dimethyl malonate.

110

4 Slim Substrates: Big Problems and their Solutions

The front view in Figure 1 shows quite clearly that the chiral ligand mainly provides interactions at its wings. It appears likely that allylic systems with big substituents, such as phenyl, should display high *exo-endo*-ratios and enantioselectivity, but narrow systems, with small substituents or cyclic compounds, might give low selectivity. This is exactly what was found. In Scheme 6 substrates are ordered according to their "broadness" and it is quite remarkable how closely the ee values parallel the steric extension (the isopropyl case is taken from ref. 7). The importance of this parameter is further underlined by NMR data of the corresponding π-complexes: ratios of 1.8:1, 4:1, and 9:1 for the cyclohexenyl, the 1,3-dimethyl- and the 1,3-diphenylallyl derivative (CDCl₃ solution), respectively. The cause of enantioselectivity though is a kinetic phenomenon, *i.e.*, a function of differing reaction rates at the allylic termini in *exo* and *endo* complexes. Recent results indicate a significant difference of relative reaction rates in acyclic and cyclic substrates with respect to *exo* and *endo* isomers.

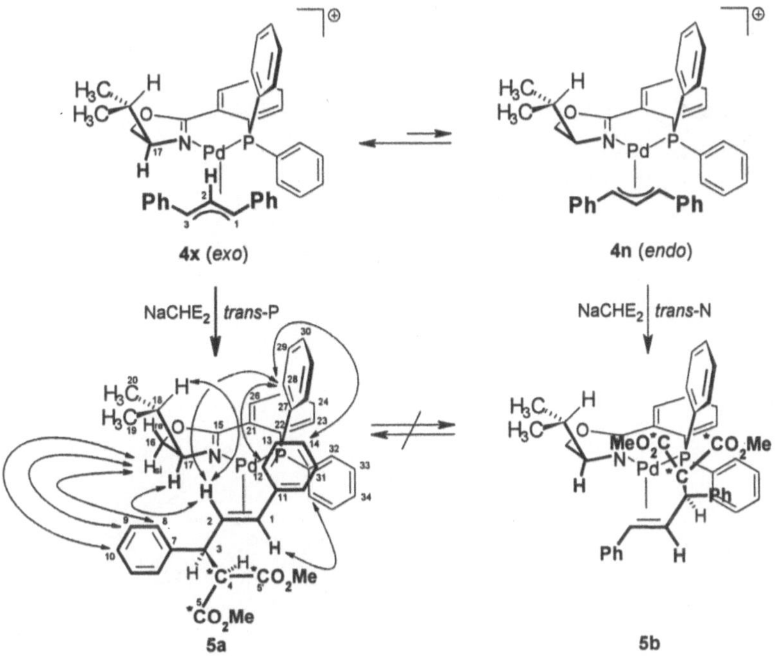

Scheme 5

The rather clear relationship between the size of the π-allyl moiety and enantioselectivity was very satisfactory, as it was in excellent agreement with our mechanistic assumptions. However, the production of racemic product from the cyclic substrate (cf. Scheme 6) was somewhat unsatisfactory from a preparative point of view. Clearly, a ligand was required that would reach into the narrow area directly above or below the allylic sp^2 centers. As such ligands the biphenyl derivatives **A-C** (Scheme 3) were conceived (ref. 22). We were able to obtain a high resolution x-ray crystal structure of the complex $[Pd(\eta^3\text{-}C_6H_9)(\mathbf{A})]SbF_6$ derived from ligand **A** (R = *i*-Pr), and indeed, in the crystal conformer α of the cyclohexenyl π-allyl Pd complex is found in which the phenyl of the 2-biphenyl group is located directly above the allylic moiety as described in Scheme 7.

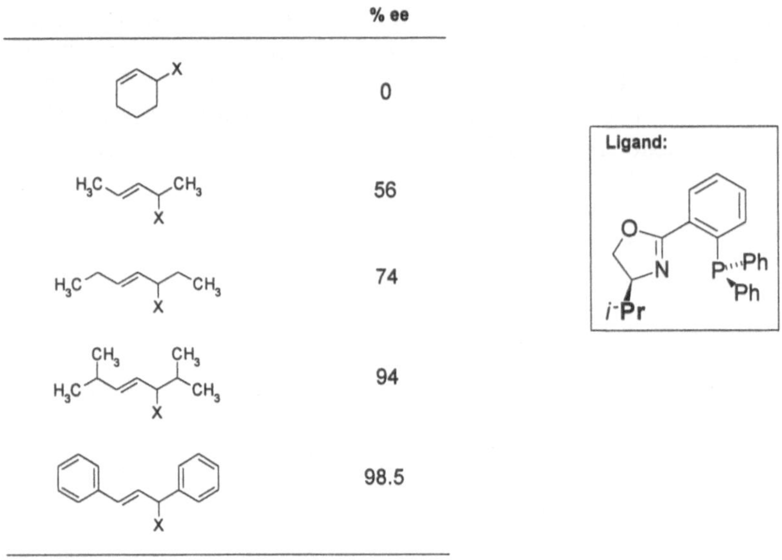

	% ee
(cyclohexenyl-X)	0
H_3C—CH_3 (X)	56
H_3C—CH_3 (X)	74
CH_3 CH_3 H_3C—CH_3 (X)	94
(Ph—Ph, X)	98.5

Ligand:

(structure of oxazoline-phosphine ligand with *i*-Pr, P"Ph, Ph)

Scheme 6

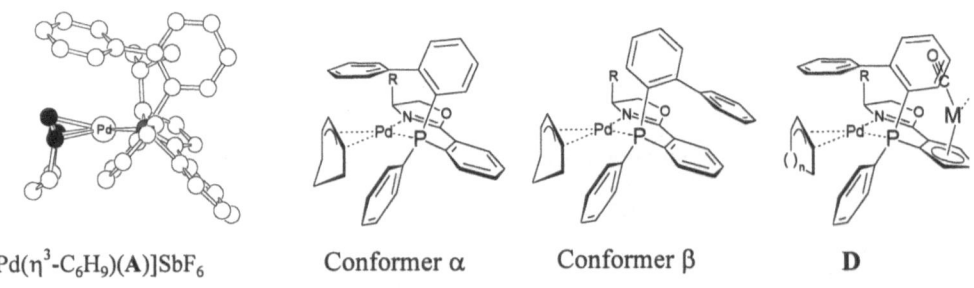

$[Pd(\eta^3\text{-}C_6H_9)(\mathbf{A})]SbF_6$ Conformer α Conformer β **D**

Scheme 7

Despite the fact that conformer α is favored in the solid state enantioselectivities resulting with the ligand **A** (cf. Scheme 3) were not satisfactory. Distinct dependence on ring size of the substrate and, to a certain extent, on reaction conditions are apparent from the data given in Scheme 8. In order to enhance electronic effects, the ligand **B** with electron withdrawing CF_3 groups was prepared. With this ligand improved enantioselectivity was obtained with methylene chloride as solvent.

Nevertheless, results were still not satisfactory. A hint towards improvement was gained by an NMR analysis of the complex $[Pd(\eta^3\text{-}C_6H_9)(A)]SbF_6$ which indicated the existence of several conformers in solution, including the unfavorable conformer β with the crucial phenyl group rotated away from the allylic moiety. In order to destabilize conformers of this type, the cymantrene-based ligand **D** (Scheme 7) was conceived and could be prepared in a reasonably straightforward way (ref. 23). This ligand induces excellent catalytic activity and displays long shelf life. Conformers analogous to β are apparently destabilized by interaction with the manganese tricarbonyl group. High enantioselectivity with this new ligand was indeed obtained.

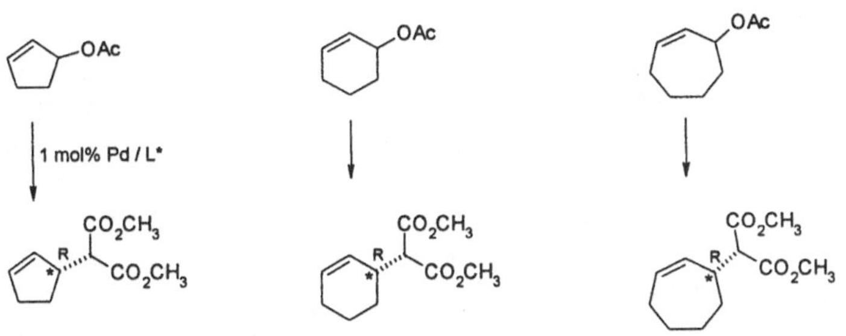

Scheme 8

Ligand	Method	5-membered ring [% ee]	6-membered ring [% ee]	7-membered ring [% ee]
A	a	56	51	83
B	a	63	53	83
B	b	64	72	85
D	c	95	93	> 99

Method a: 1.5 eq $LiCH(COOCH_3)_2$, dioxane, rt.
Method b: 2.5 eq $CH_2(COOCH_3)_2$, BSA method, methylene chloride, 0 °C.
Method c: 1.5 eq $NaCH(COOCH_3)_2$, dimethylformamide, -50 °C to 0 °C.

Prior to the development of the new ligand **D** high enantioselectivities with cyclic substrates were achieved with salts of the easily available β-phosphinocarboxylic acid **E** as chiral ligand (ref. 24): enantiomeric excess of 85, 98 and >99 % ee for the 5-, 6- and 7-membered ring derivatives, respectively [with $LiCH(COO\text{-}t\text{-}Bu)_2$ as nucleophile]. Products with S-configuration are formed with **E**.

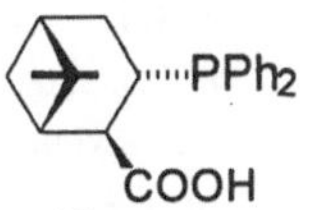

E

Acknowledgements

This work was supported by the Deutsche Forschungsgemeinschaft (SFB 247) and the Fonds der Chemischen Industrie.

5 References and Notes

1. (a) T. Hayashi in *Catalytic Asymmetric Synthesis*, I. Ojima (Ed.), VCH, Weinheim, 1993, pp 325-365; (b) T. Lübbers, P. Metz in Houben-Weyl E21, *Stereoselective Synthesis*; G. Helmchen, R.W. Hoffmann, J. Mulzer, E. Schaumann (Eds.), 1995, pp. 2371-2473 and 5643-5676; (c) B.M. Trost, D.L. Van Vranken, *Chem. Rev.* **1996**, *96*, 395-422.

2. (a) D. Müller, G. Umbricht, B. Weber, A. Pfaltz, *Helv. Chim. Acta* **1991**, *74*, 722-240; (b) U. Leutenegger, G. Umbricht, C. Fahrni, P. von Matt, A. Pfaltz, *Tetrahedron* **1992**, *48*, 2143; (c) P. von Matt, G.C. Lloyd-Jones, A.B.E. Minidis, A. Pfaltz, L. Macko, M. Neuburger, M. Zehnder, H. Rüegger, P.S. Pregosin, *Helv. Chim. Acta* **1995**, *78*, 265-284.

3. B.M. Trost, D.L. Van Vranken, C. Bingel, *J. Am. Chem. Soc.* **1992**, *114*, 9327-9343.

4. J.W. Faller, K.-H. Chao, H.H. Murray, *Organometallics* **1984**, *3*, 1231-1240, and earlier work cited therein.

5. E. Cesarotti, F. Demartin, M. Grassi, L. Prati, *J. Chem. Soc., Dalton Trans.* **1991**, 2073-2082.

6. J. Sprinz, G. Helmchen, *Tetrahedron Lett.* **1993**, 1769-1772.

7. P. von Matt, A. Pfaltz, *Angew. Chem. Int. Ed. Engl.* **1993**, *32*, 566-567.

8. (a) G.J. Dawson, C.G. Frost, J.M.J. Williams, S.J. Coote, *Tetrahedron Lett.* **1993**, *34*, 3149-3150; (b) S.J. Coote, G.J. Dawson, C.G. Frost, J.M.J. Williams, *Synlett* **1993**, 509-510.

9. (a) J.M. Brown, D.I. Hulmes, P.I. Guiry, *Tetrahedron* **1994**, *50*, 4493-4506; (b) P.B. Mackanzie, J. Whelan, B. Bosnich, *J. Am. Chem. Soc.* **1985**, *107*, 2046-2054.

10. T.G. Gant, A.I. Meyers, *Tetrahedron* **1994**, *50*, 2297-2360.

11. G. Koch, G.C. Lloyd-Jones, O. Loiseleur, A. Pfaltz, R. Prétôt, S. Schaffner, P. Schnider, P. von Matt, *Recl. Trav. Chim. Pays-Bas* **1995**, *114*, 206-210.

12. J.V. Allen, S.J. Coote, G.J. Dawson, C.G. Frost, C.J. Martin, J.M.J. Williams, *J. Chem. Soc., Perkin Trans. 1* **1994**, 2065-2072, and literature cited therein.

13. H. Vorbrüggen, K. Krolikiewics, *Tetrahedron* **1993**, *49*, 9353-9372, and earlier work cited therein.

14. Full paper with experimental details: M. Peer, J.C. de Jong, M. Kiefer, Th. Langer, H. Rieck, H. Schell, P. Sennhenn, J. Sprinz, H. Steinhagen, B. Wiese, G. Helmchen, *Tetrahedron* **1996**, *51*, 7547-7583.

15. P. von Matt, O. Loiseleur, G. Koch, A. Pfaltz, C. Lefeber, T. Feucht, G. Helmchen, *Tetrahedron: Asymmetry* **1994**, *5*, 573-584.

16. H. Rieck, G. Helmchen, *Angew. Chem. Int. Ed. Engl.* **1995**, *34*, 2687-2689.

17. H. Eichelmann, H.-J. Gais, *Tetrahedron: Asymmetry* **1995**, *6*, 643-646.

18. B. Bosnich, P.B. Mackenzie, *Pure Appl. Chem.* **1982**, *54*, 189-195.

19. Arguments in favor of a late transition state have been presented for allylic substitutions carried out with QUINAP as chiral ligand (ref. 9). Assumption of a late transition state also leads to preference for substitution *trans* to phosphorus.

20. (a) J. Sprinz, M. Kiefer, G. Helmchen, M. Reggelin, G. Huttner, O. Walter, L. Zsolnai, *Tetrahedron Lett.* **1994**, 1523-1526; (b) G. Helmchen, J. Janssen, M. Kiefer, M. Peer, H. Rieck, P. Sennhenn, J. Sprinz in *Electronic Conference on Trends in Organic Chemistry (ECTOC-1)*, June 12-July 7, 1995, Eds. H. S. Rzepa, C. Leach, J. M. Goodman (CD-ROM), The Royal Society of Chemistry, **1996**; ISBN 0 85404 899 5; (c) C. Breutel, P.S. Pregosin, R. Salzmann, A. Togni, *J. Am. Chem. Soc.* **1994**, *116*, 4067-4068.

21. H. Steinhagen, M. Reggelin, G. Helmchen, *Angew. Chem. Int. Ed. Engl.* **1997**, in print.

22. P. Sennhenn, B. Gabler, G. Helmchen, *Tetrahedron Lett.* **1994**, 8595-8598.

23. Corresponding ligands with a ferrocene or a benzenechromium tricarbonyl moiety were also prepared and tested in catalytic allylic alkylations. A report (S. Kudis, G. Helmchen) on preparations and catalysis data will soon be published.

24. G. Knühl, P. Sennhenn, G. Helmchen, *J. Chem. Soc., Chem. Commun.* **1995**, 1845-1846.

New Preparations and Reactions of Organozinc Reagents

Paul Knochel

Fachbereich Chemie der Philipps-Universität Marburg, Hans-Meerwein-Straße, D-35032 Marburg, Germany

1 Preparation methods of zinc organometallics

Polyfunctional organozinc halides are best prepared by the direct insertion of zinc dust to alkyl iodides [1,7]. This method allows the preparation of organozinc iodides bearing almost all possible organic functionalities with the exception of nitro, azido and hydroxy groups. With primary alkyl iodides, the insertion reaction is usually performed by adding a concentrated solution (3 M) of the alkyl iodide in THF to a suspension of zinc dust (-325 mesh, ca. 3 equiv.) in THF at 40 °C. The zinc dust is treated with 1,2-dibromoethane and TMSCl prior the halide addition [5, 8, 9]. Secondary alkyl iodides react even with zinc dust at 25 °C [5, 9, 10] whereas benzylic bromides undergo an optimal insertion reaction at 0 °C [11]. The insertion of zinc into Csp^2-I bonds is usually less straightforward and may require longer reaction times, higher temperature or the use of a polar solvent [12]. Thus the polyfunctional zinc reagents 1-5 have been prepared in high yield using these procedures (Figure 1).

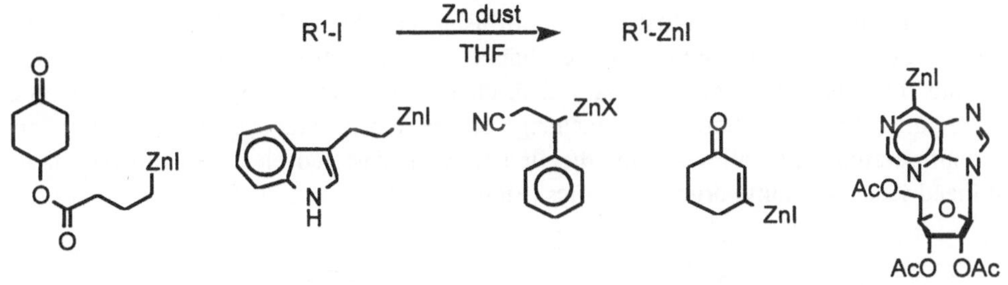

Figure 1

Alternatively, more activated zinc prepared by the reduction of zinc halides (Rieke-zinc) can be used starting with the less reactive aryl iodides or bromides, but also with secondary and tertiary alkyl bromides [16, 17, 18].

Diorganozincs (R^1_2Zn) are more reactive than organozinc halides [7, 19] and undergo transmetalation reactions more readily. They are important for the performance of catalytic asymmetric additions to various aldehydes allowing a general preparation of polyfunctional secondary alcohols in high enantioselectivity [20, 21]. Two general methods have been developed in our laboratory for the preparation of these zinc reagents (Figure 2). The first method involves an iodine-zinc exchange reaction [22] and is suited for primary or secondary alkyl iodides. It requires a catalytic amount of CuX and Et_2Zn or i-Pr_2Zn as starting materials. The second method involves a boron-zinc exchange and starts with functionalized olefins which are hydroborated with Et_2BH [23] and treated in a second step with Et_2Zn or i-Pr_2Zn [23, 24].

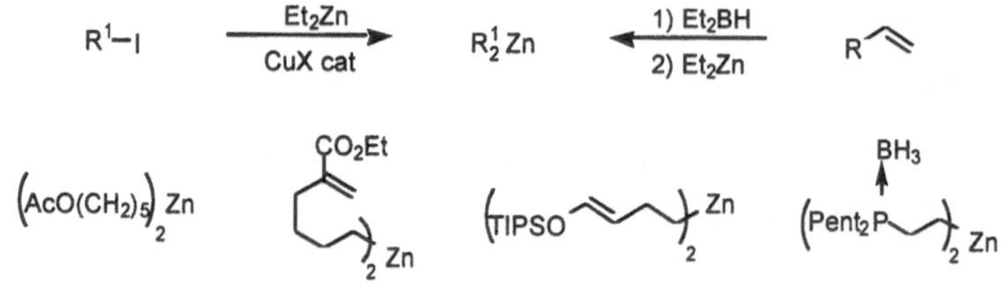

Figure 2

Under these conditions, a broad range of new polyfunctional dialkylzincs like **6-9** have been prepared. The boron-zinc exchange proceeds under significantly milder conditions (0 °C instead of 50 °C) for primary alkyl derivatives and requires only a few minutes compared to several hours which are necessary in the case of the iodine-zinc exchange. Nevertheless the boron-zinc exchange reaction using Et_2Zn has some drawbacks and does not proceed rapidly with hindered secondary alkyldiethylboranes, also a mixture of diastereomeric zinc reagents is produced with diastereomerically pure alkylboranes [23c].

Remarkably, it was found that the use of diisopropylzinc (i-Pr_2Zn) facilitates considerably the boron-zinc and iodine-zinc exchange with *primary and secondary* substrates allowing the preparation of secondary dialkylzincs. By performing the reaction at low temperature (-10 °C) the boron-zinc exchange occurs with several systems with retention of the configuration [27], allowing for the first time, a diastereoselective synthesis of non-stabilized secondary alkyl organometallics (Figure 3) [28].

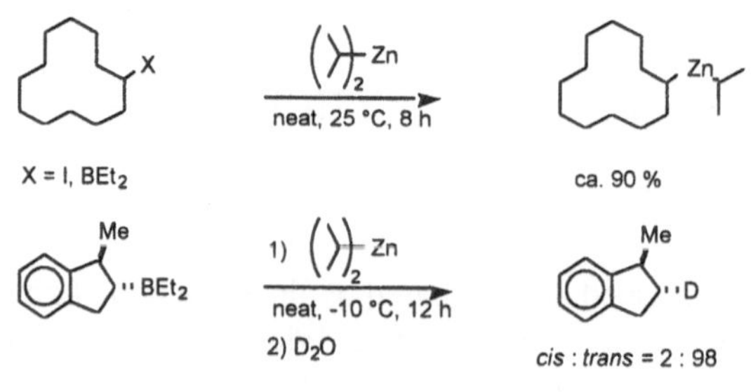

Figure 3

118

A few less general methods are available for the preparation of zinc organometallics. Thus organozinc halides can also be obtained from the alkyl mesylates in the presence of sodium bromide and iodide in *N,N*-dimethylpropyleneurea (DMPU) [29] using zinc dust [30] or by an iodine-lithium exchange performed on functionalized alkenyl or aryl iodides followed by a transmetalation with zinc bromide [31] (Figure 4). A less general preparation of organozinc halides involves the treatment of diorganomercurials with zinc dust in the presence of zinc bromide [32].

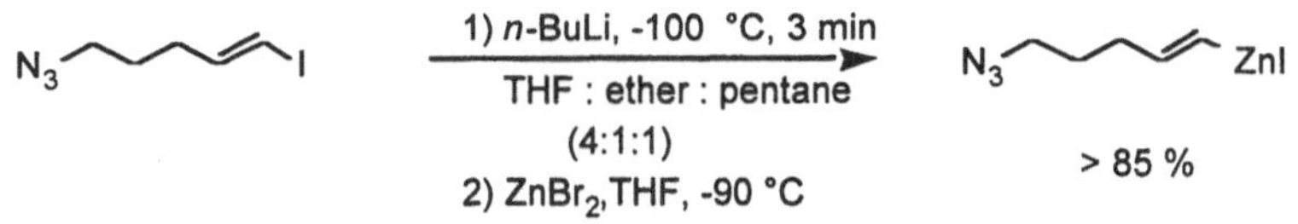

Figure 4

Finally some transition metal (Ni, Mn) catalyzed iodine- or bromine-zinc exchange reactions using Et$_2$Zn constitute a convenient approach to organozinc halides under mild reaction conditions [32, 33].

Diorganozincs can be obtained by the hydrozincation of olefins using diethylzinc and catalytic amounts of Ni(acac)$_2$. The resulting diorganozincs are well suited for applications in asymmetric synthesis [33, 34] (Figure 5).

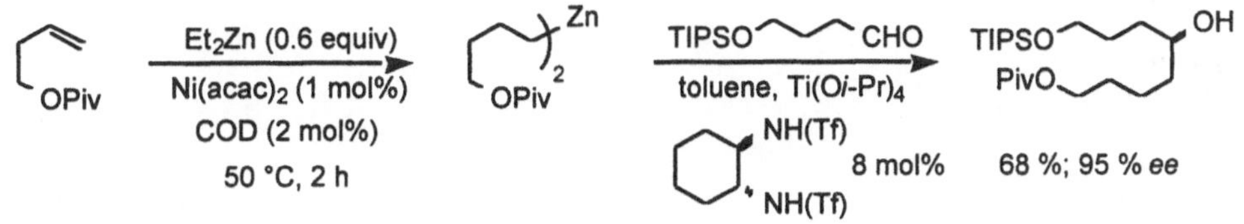

Figure 5

The presence of a heteroatom in the olefin at an allylic or homoallylic position considerably stabilizes the zinc organometallic obtained after the hydrozincation of the olefin. Thus allylic alcohols and amines are especially well suited as substrates for this hydrozincation procedure.

2 Uncatalyzed reactions of zinc reagents

Many organometallic zinc species are too unreactive to undergo reactions with carbon nucleophiles. This general statement cannot be applied to allylic organometallic species which react smoothly with several electrophiles like carbonyl compounds [35, 36], nitriles [37, 38] or triple bonds [39, 40]). Allylic zinc reagents undergo cross-coupling reactions with reactive halides leading to 1,5-dienes. Usually the new carbon-carbon bond is formed from the more substituted end of the allylic system [30]. A very reactive and selective carbon electrophile, tosyl cyanide (TosCN) [41] reacts with various organozinc halides affording polyfunctional nitriles in good yields (Figure 6) [42].

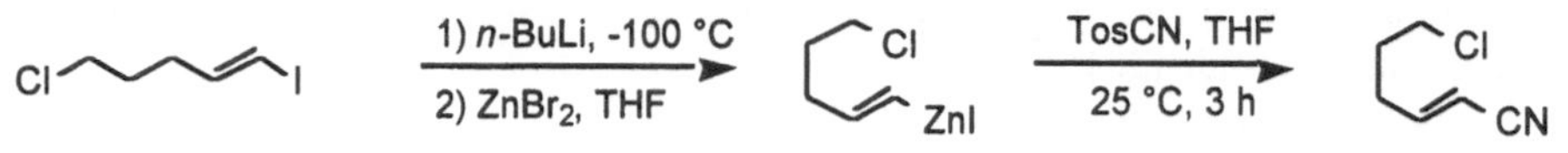

Cl~~~~~I
1) *n*-BuLi, -100 °C
2) ZnBr$_2$, THF
→
Cl
ZnI
TosCN, THF
25 °C, 3 h
→
Cl
CN

10: 72 %

Figure 6

The reaction can be used to convert 1,1-bimetallic reagents of zinc and magnesium like **11** to unsaturated nitriles such as **12** (Figure 7) [42]. The regioselectivity of the reaction of benzylzinc bromide with TosCN affords *o*-methylbenzonitrile **13** whereas the reaction of the corresponding zinc-copper derivative provides the benzyl cyanide **14** (Figure 8).

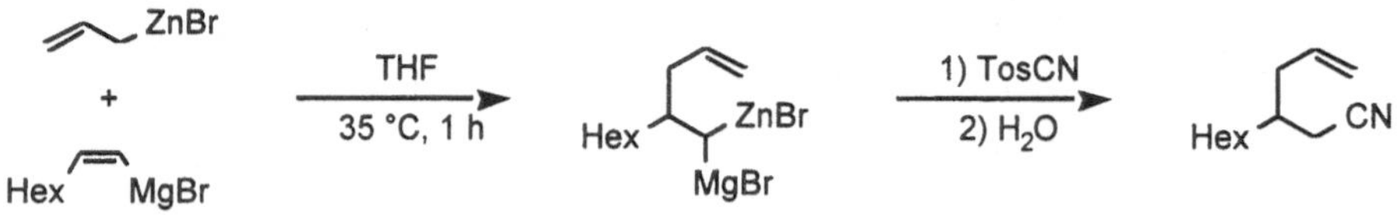

ZnBr
+
Hex—MgBr
THF
35 °C, 1 h
→
Hex
ZnBr
MgBr

11

1) TosCN
2) H$_2$O
→
Hex
CN

12: 93 %

Figure 7

Me
CN
←
TosCN
ZnBr
1) CuCN·2 LiCl
2) TosCN
→
CN

13: 76 %   **14**: 80 %

Figure 8

The reactivity of diorganozincs can be increased by performing the reaction in a polar solvent. Thus, it was found that *N*-methylpyrrolidinone (NMP) is especially well suited and allows the performance of Michael-addition reactions with a variety of double bonds bearing electron-withdrawing groups like enones **15**, alkylidenemalonates or nitroolefins (Figure 9) [43]. One equivalent of NMP is sufficient to promote the addition reaction but the presence of Me$_3$SiCl is mandatory. Concerning the addition to enones, a range of b-monosubstituted enones can be used, however b-disubstituted enones undergo only reluctantly the Michael-addition.

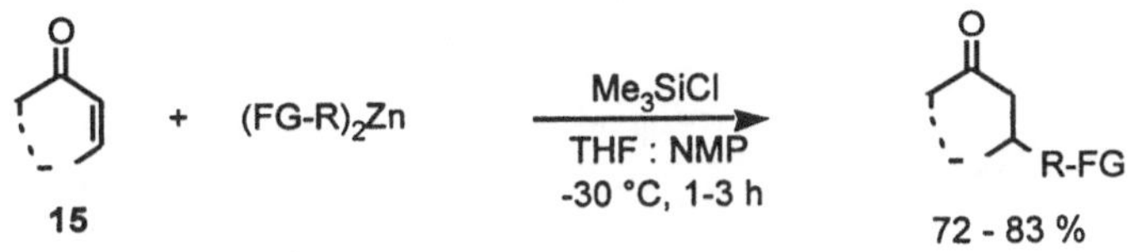

Figure 9

3 Transition metal catalyzed reactions of zinc reagents

Organozinc can react with various electrophiles in the presence of copper(I)salts. Herein we wish to focuss on some new palladium and nickel catalyzed reactions. The iodine-zinc exchange of an alkyl iodide with Et_2Zn is promoted by the addition of small amounts of copper(I) salts like CuCN or CuI. Although the exact reason for this copper catalysis is not known, it has been speculated that the presence of copper(I) salts promotes a radical chain-reaction resulting in the formation of a dialkylzinc species [22b]. Similarly, the addition of other transition metals like nickel and palladium salts promotes radical reactions. These reactions turned out to be preparatively very useful since they allow the performance of radical cyclization reactions but produce an organozinc halide as a final product (Figure 10) [44-49]. The treatment of an unsaturated alkyl halide (X = Br, I) **16** with a palladium(0) or nickel(0) complex produces via a mono-electron transfer [51] a paramagnetic nickel(I) or palladium(I) salt $ML_n(X)$ (M=Ni, Pd) and a radical **17** which undergoes a smooth cyclization reaction and produces after recombination with the transition metal moiety the nickel(II) or palladium(II) species **18**. After transmetalation with a zinc(II) salt, a stable organozinc cyclopentylmethyl derivative of type **19** is produced.

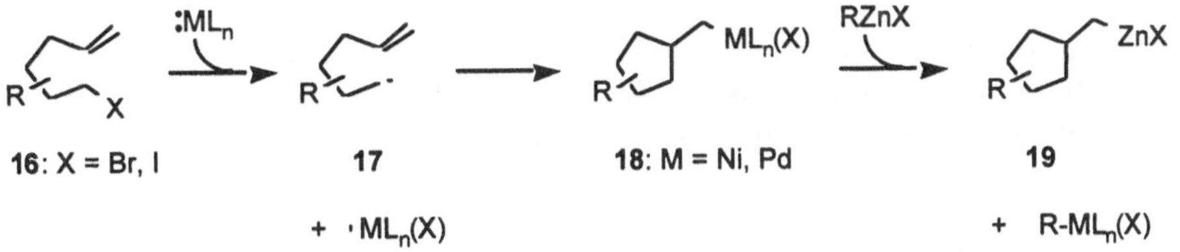

Figure 10

The overall reaction allows to perform intramolecular carbozincation [44-50] via a radical cyclization. This useful preparation of cyclopentylmethylzinc derivatives proceeds with excellent stereoselectivity and allows to prepare quaternary centers. After cyclization, the zinc organometallic can be transmetalated with CuCN·2LiCl and reacted with a broad range of electrophiles like acid chlorides, allylic and alkynyl halides, ethyl propiolate, 3-iodo-2-cyclohexen-1-one and nitroolefins such as nitrostyrene leading to products of type **20** (Figure 11) [44, 49].

121

Figure 11

The cyclization is highly stereoselective according to Beckwith radical cyclization rules [52]. Thus, the allylic benzylether **21** (1:1 mixture of diastereoisomers) undergoes a smooth stereoconvergent cyclization in the presence of Et$_2$Zn (2 equiv) and PdCl$_2$ (dppf)$_2$ (2 mol %) through a radical intermediate **22** which adopts a conformation such as all its substituents are in pseudo-equatorial positions. After allylation with ethyl 2-(bromomethyl)acrylate, the cyclopentane derivative **23** is obtained with > 99:1 *trans* selectivity between substituents at positions 1 and 2 and 95:5 *cis* selectivity between the substituents at positions 2 and 3 (Figure 12).

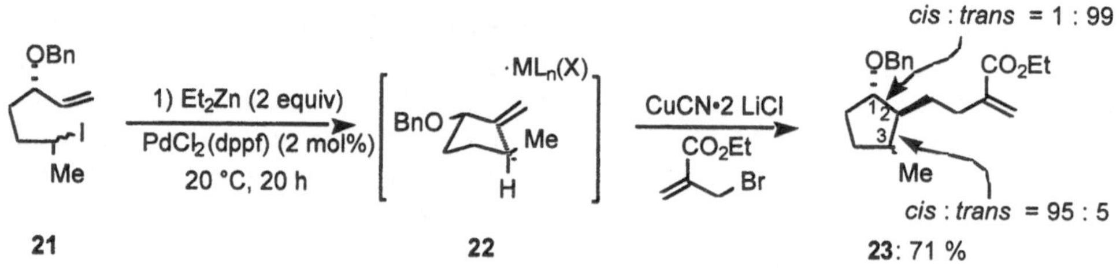

Figure 12

Multiple cyclization reactions are possible as well as preparation of heterocycles (Figure 13) [44-49]. Several natural products like (+)-methyl epijasmonate (**24**) [48] and the antitumor antibiotic (-)-methylenolactocin **25** (Figure 14) [47, 53].

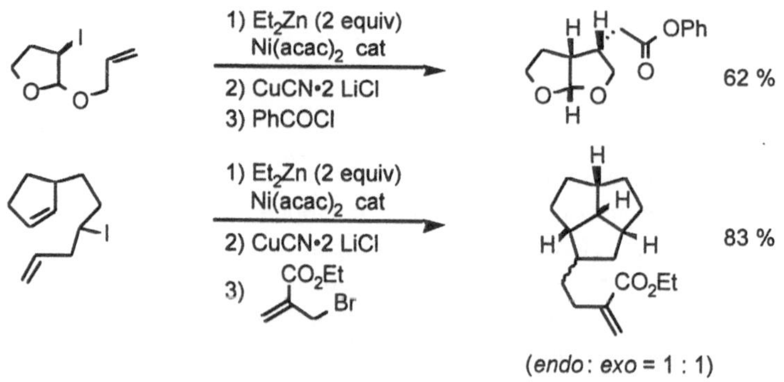

Figure 13

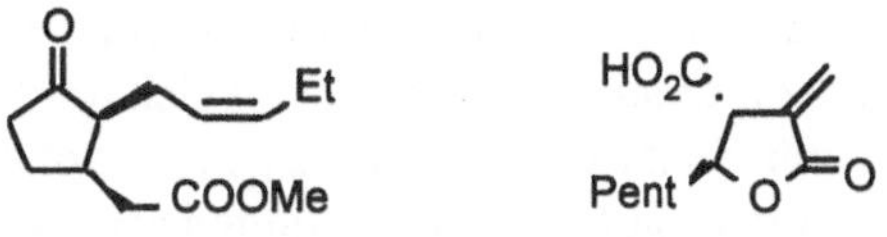

24: (+)-methyl epijasmonate   **25**: (-)- methylenolactocin

Figure 14

The addition of organometallics to internal alkynes is seldom [54-56] and proceeds with moderate stereoselectivity [55]. In the presence of a nickel catalyst like Ni(acac)$_2$ it is possible to add diorganozincs to substituted phenylacetylenes. The reaction proceeds with high stereoselectivity (> 99 % *syn*-addition) and is often highly regioselective. The resulting alkenylzinc organometallics can be quenched with several types of electrophiles (Figure 15) [57].

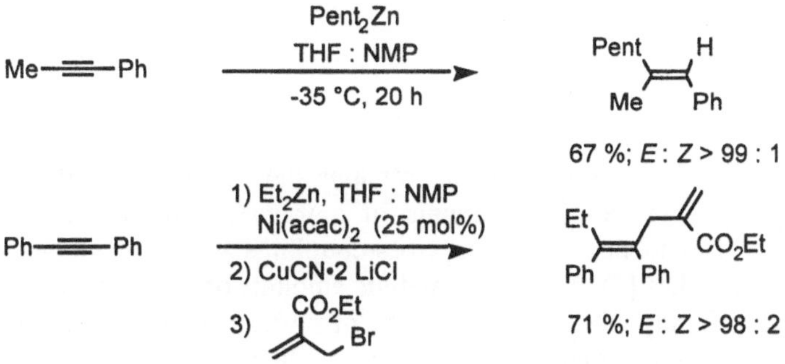

Me—≡—Ph   →   Pent$_2$Zn / THF : NMP / -35 °C, 20 h   →   67 %; *E* : *Z* > 99 : 1

Ph—≡—Ph   →   1) Et$_2$Zn, THF : NMP Ni(acac)$_2$ (25 mol%) 2) CuCN·2 LiCl 3) CO$_2$Et / Br   →   71 %; *E* : *Z* > 98 : 2

Figure 15

Remarkably, this carbozincation procedure works well at low temperature and allows even an efficient addition of the relatively unreactive Me$_2$Zn. It was soon found that diarylzincs undergo the addition even more readily and this reaction has been used to prepare Z-Tamoxifen **26** which is an commercial antitumor drug; (Figure 16) [57].

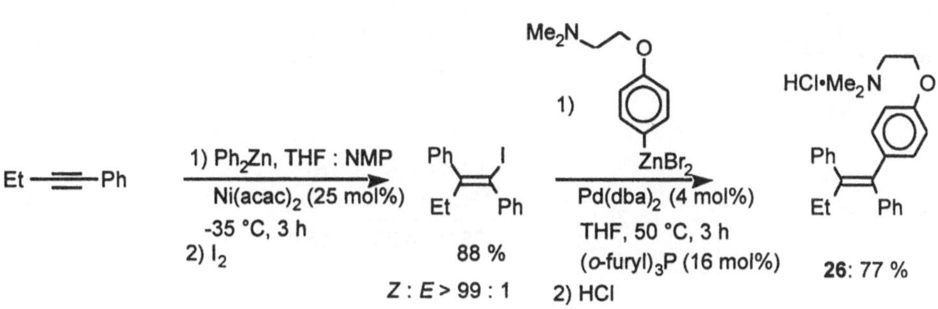

Et—≡—Ph   →   1) Ph$_2$Zn, THF : NMP Ni(acac)$_2$ (25 mol%) -35 °C, 3 h 2) I$_2$ →   88 % *Z* : *E* > 99 : 1   →   1) ZnBr$_2$ Pd(dba)$_2$ (4 mol%) THF, 50 °C, 3 h (*o*-furyl)$_3$P (16 mol%) 2) HCl   →   **26**: 77 %

Figure 16

123

The carbometalation can be performed in an intramolecular way leading in this case to an *unsaturated* diorganonickel species of type **27** which undergoes a rapid reductive elimination furnishing polyfunctional cyclopentylalkylidenes of type **28** (Figure 17). The *syn*-addition is demonstrated by using a phenyl substituted acetylenic iodide like **29** [57].

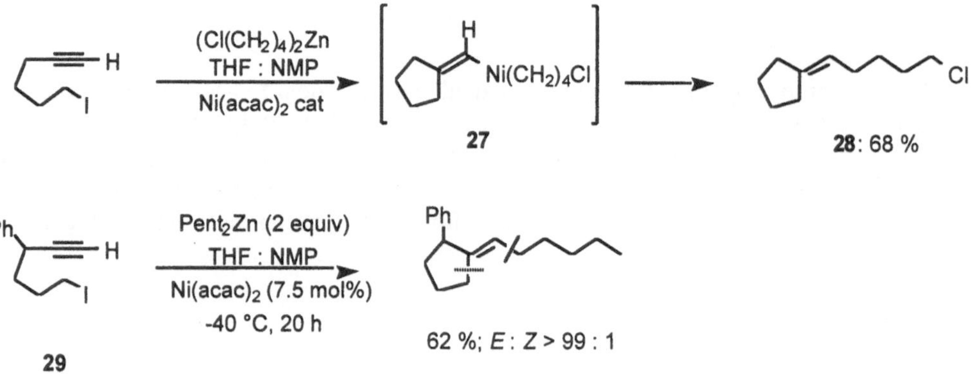

Figure 17

A major problem in organic synthesis remains the performance of Csp^3-Csp^3 cross-coupling reactions using catalytic amounts of a transition metal. Whereas the use of organocuprates allows to perform these cross-couplings with stoichiometric amounts of copper(I) organometallics [2], the use of catalytic amounts of a transition metal salt remains a challenge. The difficulty with the performance of such a reaction is the reductive coupling which is slow with saturated diorganometallic intermediates such as R^1-M-R^2.

By removing electron density from the metal center, such reductive eliminations should be easier [58-60]. Thus, it was observed that whereas the unsaturated alkyl bromide **30** undergoes a smooth cross-coupling reaction with Et_2Zn in the presence of $Ni(acac)_2$ leading to **32**, the corresponding saturated alkyl bromide **31** does not undergo the cross-coupling reaction but instead produces the bromide-zinc exchange product **33**; (Figure 18) [61].

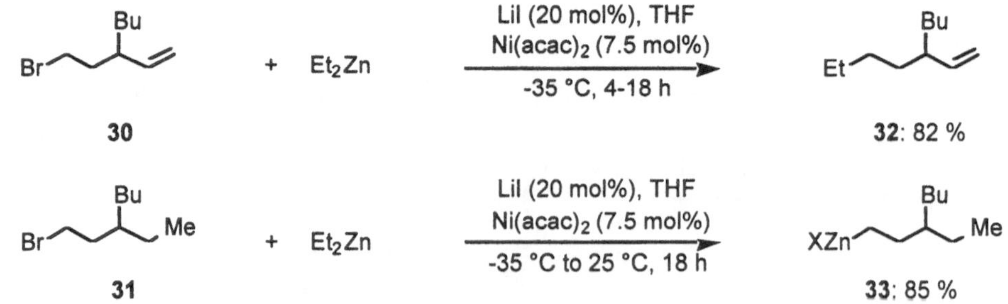

Figure 18

This behavior can be rationalized by assuming that the remote double bond coordinates to the nickel center. Although a double bond coordinated to a metal center acts as a s-donor, it is also a p-acceptor and therefore removes electron density from the metal center and facilitates the reductive coupling reaction. If the coordination to the double bond is too loose or prohibited due to steric hindrance, the reductive elimination does not occur, but instead a ligand exchange reaction occurs leading to the transmetalation product (Figure 19) [61].

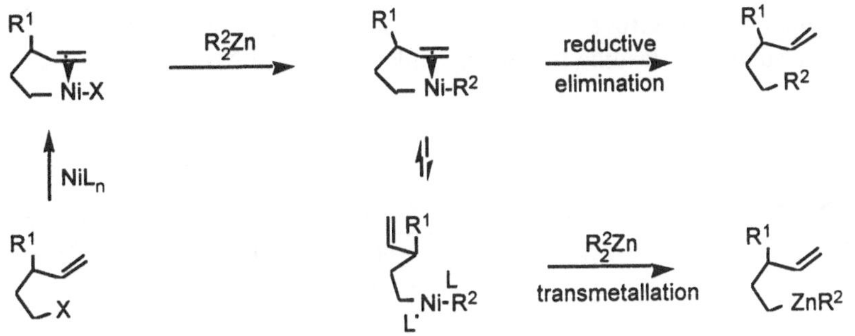

Figure 19

The synthetic applications can be greatly extended by using the polar cosolvent *N*-methylpyrrolidinone (NMP) which allows to couple a variety of polyfunctional zinc organometallics with alkyl iodides bearing a remote double bond like **34** leading to cross-coupling products such as **35** (Figure 20) [61].

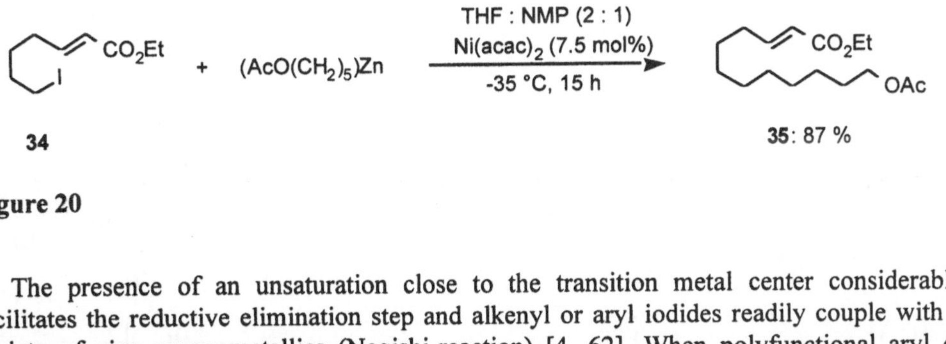

34   **35**: 87 %

Figure 20

The presence of an unsaturation close to the transition metal center considerably facilitates the reductive elimination step and alkenyl or aryl iodides readily couple with a variety of zinc organometallics (Negishi-reaction) [4, 62]. When polyfunctional aryl or alkenyl zinc reagents are used, optimal reaction conditions are obtained by using *bis*(dibenzylidene-acetone)palladium(0) (Pd(dba)$_2$) [63] (4 mol %) and *tris-o-*furylphosphine (TFP) [64] or triphenylphosphine (TPP) as ligand. Under these conditions, the reaction is completed within a few hours at rt. Thus, the aryl bromide **36** undergoes a cross-coupling reaction with a functionalized alkenyl iodide furnishing the polyfunctional styrene **37** (Figure 21) [65].

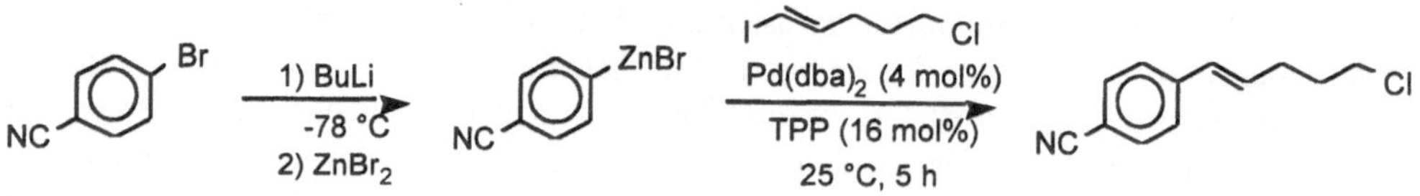

Figure 21

Attempts to apply these reaction conditions to the cross-coupling between alkenyl or aryl zinc derivatives with aryl triflates were disappointing. However, in the presence of 1,1-(diphenylphosphino)ferrocene (dppf) [66], the cross-coupling reaction occurs at 60 °C in satisfactory conditions leading to biphenyls like **38** (Figure 22) [65].

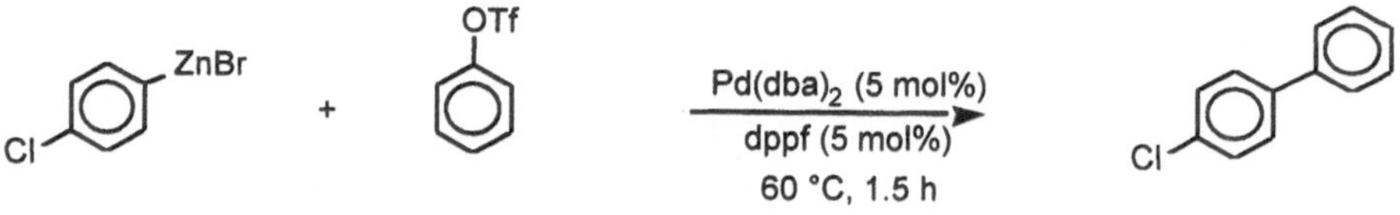

Figure 22

Selective Pd(0)-catalyzed arylations can be performed with aryl iodides bearing a triflate function using an appropriate palladium catalyst. Under these conditions, aromatic iodo-triflates like **39** can play the role of *multi-coupling* reagents [67]. Thus, the reaction of **39** with a functional arylzinc bromide provides the functionalized biphenyl triflate **40** [68]. By using dppf as ligand, these biphenyl triflates can be selectively converted to a terphenyl like **41** (Figure 23) [68].

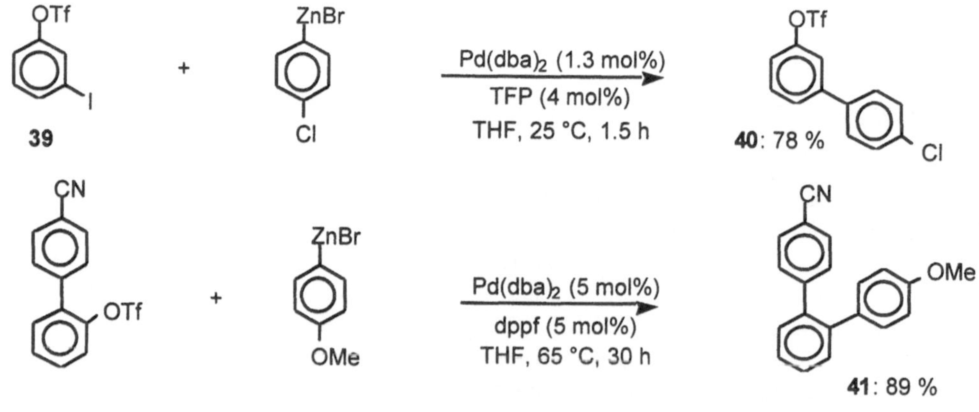

Figure 23

126

4 Conclusion

The reactions between various polyfunctional organozinc derivatives and a range of organic electrophiles provides an expedite access to numerous polyfunctional molecules. The functional group compatibility allows a unique diversity for the organometallic reagent. After transmetalation by the addition of often catalytic quantities of transition metal salts (Cu(I), Pd(II), Ni(II)) smooth cross-coupling reaction can usually be performed in high yields. The applications of organozinc compounds range from asymmetric synthesis, the preparation of biological relevant molecules, new materials to combinatorial chemistry and it can be predicted that broader applications will be developed in the future.

References

[1]   K. Nützel in *Methoden der Organischen Chemie: Metallorganische Verbindungen Be, Mg, Ca, Sr, Ba, Zn, Cd,* Thieme, Stuttgart, **1973**, Vol. 13, 2a, p. 552-858.

[2]   B. H. Lipshutz, S. Sengupta, *Org. React.* **1992**, *41*, 135-631.

[3]   (a) B. Weidemann, D. Seebach, *Angew. Chem., Int. Ed. Engl.* **1983**, *22*, 31-45. (b) M. T. Reetz, *Top. Curr. Chem.* **1982**, *106*, p. 1-53.

[4]   E. Negishi, *Acc. Chem. Res.* **1982**, *15*, 340-348.

[5]   P. Knochel, M. C. P. Yeh, S. C. Berk, J. Talbert, *J. Org. Chem.* **1988**, *53*, 2390-2392.

[6]   P. Knochel, M. J. Rozema, C. E. Tucker, C. Retherford, M. Furlong, S. AchyuthaRao, *Pure and Appl. Chem.* **1992**, *64*, 361-369.

[7]   P. Knochel, R. Singer, *Chem. Rev.* **1993**, *93*, 2117-2188.

[8]   M. C. P. Yeh, H. G. Chen, P. Knochel, *Org. Synth.* **1991**, *70*, 195-203.

[9]   P. Knochel, M. J. Rozema, C. E. Tucker in *A Practical Approach-Organocopper Reagents* (Ed. R. J. K. Taylor), Oxford University Press **1993**, 85-104.

[10]   H. Stadtmüller, B. Greve, K. Lennick, A. Chau, P. Knochel, *Synthesis* **1995**, 69-72.

[11]   S. C. Berk, M. C. P. Yeh, N. Jeong, P. Knochel, *Organometallics* **1990**, *9*, 3053-3064.

[12]   T. N. Majid, P. Knochel, *Tetrahedron Lett.* **1990**, *31*, 4413-3316.

[13]   H. P. Knoess, M. T. Furlong, M. J. Rozema, P. Knochel, *J. Org. Chem.* **1991**, *56*, 5974-5978

[14]   (a) C. JanikaramRao, P. Knochel, *J. Org. Chem.* **1991**, *56*, 4593-4596. (b) C. JanikaramRao, P. Knochel, *Tetrahedron* **1993**, *49*, 29-48.

[15]   T. Stevenson, B. Prasad, J. Citineni, P. Knochel, *Tetrahedron Lett.* **1996**, *37*, in press.

[16]   L. Zhu, R. M. Wehmeyer, R. D. Rieke, *J. Org. Chem.* **1991**, *56*, 1445-1453.

[17]   R. D. Rieke, *Science* **1989**, *246*, 1260-1264.

[18]   M. V. Hanson, R. D. Rieke, *J. Am. Chem. Soc.* **1995**, *117*, 10775-10776.

[19]   P. Knochel, *Synlett* **1995**, 393-403.

[20]   K. Soai, S. Niwa, *Chem. Rev.* **1992**, *92*, 833-856.

[21]   P. Knochel, S. Vettel, C. Eisenberg, *Appl. Organomet. Chem.* **1995**, *9*, 175-188.

[22]   (a) M. J. Rozema, S. AchyuthaRao, P. Knochel, *J. Org. Chem.* **1992**, *57*, 1956-1958. (b) M. J. Rozema, C. Eisenberg, H. Lütjens, R. Ostwald, K. Belyk, P. Knochel, *Tetrahedron Lett.* **1993**, *34*, 3115-3118.

[23] (a) F. Langer, J. R. Waas, P. Knochel, *Tetrahedron Lett.* **1993**, *34*, 5261-5264. (b) F. Langer, A. Devasagayaraj, P.-Y. Chavant, P. Knochel, *Synlett* **1994**, 410-412. (c) F. Langer, L. Schwink, P.-J. Chavant, P. Knochel, *J. Org. Chem.* **1996**, *61*, in press.

[24] L. Schwink, P. Knochel, *Tetrahedron Lett.* **1994**, *35*, 9007-9010.

[25] A. Devasagayaraj, L. Schwink, P. Knochel, *J. Org. Chem.* **1995**, *60*, 3311-3317.

[26] A. Longeau, F. Langer, P. Knochel, *Tetrahedron Lett.* **1996**, *37*, 2209-2212.

[27] R. Duddu, M. Eckhardt, M. Furlong, H. P. Knoess, S. Berger, P. Knochel, *Tetrahedron* **1994**, *50*, 2415-2432.

[28] M. Oestreich, L. Micouin, P. Knochel, *Angew. Chem.* in press.

[29] (a) T. Mukhopadhyay, D. Seebach, *Helv. Chim. Acta* **1982**, *65*, 385-391. (b) D. Seebach, A. K. Beck, T. Mukhopadhyay, E. Thomas, *Helv. Chim. Acta* **1982**, *65*, 1101-1133.

[30] C. Jubert, P. Knochel, *J. Org. Chem.* **1992**, *57*, 5425-5431.

[31] C. E. Tucker, T. N. Majid, P. Knochel, *J. Am. Chem. Soc.* **1992**, *114*, 3983-3985.

[32] I. Klement, K. Chau, G. Cahiez, P. Knochel, *Tetrahedron Lett.* **1994**, *35*, 1177-1180.

[33] S. Vettel, A. Vaupel, P. Knochel, *J. Org. Chem.* **1996**, in press.

[34] S. Vettel, A. Vaupel, P. Knochel, *Tetrahedron Lett.* **1995**, *36*, 1023-1026.

[35] L. Miginiacin *The chemistry of the metal-carbon bond* (Eds. F. R. Hartley, S. Patai), Wiley, New York, **1985**, Vol. 3, 99-141.

[36] Y. A. Dembélé, C. Belaud, R. Hichcock, J. Villieras, *Tetrahedron: Asymmetry* **1992**, *3*, 351-351

[37] G. Rousseau, J. Drouin, *Tetrahedron* **1983**, *39*, 2307-2310

[38] P. Knochel, J. F. Normant, *Tetrahedron Lett.* **1984**, *25*, 4383-4386.

[39] P. Knochel, J. F. Normant, *Tetrahedron Lett.* **1984,** *25*, 1475-1478.

[40] P. Knochel, J. F. Normant, *J. Organomet. Chem.* **1986**, *309*, 1-23.

[41] (a) J. M. Cox, R. Gosh, *Tetrahedron Lett.* **1969**, 3351-3354. (b) D. Kahne, D. B. Collum, *Tetrahedron Lett.* **1981**, *22*, 5011-5014.

[42] I. Klement, K. Lennick, C. E. Tucker, P. Knochel, *Tetrahedron Lett.* **1993**, *34*, 4623-4626.

[43] C. K. Reddy, A. Devasagayaraj, P. Knochel, *Tetrahedron Lett.* **1996**, *37*, 4495-4498.

[44] H. Stadtmüller, R. Lentz, W. Dörner, T. Stüdemann, C. E. Tucker, P. Knochel, *J. Am. Chem. Soc.* **1993**, *115*, 7027-7028.

[45] H. Stadtmüller, C. E. Tucker, A. Vaupel, P. Knochel, *Tetrahedron Lett.* **1993**, *34*, 7911-7914.

[46] A. Vaupel, P. Knochel, *Tetrahedron Lett.* **1994**, *35*, 8349-8352.

[47] A. Vaupel, P. Knochel, *Tetrahedron Lett.* **1995**, *36*, 231-232.

[48] H. Stadtmüller, P. Knochel, *Synlett* **1995**, 463-464.

[49] H. Stadtmüller, A. Vaupel, C. E. Tucker, T. Stüdemann, P. Knochel, *Chem. Eur. J.* **1997**, *2*, 1204-1220.

[50] C. Meyer, I. Marek, G. Courtemanche, J.-F. Normant, *Synlett* **1993**, 266-268.

[51] (a) A. V. Kramer, J. A. Labinger, J. S. Bradley, J. A. Osborn, *J. Am. Chem. Soc.* **1974**, *96*, 7145-7147. (b) A. V. Kramer, J. A. Osborn, *J. Am. Chem. Soc.* **1974**, *96*, 7832-7833.

[52] A. L. J. Beckwith, C. H. Schiesser, *Tetrahedron* **1985**, *41*, 3925-3941.

[53] M. B. M. de Azevedo, M. M. Murta, A. E. Greene, *J. Org. Chem.* **1992**, *57*, 4567-4568.

[54] P. Knochel in *Comprehensive Organic Chemistry* (Eds.: B. M. Trost), Pergamon Press, New York, **1991**, Vol. 4, p. 865.

[55] J. F. Normant, A. Alexakis, *Synthesis* **1981**, 841-870.

[56] S. AchyuthaRao, P. Knochel, *J. Am. Chem. Soc.* **1992**, *114*, 7579-7581.

[57] T. Stüdemann, P. Knochel, *Angew. Chem.* **1996**, *37*, in press

[58] T. Yamamoto, A. Yamamoto, S. Ikeda, *J. Am. Chem. Soc.* **1971** *93*, 3350-3359.

[59] (a) R. Sustmann, J. Lau, M. Zipp, *Tetrahedron Lett.* **1986**, *27*, 5207-5210. (b) R. Sustmann, J. Lau, *Chem. Ber.* **1986**, *119*, 2531-2541. (c) R. Sustmann, J. Lau, M. Zipp, *Recl. Trav. Chim. Pays-Bas* **1986**, *105*, 356-359. (d) R. Sustmann, P. Hopp, P. Holl, *Tetrahedron Lett.* **1989**, *30*, 689-692.

[60] R. van Asselt, C. J. Elsevier, *Tetrahedron* **1994**, *50*, 323-334.

[61] A. Devasagayaraj, T. Stüdemann, P. Knochel, *Angew. Chem. Int. Ed. Engl.* **1995**, *34*, 2723-2725.

[62] (a) E. Negishi, L. F. Valente, M. Kobayashi, *J. Am. Chem. Soc.* **1980**, *102*, 3298-3299. (b) M. Kobayashi, E. Negishi, *J. Org. Chem.* **1980**, *45*, 5223-5225.

[63] M. F. Rettig, P. M. Maitlis, *Inorg. Synth.* **1977**, *17*, 134-137.

[64] (a) V. Farina, B. Krishnan, *J. Am. Chem. Soc.* **1991**, *113*, 9585-9595. (b) V. Farina, S. Kapadia, B. Krishnan, C. Wang, L. S. Liebeskind, *J. Org. Chem.* **1994**, *59*, 5905-9511.

[65] I. Klement, M. Rottländer, C. E. Tucker, T. N. Majid, P. Knochel, P. Venegas, G. Cahiez, *Tetrahedron* **1996**, *52*, 7201-7220.

[66] T. Hayashi, M. Konishi, Y. Kobori, M. Kumada, T. Higuchi, K. Hirotsu *J. Am. Chem. Soc.* **1984**, *106*, 158-163.

[67] P. Knochel, D. Seebach, *Tetrahedron Lett.* **1982**, *32*, 3897-3900.

[68] M. Rottländer, N. Palmer, P. Knochel, *Synlett* **1996**, 573-575.

Stereoselective Synthesis of Axially Chiral Biaryls Utilizing Planar Chiral (Arene)chromium Complexes

Motokazu Uemura

Department of Chemistry, Faculty of Integrated Arts and Sciences,
Osaka Prefecture University, Sakai, Osaka 593, Japan

Summary

Tricarbonyl(2,6-disubstituted 1-bromobenzene)chromium complexes were treated with *ortho*-substituted arylboronic acids in the presence of Pd(0) catalyst to give mono-Cr(CO)$_3$ complexes of biaryl compounds with complementary axial chirality depending upon the steric bulkiness of the *ortho* substituents. Thus, cross-coupling of *o*-alkyl phenylboronic acids with (arene)chromium complexes gave diastereoselectively Cr(CO)$_3$-complexed biaryls in which the *ortho* substituents are in a *syn*-orientation to the tricarbonylchromium fragment. With *o*-formyl phenylboronic acids, diastereoisomeric *anti*-coupling products were stereoselectively obtained. The kinetically controlled *syn*-coupling products were easily isomerized to thermodynamically more stable mono-Cr(CO)$_3$-complexed *anti*-biaryls by modification of the *o*-substituent to less hindered ones, or the thermal conditions assisted the axial isomerization. The overall process can be considered to be an enantioselective preparation of both axially chiral biaryls starting from a single planar chiral (arene)chromium complex.

Introduction

Biaryl compounds with axial chirality are of potential importance not only as chiral ligands for asymmetric reactions but also intermediates for the synthesis of biologically active natural products [1]. (η6-Arene)chromium complexes exist in two enantiomeric forms based on planar chirality, when the arene ring is substituted at the *ortho*- or *meta*-positions with different substituents. In this paper, we describe the stereoselective synthesis of axially chiral biaryls utilizing the planar chiral (arene)chromium complexes by Pd(0) catalyzed cross-coupling or nucleophilic addition with aryl Grignard reagents.

Diastereoselective Cross-Coupling of (Arylhalide)Cr(CO)$_3$ with Arylboronic Acids.

Cross-coupling reactions of aryl halides or aryl triflates with arylmetals catalyzed by palladium(0) are commonly used for the preparation of biaryls [2]. An oxidative addition of the carbon-halogen bond of the aryl halide to the palladium(0) is accelerated by a coordination of an electron withdrawing tricarbonylchromium fragment to the arene ring [3]. Even chlo-

131

robenzene can be made susceptible to oxidative addition by utilizing the corresponding tricarbonylchromium complex to give cross-coupling products. Thus, the cross-coupling of tricarbonylchromium complexes of *o*-substituted aryl halides with phenylmetals such as phenylboronic acid, Grignard reagent, and phenylzinc chloride, in the presence of palladium(0) catalyst, afforded the hetero-cross-coupling products in good yields [4].

Therefore, we next examined the axial stereochemistry in the cross-coupling reaction of tricarbonyl(aryl halide)chromium complexes with *o*-substituted arylboronic acids (eq. 1, Table 1) [5, 6]. Tricarbonyl(2-methoxy-6-methyl-1-bromobenzene)chromium (**1a**) was allowed to react with *o*-methyl phenylboronic acid (**2a**) catalyzed by 5 mol % of Pd(PPh$_3$)$_4$ in the presence of sodium carbonate in aqueous methanol at 75° C to give cross-coupling product **3** in 96% yield without formation of the corresponding atropisomer. The methyl group on the B-ring of complex **3** is directed toward the tricarbonylchromium fragment of the chromium-complexed A-ring in spite of a severe steric interaction between the methyl and Cr(CO)$_3$ groups. Similarly, the cross-coupling reaction of *o*-methyl phenylboronic acid with other chromium complexes gave the products **3** with the same axial chirality. With *o*-methoxy phenylboronic acid, the axial chirality of the coupling products was dependent upon the steric bulkiness of the *ortho* substituents of (aryl halide)chromium complexes (entries 9,10). On the other hand, cross-coupling of *o*-formyl phenylboronic acid **2b** with chromium complexes gave diastereoisomeric (S_p*,R_a*)-complexes **4** as the only isolated coupling products regardless of the *ortho* substituents on the (arene)chromium complexes (entries 5-7). The formation of (S_p*,R_a*)-products **4** can be attributed to a thermodynamically controlled reaction *via* an axial isomerization of the initially formed (S_p*,S_a*)-products (*vide infra*). The coupling of *o*-hydroxymethyl phenylboronic acid with (arene)chromium complexes gave the (S_p*,S_a*)-biphenyl complex **3** and CO-inserted product **5** in various ratio depending upon the *ortho* substituent of chromium complexes. It is obvious from the above results that the axial stereochemistry of the coupling products is strictly controlled by the steric bulkiness of the *ortho* substituents adjacent to the coupling positions.

Table 1 Palladium(0)-Catalyzed Cross-Coupling of (Arene)chromium Complexes 1 with Arylboronic Acids 2.

entry	complex	phenyl-boronic acid	R^1	R^2	ratio 3 : 4 : 5	yield (%)
1	**1a**	**2a**	Me	Me	100:0:0	96
2	**1b**	**2a**	CHO	Me	92:0:8	89
3	**1c**	**2a**	CHO(CH$_2$)$_2$	Me	100:0:0	81
4	**1d**	**2a**	CH$_2$OH	Me	100:0:0	77
5	**1a**	**2b**	Me	CHO	0:100:0	95
6	**1b**	**2b**	CHO	CHO	0:100:0	43
7	**1c**	**2b**	CHO(CH$_2$)$_2$	CHO	0:100:0	52
8	**1a**	**2c**	Me	CH$_2$OH	81:0:19	68
9	**1a**	**2d**	Me	OMe	97:3:0	94
10	**1b**	**2d**	CHO	OMe	4:96:0	85

132

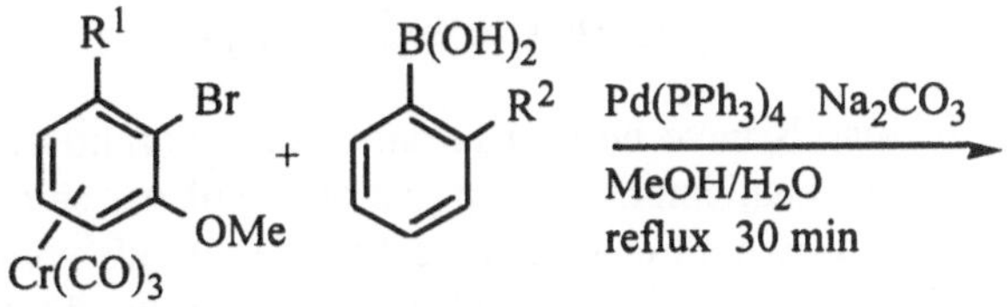

1

a: $R^1 = Me$
b: $R^1 = CHO$
c: $R^1 = $ (dioxolane)
d: $R^1 = CH_2OH$

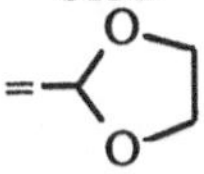

2

a: $R^2 = Me$
b: $R^2 = CHO$
c: $R^2 = CH_2OH$
d: $R^2 = OMe$

(eq 1)

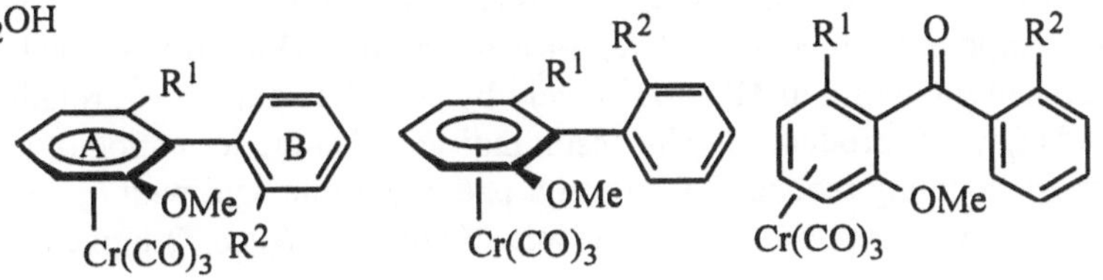

3   **4**   **5**

We further studied the stereochemistry in the palladium(0)-catalyzed cross-coupling of (arene)chromium complexes with naphthylboronic acids, as directed toward the total synthesis of naphthyltetrahydroisoquinoline alkaloids (eq 2) [5, 7]. Tricarbonyl(2-methoxy-6-methyl-1-bromobenzene)chromium **6** ($R^1 = OMe$, $R^2 = Me$) coupled with 1-naphthyl-boronic acid **7** ($R^3 = H$) to give the (S_p*,S_a*)-complex **8** ($R^1 = OMe$, $R^2 = Me$, $R^3 = H$) in 88% yield without formation of diastereoisomeric compound **9**. In other cases, the products **8** were obtained as major product in a various ratio depending upon the steric bulk of substituent.

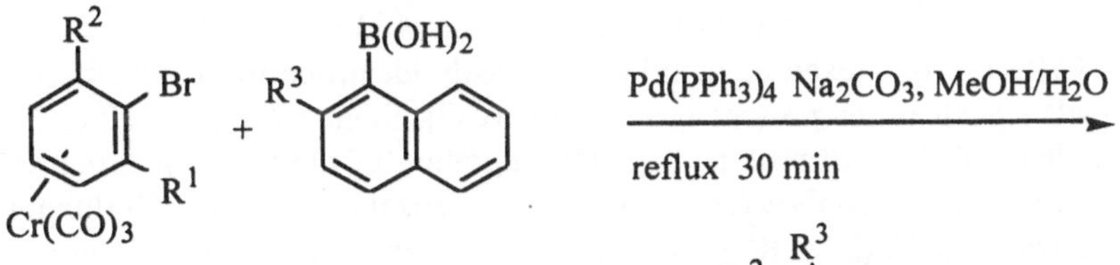

6   **7**

$R^3 = H, OMe, Me$

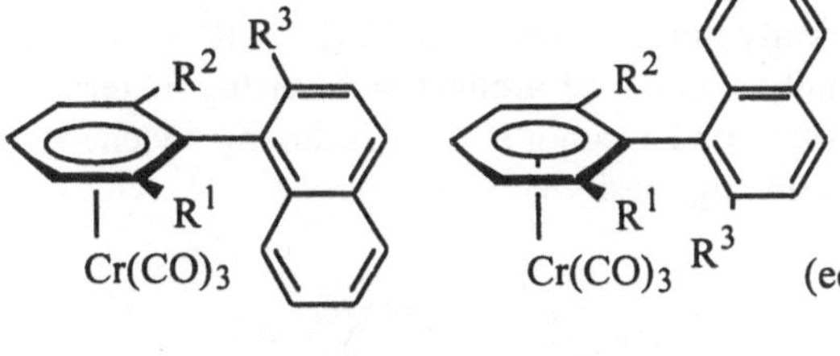

(eq 2)

8   ratio   **9**

$(100\sim71 : 0 \sim 29)$

133

Axial Isomerization of Mono-Cr(CO)$_3$-Complexed Biaryls

Since the mono-Cr(CO)$_3$ complexes of biaryls have both axial and planar chiralities, the chromium-complexed biaryls can exist in an enatiomeric form based on the planar chirality even when the central bond rotates. Therefore, both enantiomerically pure biaryls with an axial chirality could be obtainable by the cross-coupling and subsequnt central bond rotation, starting from a single planar chiral arene-chromium complex. The axial isomerization of the chromium-complexed biaryls was achieved by following two procedures: (1) modification of both *ortho* substituents to one less hindering and (2) the thermal conditions to assist the central bond isomerization [8].

(S_p*,S_a*)-[(1,2,3,4,5,6-η)-2-methoxy-2'-hydroxymethyl-6-methylbiphenyl]chromium **10**, obtained by the palladium(0)-catalyzed cross-coupling, was oxidized with DMSO/Ac$_2$O at room temperature to give (S_p*,R_a*)-tricarbonyl[[(1,2,3,4,5,6-η)-2-methoxy-2'-formyl-6-methylbiphenyl]chromium **11** in 53% yield by complete axial bond rotation of the initially produced (S_p*,S_a*)-product **12**. This result indicates that the formation of (S_p*,R_a*)-biaryls **4** in the cross-coupling reaction of *o*-formyl phenylboronic acid with (arene)chromium complexes is due to the isomerization of the initially formed (S_p*,S_a*)-biphenyls **3**.

Several Cr(CO)$_3$-complexed (S_p*,S_a*)-biaryls obtained by the palladium(0)-catalyzed cross-coupling would be expected to isomerize to the thermodynamically more stable (S_p*,R_a*)-complexes upon refluxing in high boiling solvent [8]. When the complex **13** (R^1 = CHO, R^2 = Me) in toluene was refluxed for 2h, the axially isomerized biphenyl **14** was obtained along with a small amount of the starting material (eq. 4, Table 2). The axial stereochemistry of the Cr(CO)$_3$-complexed biphenyl can be easily identified by ^{1}H NMR analysis. Generally, NMR signals of the protons *syn* to the Cr(CO)$_3$ fragment are shifted to far lower field than are those of the *anti*-protons [9]. No migration of the Cr(CO)$_3$ group to another arene ring or to solvent was observed during the isomerization reaction. Refluxing of sterically hindered complex **13** (R^1 = R^2 = Me) in xylene for 2h gave a 20:80 mixture of the axially isomerized product and starting material. Although this axial isomerization was incomplete under conditions of refluxing xylene, the rotation of the axial bond increased upon refluxing mesitylene.

Table 2 Axial Isomerization of Biaryls under Thermal Conditions

complex 13	solvent	ratio 13 : 14	yield (%)
R^1 = CHO, R^2 = Me	toluene	2:98	98
R^1 = CHO, R^2 = Me	xylene	1:>99	98
R^1 = Me, R^2 = OMe	toluene	42:58	98
R^1 = Me, R^2 = OMe	xylene	9:91	98
R^1 = R^2 = Me	xylene	80:20	98
R^1 = R^2 = Me	mesitylene	2:98	79

Synthesis of Both Enantiomers of Axial Chiral Biphenyls

The diastereoselective cross-coupling reaction provides a promising approach to the synthesis of both optically pure atropisomers starting from a single chiral arene chromium complex [5]. Enantiomerically pure (+)-chromium complex **15** was coupled with *o*-methyl phenylboronic acid to give (+)-(*R*,*R*)-complex **16**, which was converted into (–)-(*R*)-2-methoxy-2'-methyl-6-(1,3-dioxolanyl)biphenyl **17** upon exposure to sunlight.

Synthesis of Both Atropisomers of Axial Biphenyl by Cross-Coupling

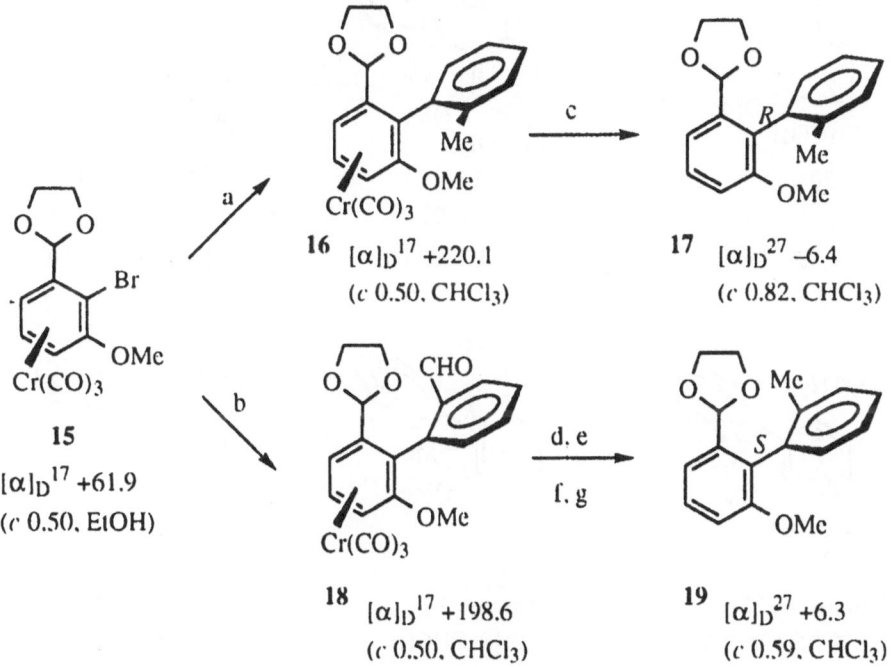

(a) *o*-methyl phenylboronic acid, Pd(PPh₃)₄, Na₂CO₃, McOH, H₂O, 75°C, 30 min, 81%;
(b) *o*-formyl phenylboronic acid, Pd(PPh₃)₄, Na₂CO₃, McOH, H₂O, 75°C, 30 min, 52%;
(c) *hμ*-O₂, 90%; (d) LAH, 62%; (e) MsCl, py, 38%; (f) LAH, 51%; (g) *hμ*-O₂, 90%

On the other hand, diastereomeric (+)-(*R*,*S*)-chromium complex **18** was stereoselectively obtained by the cross-coupling of the complex **15** with *o*-formyl phenylboronic acid. The (*R*,*S*)-biphenyl complex **18** was converted to an antipode (+)-(*S*)-2-methoxy-2'-methyl-6-(1,3-dioxolanyl)biphenyl **19** by reduction of the formyl group to a methyl group.

Combination of the axial isomerization described above and the diastereoselective cross-coupling reaction can also utilized for the preparation of both enantiomers of axial biphenyls. (–)-(2-Bromo-3-methoxybenzaldehyde)chromium complex **20** was reacted with 2-methyl phenylboronic acid in the presence of palladium(0) catalyst to give (+)-(*R*,*R*)-tricarbonyl[(1,2,3,4,5,6-η)-2-methoxy-2'-methyl-6-formylbiphenyl]chromium **21**, which was successively reduced with $NaBH_4$ acetylated to give (+)-(*R*,*R*)-complex **22**. Photooxidative demetalation of the complex **22** gave (–)-(*R*)-2-methoxy-2'-methyl-6-acetoxymethylbiphenyl **23**. On the other hand, the axial isomerization of **21**, upon refluxing in xylene for 2h, gave the diastereisomer, (*R*,*S*)-tricarbonyl[(1,2,3,4,5,6-η)-2-methoxy-2'-methyl-6-formylbiphenyl]chromium **24**. The axially isomerized product **24** was converted into the corresponding antipode (+)-(*S*)-2-methoxy-2'-methyl-6-acetoxymethylbiphenyl **25** by a similar reaction sequence.

Synthesis of Both Atropiomers by Cross-Coupling and Axial Isomerization

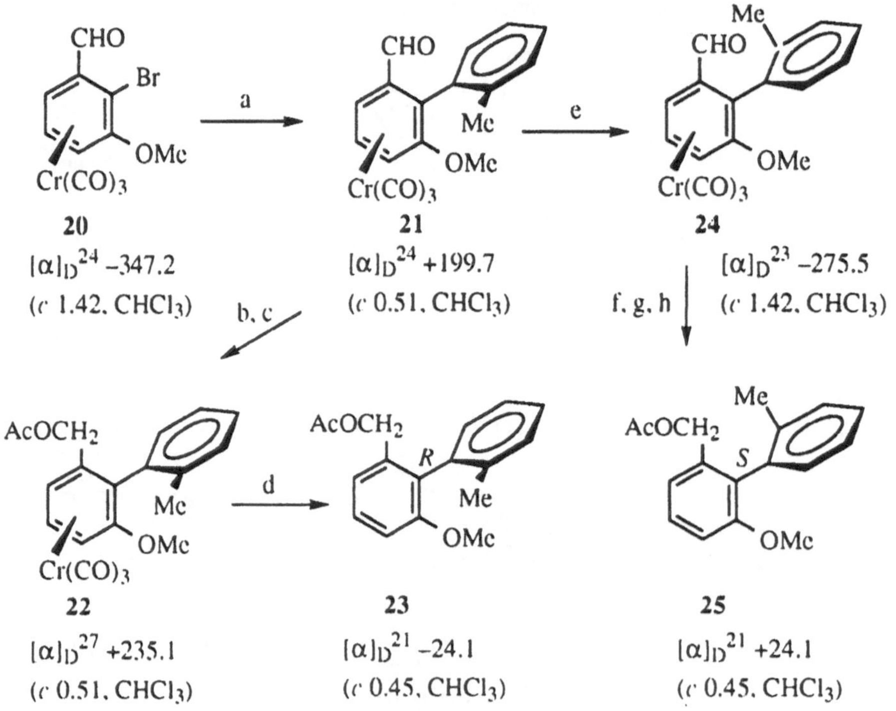

CHO
Br
OMe
Cr(CO)3
20
[α]D24 –347.2
(c 1.42, CHCl3)
a
CHO
Me
OMe
Cr(CO)3
21
[α]D24 +199.7
(c 0.51, CHCl3)
e
Me
CHO
OMe
Cr(CO)3
24
[α]D23 –275.5
(c 1.42, CHCl3)
b, c
f, g, h
AcOCH2
Me
OMe
Cr(CO)3
22
[α]D27 +235.1
(c 0.51, CHCl3)
d
AcOCH2
R
Me
OMe
23
[α]D21 –24.1
(c 0.45, CHCl3)
Me
AcOCH2
S
OMe
25
[α]D21 +24.1
(c 0.45, CHCl3)

(a) *o*-methyl phenylboronic acid, $Pd(PPh_3)_4$, Na_2CO_3, MeOH, H_2O, 75°C, 30 min, 80%; (b) $NaBH_4$, MeOH, 98%; (c) Ac_2O, py, 91%; (d)$h\mu$-O_2, 60%; (e) reflux in xylene, 98%; (f) $NaBH_4$, MeOH, 96%; (c) Ac_2O, py, 90%; (d)$h\mu$-O_2, 52%

Reaction Mechanism of Diastereoselective Cross-Coupling

Two *cis* diorganopalladium(II) intermediates **26** and **27**, having a square configuration, would be the transient species prior to the biaryl carbon-carbon bond formation. The intermediate **26** depicts a crowded system in which the substituent R on the B-ring and the sterically bulky L substituent are face to face. The altenative complex **27** appears to be free of any severe nonbonded interactions. The R substituent of **27** is oriented *syn* to the less bulky S group, and the tricarbonylchromium fragment is farther away from bulky triphenylphosphine. Both arene rings are coupled avoiding severe nonbonding interactions between the R and triphenylphosphine groups from **27**, and R substituent rotates toward the Cr(CO)₃ moiety to give (S*,S*)-complexes **3**. The formation of (S*,R*)-complexes **4** by coupling with *o*-formyl phenylboronic acid can be attributed to an axial isomerization of the kinetically controlled (S*,S*)-complexes, as mentioned above.

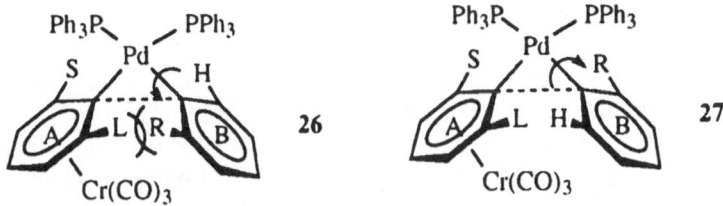

Synthesis of (–)-Steganone

(–)-Steganone **28**, an antileukaemic bisbenzocyclooctadiene lignan lactone with axial chirality, isolated from *Steganotaenia araliacea*, has attracted considerable synthetic interest. The novel feature of our strategy for the total synthesis of (–)-steganone **27** is initially to prepare an enantiomerically pure (aryl halide)chromium complex with planar chirality, and subsequent to form the axial asymmetry of the biaryl bond by the palladium(0)-catalyzed Suzuki cross-coupling reaction [10].

3,4,5-Trimethoxybenzaldehyde was converted to the corresponding enantiomerically pure bromo benzaldehyde chromium complex **29** by diastereoselective *ortho* lithiation of chiral chromium-complexed benzaldehyde acetal. Stereoselective cross-coupling with *o*-formylphenylboronic acid gave **30** with (R)-axial chirality, which was converted to (–)-steganone **28** by several steps.

OCH₂CH=CH
Me
MeO
MeO
OMe
CH₂CH(CO₂Me)₂
O
MeO
MeO
OMe

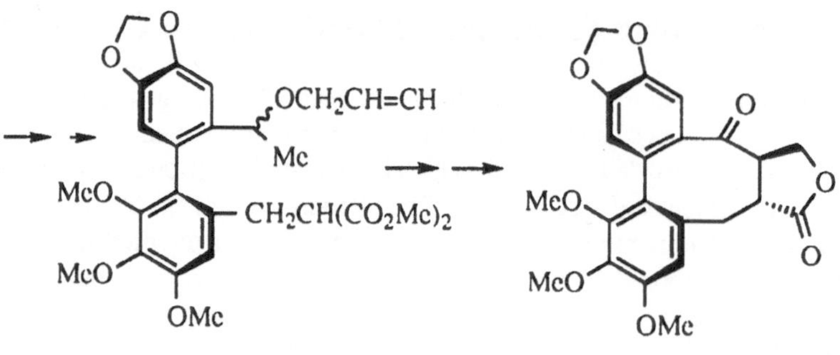

31   28 (−)-Steganone

Nucleophilic Substitution of (Arene)chromium Complexes with Aryl Grignard Reagents

Nucleophilic substitution of an *o*-alkoxy group of activated arenes having electron-withdrawing substituents such as oxazoline, ester, sulfinyl, sulfonyl and diphenylphosphinyl groups with aryl Grignard reagents produces the biaryl compounds. It is well known that the chiral *o*-methoxyphenyloxazolines are reacted with aryl Grignard reagents to give axial biaryls with high enantiomerically purity [1]. The strong electron-withdrawing ability of tricarbonylchromium fragment accelerates the nucleophilic addition [11].

Enantiomerically pure (*R*)-axial biaryl **35** was synthesized from the planar chiral (arene)chromium complex **33** having a sterically bulky 2,4,6-trimethylphenyl ester group by stereoselective nucleophilic substitution with aryl Grignard reagent as follows [12].

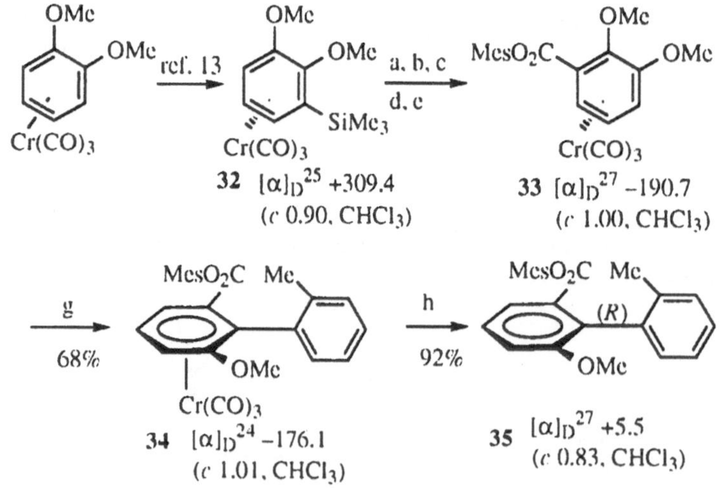

OMe
OMe
Cr(CO)₃
ref. 13
SiMe₃
a, b, c
d, e
MesO₂C
32 [α]D25 +309.4
(c 0.90, CHCl₃)
33 [α]D27 −190.7
(c 1.00, CHCl₃)
g
68%
h
92%
Me
(R)
34 [α]D24 −176.1
(c 1.01, CHCl₃)
35 [α]D27 +5.5
(c 0.83, CHCl₃)

(a) n-BuLi, THF, TMEDA, −78°; (b) $ClCO_2Et$, THF; (c) n-Bu_4NF, THF (76% from **32**; (d) 1M-KOH, THF, H_2O; (e) 2,4,6-trimethylphenol, $(CF_3CO)_2O$, reflux, 34%; (g) *o*-tolylmagnesium bromide, benzene, reflux; (h) $h\mu$-O_2

Acknowledgment

This work was supported by a Grant-in-Aid for Scientific Research from the Ministry of Education, Science and Culture of Japan. The author thanks CIBA GEIGY Foundation (Japan) and The Asahi Glass Foundation for the financial support. I acknowledge the many enthusiastic co-workers who contributed over the years to the development of this fascinating chemistry.

References

[1] G. Bringmann, R. Walter, R. Weirich, *Angew. Chem. Int. Ed. Engl.* **1990**, *29*, 977. T. G. Gant, A. I. Meyers, *Tetrahedron*, **1994**, *50*, 2297.

[2] A. Suzuki, *Pure Appl. Chem.* **1991**, *63*, 419; N. Miyaura, T. Yanagi, A. Suzuki, *Synth. Commun.* **1981**, *11*, 513; A. Suzuki, N. Miyaura, *Chem. Rev.* **1995**, *95*, 2457; J. K. Stille, *Angew. Chem. Int. Ed. Engl.* **1986**, *25*, 508; E-I. Negishi, F. T. Luo, R. Frisbee, H. Matsushita, *Heterocycles*, **1982**, 18, 117.

[3] W. J. Scott, *J. Chem. Soc. Chem. Commun.* **1987**, 1755; J. M. Clough, I. S. Mann, D. A. Widdowson, *Tetrahedron Lett.* **1987**, *28*, 2645.

[4] M. Uemura, H. Nishimura, K. Kamikawa, K. Nakayama, Y. Hayashi, *Tetrahedron Lett.* **1994**, *35*, 1909.

[5] K. Kamikawa, T. Watanabe, M. Uemura, *J. Org. Chem.* **1996**, *61*, 1375.

[6] M. Uemura, K. Kamikawa, *J. Chem. Soc. Chem. Commun.* **1994**, 2697.

[7] T. Watanabe, K. Kamikawa, M. Uemura, *Tetrahedron Lett.* **1995**, *36*, 6695.

[8] K. Kamikawa, T. Watanabe, M. Uemura, *Synlett*, **1995**, 1040.

[9] M. Uemura, H. Nishimura, K. Kamikawa, M. Shiro, *Inorg. Chim. Acta.* **1994**, *222*, 63; G. Bringmann, L. Göbel, K. Peters, E.-M. Perters, H. G. Schnering, *Inorg. Chim. Acta.* **1994**, *222*, 255.

[10] M. Uemura, A. Daimon, Y. Hayashi, *J. Chem. Soc. Chem. Commun.* **1995**, 1943.

[11] M. F. Semmelhack, Comprehensive Organometallic Chemistry II, Eds. by E. W. Abel, F. G. A. Stone, G. Wilkinson, Pergamon, 1995, Vol. 12, p. 979; J. P. Collman, L. S. Hegedus, J. R. Norton, R. G. Finke, Principles and Applications of Organotransition Metal Chemistry, University Science Books, Mill Valley, Calf., 1987, p. 921.

[12] K. Kamikawa, M. Uemura, *Tetrahedron Lett.* **1996**, *37*, 6359.

[13] D. A. Price, N. S. Simpkins, A. M. MacLeod, A. P. Watt, *J. Org. Chem.* **1994**, *59*, 1691.

Metal-Assisted Synthesis and Application of Axially Chiral Biaryl Systems

G. Bringmann*[a], M. Breuning[a], S. Busemann[a], J. Hinrichs[a], T. Pabst[a], R. Stowasser[a], S. Tasler[a], A. Wuzik[a], W. A. Schenk*[b], J. Kümmel[b], D. Seebach*[c], G. Jaeschke[c]

[a] Institute of Organic Chemistry, Am Hubland, D-97074 Würzburg, Germany
[b] Institute of Inorganic Chemistry, Am Hubland, D-97074 Würzburg, Germany
[c] Laboratory of Organic Chemistry, ETH-Zentrum, Universitätsstr. 16, CH-8092 Zürich, Switzerland

1   The Basic Principle of the 'Lactone Methodology'

Axially chiral biaryl compounds have been applied successfully as auxiliaries in asymmetric synthesis and several of them are pharmacologically active natural products [1,2]. For the efficient regio- and stereoselective access to stereochemically homogeneous biaryls, we have developed a methodology [1-3] in which the two aromatic moieties are prefixed *via* an ester bridge followed by an intramolecular aryl coupling to give configuratively labile lactones like **1** as a mixture of rapidly interconverting helimers (*M*)-**1** and (*P*)-**1**. Out of this equilibrium, **1** can be opened stereoselectively by chiral reagents, with the faster reacting isomer being permanently supplied out of the helimerization equilibrium. Thus the configuratively stable ring opening product **3** can be obtained in excellent enantiomeric or diastereomeric ratios and high chemical yields.

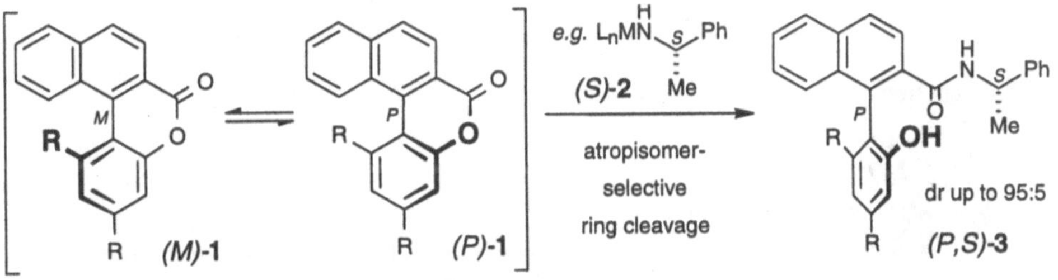

Figure 1 The lactone concept for the stereoselective synthesis of axially chiral biaryls

2   Metal-Assisted Atroposelective Cleavage of Biaryl Lactones

2.1   Principal Possibilities for the Asymmetric Induction

Three different metal-assisted variations of a chiral induction have been elaborated: by metallated chiral nucleophiles (method A), by η^6-bonded transition metal fragments with the

141

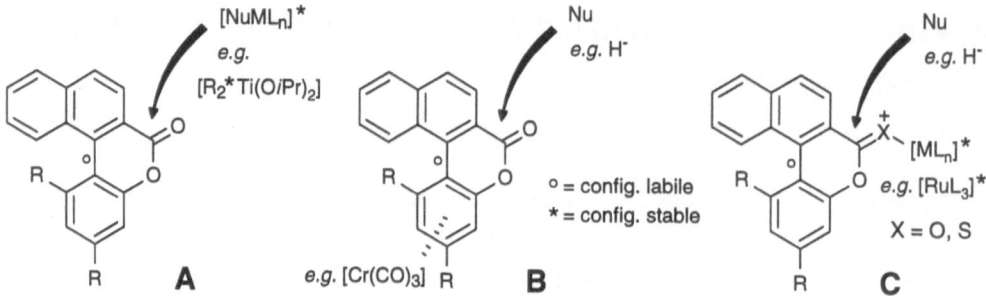

Figure 2 Possibilities for the metal-assisted ring opening reaction

additional element of planar chirality (method **B**), or by activation of the carbonyl group or its thiono analog with chiral Lewis acids (method **C**).

2.2 Realization of the Approaches A-C

The cleavage of the lactone bridge with metallated amides (method **A**) [1] has recently been further optimized: thus, reaction of **1** with (*S*)-**2** (s. Fig. 1; M = Li, Na, K; R = Me, OMe) leads to diastereomeric ratios of up to 95:5 (M = K, R = Me) and almost quantitative yields. The major isomer can be enriched further by recrystallization.

Activation of the phenolic leaving group was achieved by a chromium tricarbonyl fragment (method **B**). Ring cleavage with NaBH$_4$ yielded only one diastereomer (s. Fig. 2, **B**; R = Me) [4] (for regiochemistry and dynamics of the Cr-rotor, see our second contribution). For (η^6-arene)RuCp* complexes, the relative reactivity towards different nucleophiles was predicted using Density Functional calculations (Fukui function) and confirmed by experimental data. Thus, cleavage of the lactone unit of **4** can be attained with hard nucleophiles, *e.g.* methanolate, but fails with softer nucleophiles like superhydride (LiBHEt$_3$).

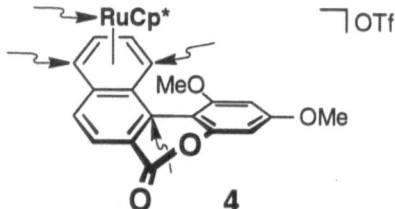

Figure 3 Ru-complex with positions of maximum softness

Method **C** was realized by activation of thionolactones **5** with ruthenium complexes [5]. Transformation of **5** to the chiral complex **6**, stereoselective ring cleavage with an achiral hydride transfer reagent and subsequent demetallation gave enantiomeric ratios of up to 88:12 (see also contribution by W. Schenk *et al.*).

In the atropo-enantioselective ring opening reaction with Ti-TADDOLates like **8**, the reagent acts as a chiral Lewis acid *and* as a metal-activated nucleophile at the same time (methods **A** + **C**) [6]. After reduction of the resulting esters **9**, the er of 79:21 could be enhanced to 99:1 by fractionated crystallization

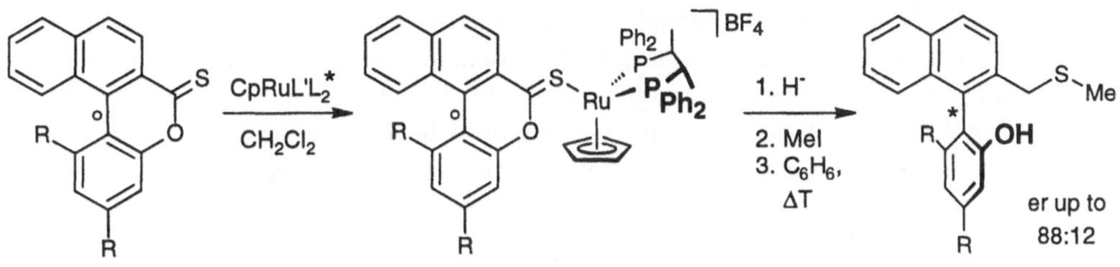

Figure 4 Asymmetric induction by chiral Ru-complexes

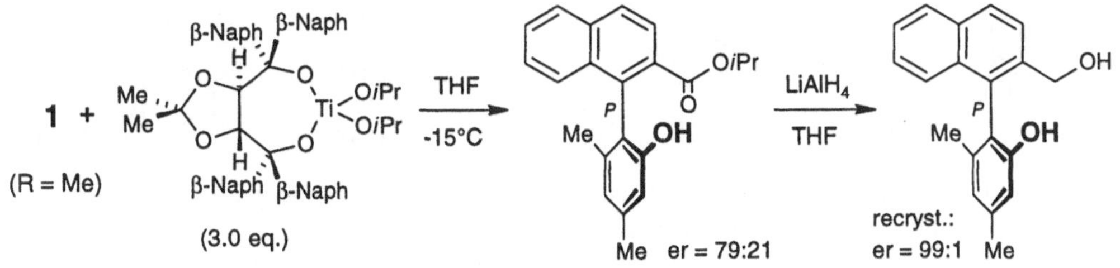

Figure 5 Atropo-enantioselective Ti-TADDOLate-assisted ring opening of **1**

3 Application of the Lactone Concept in the Preparation of Ligands for Asymmetric Synthesis and Natural Products

A first successful application of modified ring opening products as catalysts in stereoselective synthesis is the highly enantioselective addition of diethyl zinc to aldehydes **11**. Using aminoalcohol **12**, enantiomeric ratios of up to 99:1 were achieved.

With the prospect of a potential application as a chiral ligand *e.g.* in Pd-complexes for asymmetric Heck-type reactions, we have developed a synthesis that enables us to build up the phosphine **14** or optionally its enantiomer stereoselectively using the lactone concept.

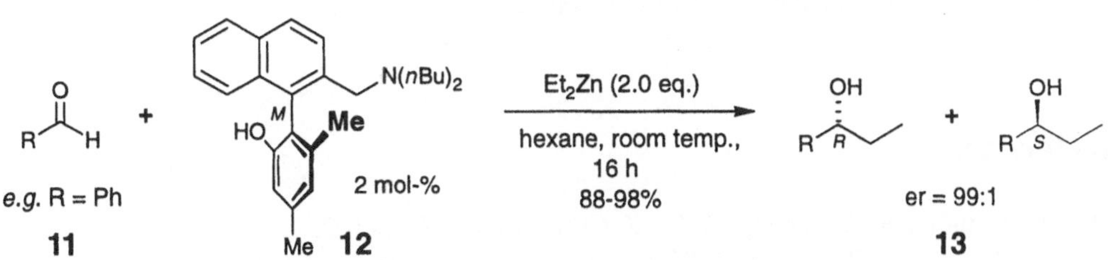

Figure 6 Biaryl-catalyzed enantioselective addition of diethyl zinc to aldehydes

The lactone methodology has already been used in numerous efficient syntheses of natural products [7]. More recently, we have prepared compound **15**, a simplified analogue of the natural nerve growth factor mastigophorene A [8]. The absolute configuration of **15** was elucidated by CD calculations. After ring opening of its lactone precursor with lithium mentholate (dr = 87:13), the minor isomer can be recycled in a single step and submitted to the dynamic kinetic deracemization once again (chiral economy).

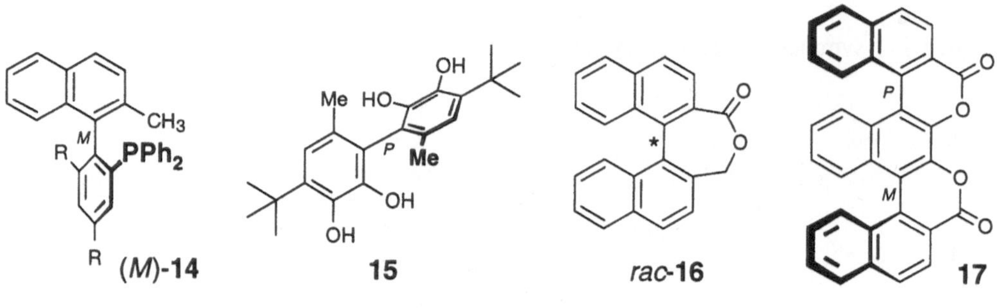

Figure 7 Compounds **14** and **15**, synthesized using the lactone-concept, and newsubstrates **16** and **17** for stereoselective ring opening reactions

4   Further New Perspectives of the Lactone Methodology

The lactone concept has recently been extended to 7-membered biaryl lactones like **16** (s. Fig. 7). The kinetic resolution of the configuratively stable racemic **16** with chiral hydride transfer reagents gives high enantiomeric ratios, *e.g.* 96:4 for **16** after 51% conversion or 97:3 after 41% conversion for the resulting diol.

Furthermore the teraryl-bislactone **17** (s. Fig. 7), a substrate for *double* atropisomer-selective metal-assisted ring opening reactions, was synthesized for the first time.

Acknowledgements

This work was supported by the DFG (project B-1 of the SFB 347 "Selektive Reaktionen Metall-aktivierter Moleküle") and by the Fonds der Chemischen Industrie.

References

[1]   G. Bringmann, O. Schupp, *S. Afr. J. Chem.* **1994**, *47*, 83-102.

[2]   G. Bringmann, R. Walter, R. Weirich in *Methods of Organic Chemistry (Houben Weyl) 4th ed.*(Eds.: G. Helmchen, R. W. Hoffmann, J. Mulzer, E. Schaumann), vol. E21a, Thieme, Stuttgart, **1995**, p. 567-587.

[3] G. Bringmann, S. Harmsen, O. Schupp, R. Walter in *Stereoselective Reactions of Metal-Activated Molecules* (Eds.: H. Werner, J. Sundermeyer), Vieweg, Braunschweig, Wiesbaden, **1995**, p. 137-142.

[4] G. Bringmann, L. Göbel, K. Peters, E.-M. Peters, H. G. von Schnering, *Inorg. Chim. Acta* **1994**,*222*, 255-260.

[5] G. Bringmann, B. Schöner, O. Schupp, W. A. Schenk, I. Reuther, K. Peters, E.-M. Peters, H. G. von Schnering, *J. Organomet. Chem.* **1994**, *472*, 275-284.

[6] D. Seebach, G. Jaeschke, K. Gottwald, K. Matsuda, R. Formisano, D. A. Chaplin, M. Breuning, G. Bringmann, *Tetrahedron* **1997**, *53*, 7539-7556.

[7] G. Bringmann, F. Pokorny in *The Alkaloids* (Ed.: G. A. Cordell), vol. 46, Academic Press, New York, **1995**, p. 127-271.

[8] G. Bringmann, T. Pabst, S. Busemann, K. Peters, E.-M. Peters, *Tetrahedron*, in preparation.

Transition-Metal-Mediated Radical Reactions: A Convenient Method for the Synthesis of 2-C-Branched Carbohydrates

Torsten Linker,* Thomas Sommermann, and Michael Maurer

Institute of Organic Chemistry, University of Würzburg, Am Hubland, D-97074 Würzburg, Germany

Introduction

C-Branched sugars are of current interest in carbohydrate chemistry. During the last 20 years, many methods have been developed for the synthesis of C-glycosides in which a carbon atom substitutes the glycosidic oxygen [1]. However, C-functionalizations at other positions of the sugar ring require many steps or the use of toxic tin and mercury compounds. Herein we present a practical and general protocol for the synthesis of 2-C-analogs of carbohydrates, which is notable for easily available starting materials, high yields, and good stereoselectivities.

During the course of our investigations on manganese(III)-mediated radical reactions [2], we revealed the first application of this methodology in carbohydrate chemistry [3]. Glycals were chosen as ideal substrates for the addition of CH-acidic compounds, since these chiral building blocks can be prepared on a multigram scale and are known to serve as precursors for a broad variety of optically active products [4]. Thus, addition of dimethyl malonate (2a) to tri-O-acetyl-D-glucal (1a) proceeds with very high regioselectivity to afford the C-analogs 3a in 66% yield (Eq. 1). However, four diastereomers are formed with only moderate selectivity. Furthermore, the unsaturated carbohydrates 4 result from an acid-catalyzed Ferrier rearrangement [5] under the drastic reaction conditions.

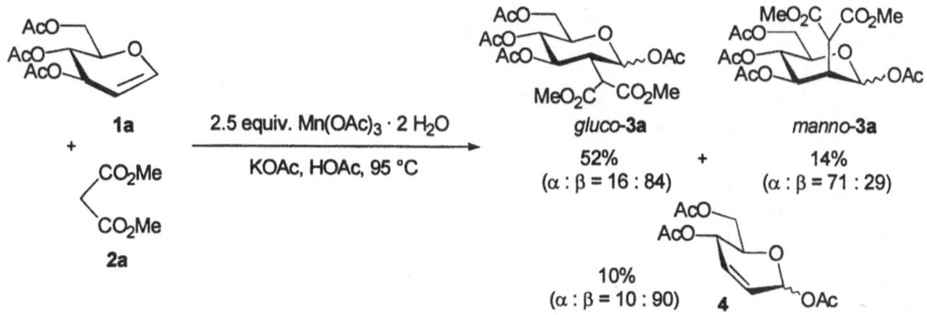

$$
\begin{array}{c}
\textbf{1a} \\
+ \\
\textbf{2a}
\end{array}
\quad
\xrightarrow[\text{KOAc, HOAc, 95 °C}]{\text{2.5 equiv. Mn(OAc)}_3 \cdot 2\,\text{H}_2\text{O}}
\quad
\begin{array}{cc}
\textit{gluco}\text{-}\textbf{3a} & \textit{manno}\text{-}\textbf{3a} \\
52\% & 14\% \\
(\alpha : \beta = 16 : 84) & (\alpha : \beta = 71 : 29)
\end{array}
\tag{1}
$$

$$
\begin{array}{c}
10\% \\
(\alpha : \beta = 10 : 90) \quad \textbf{4}
\end{array}
$$

147

Ceric(IV) Ammonium Nitrate (CAN)-Mediated Additions

To overcome the problem of acid-catalyzed Ferrier rearrangements, milder reaction conditions were next employed. Ceric(IV) ammonium nitrate (CAN) turned out to be the reagent of choice (method B), since radical generation from CH-acidic substrates takes place even in methanol at low temperatures. Thus, the addition of malonates 2 to glycals 1 afford the C-C bond-formation products 3 and 5 highly regioselectively in 86-92% yield without competing Ferrier rearrangement (Table 1). Furthermore, due to the lower reaction temperature higher stereoselectivities were obtained with cerium(IV) (method B) than with manganese(III) (method A). The methodology can be applied to various glycals 1 derived from hexoses and pentoses and provides a general and convenient entry to 2-C-branched carbohydrates [6].

Due to the similar substitution pattern of tri-O-acetyl-D-glucal (1a) and di-O-acetyl-D-xylal (1c), *gluco / manno* and *xylo / lyxo* mixtures were obtained in approximately the same ratios (Table 1). On the other hand, highest stereoselectivities were observed with tri-O-acetyl-D-galactal (1b) and di-O-acetyl-D-arabinal (1d), since two ester groups shield the same face of the carbohydrate. Furthermore, in both cases one substituent is orientated *pseudo* axial, which results in severe steric interactions with the malonyl radicals and, thus, *galacto-* and *arabino*-configurated products are formed exclusively.

Table 1 Addition of Malonates 2 to Various Glycals 1.

1	2	R	method[a] dr α:β[b]	R'	3 (%)[c]	3 (%)[c]	5 (%)[c]
1a	2a	Me	A 79:21	Ac	gluco-3a (52)[d]	manno-3a (14)[e]	-
	2b	i-Pr	A 84:16	Ac	gluco-3b (57)[f]	manno-3b (11)[g]	-
1a	2a	Me	B 85:15	Me	gluco-3a (62)	manno-3a (14)	gluco-5a (16)
	2b	i-Pr	B 91:9	Me	gluco-3b (68)	manno-3b (8)	gluco-5b (16)
1b	2a	Me	B >98:2	Me	galacto-3a (78)	-	galacto-5a (8)
	2b	i-Pr	B >98:2	Me	galacto-3b (73)	-	galacto-5b (17)
1c	2a	Me	B 93:7	Me	xylo-3a (81)[h]	lyxo-3a (6)	-
	2b	i-Pr	B 87:13	Me	xylo-3b (75)[i]	lyxo-3b (11)	-
1d	2a	Me	B <2:98	Me	arabino-3a (89)[j]	-	-
	2b	i-Pr	B <2:98	Me	arabino-3b (87)[j]	-	-

[a] Method A: 2-4 equiv. Mn(OAc)$_3$ · 2 H$_2$O, HOAc, 95 °C; method B: 3-6 equiv. CAN, MeOH, 0 °C. [b] Diastereomeric ratio (dr related to attack of malonyl radicals) determined by ^{1}H NMR analysis of the crude product (600 MHz). [c] Yield of isolated product after column chromatography. [d] α:β = 16 : 84. [e] α:β = 71 : 29. [f] α:β = 8 : 92. [g] α:β > 97 : 3. [h] α:β = 5 : 95. [i] α:β = 3 : 97. [j] α:β = 92 : 8.

148

Mechanistic Considerations

In the first step, malonyl radicals **6** are generated from malonates **2** and manganese(III) (method A) or cerium(IV) (method B) by an inner-sphere electron transfer. Such acceptor-substituted radicals are characterized by the low energy of the SOMO and exhibit electrophilic character. Thus, the interaction with the HOMO of the double bond becomes predominant, which has the largest coefficient at the 2-position of glycals (Scheme 1). This explains the highly regioselective addition of malonates **2** to afford the adduct radicals **7**, and reveals the importance of orbital interactions in radical reactions, since for steric reasons, attack at the 1-position should be favored.

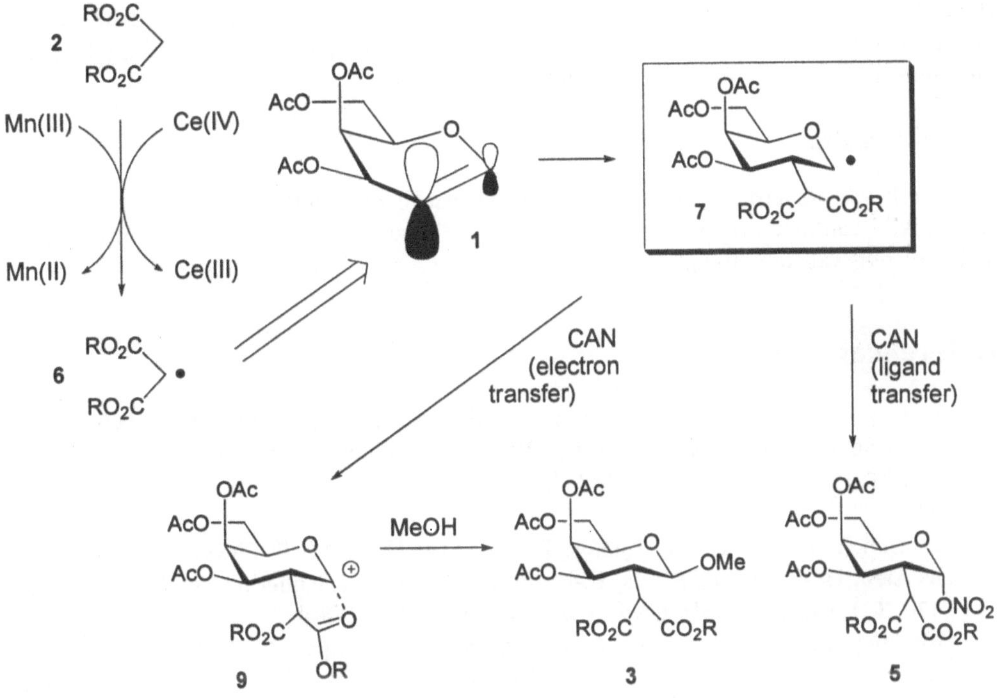

Scheme 1

The formation of methyl glycosides **3** and nitrates **5** is interesting from the mechanistic point of view. The adduct radical **7** is readily oxidized by CAN to the cation **9**, which is trapped by the solvent to afford the methyl glycoside **3**. The exclusive formation of β-galactosides, β-glucosides, and α-mannosides **3** can be rationalized by a neighboring group participation of the malonyl substituent (Scheme 1). On the other hand, the nitrates **5** are exclusively obtained as α-anomers and cannot be formed *via* the intermediate **9**. A direct ligand transfer from CAN without participation of cations is more likely, which would explain the high stereoselectivity, since carbohydrate radicals like **7** are preferentially trapped from the α-face.

Conclusion

The addition of malonates to glycals provides a general and convenient entry to carbohydrate 2-*C*-analogs. Our methodology is applicable to glycals derived from hexoses and pentoses and is characterized by easily available precursors. The generation of malonyl radicals by ceric(IV) ammonium nitrate (CAN) is superior to manganese(III)-mediated additions in terms of milder reaction conditions and yields. All reactions exhibit a very high degree of regioselectivity, since only 2-*C*-branched sugars were obtained. This result can be best rationalized by favorable orbital interactions between the SOMO of the malonyl radical and the HOMO of the double bond.

The substitution pattern of the glycals strongly alters the diastereomeric ratios. Thus, the addition of malonates to tri-*O*-acetyl-D-galactal and di-*O*-acetyl-D-arabinal occurs exclusively from one face of the carbohydrate. Strong evidence was found for a ligand transfer rather than electron transfer during the formation of nitrates, which sheds light on the mechanism of transition-metal-mediated radical reactions.

Acknowledgment

This work was generously supported by the Deutsche Forschungsgemeinschaft (SFB 347 "Selektive Reaktionen Metall-aktivierter Moleküle" and a Heisenberg fellowship for T. L.) and the Volkswagen-Stiftung.

References

[1] D. E. Levy, C. Tang, *The Chemistry of C-Glycosides*; Pergamon: Oxford, **1995**; M. H. D. Postema, *C-Glycoside synthesis*, CRC Press: London, **1995**; J.-M. Beau, T. Gallagher, *Top. Curr. Chem.* **1997**, *187*, 1-54; F. Nicotra, *Top. Curr. Chem.* **1997**, *187*, 55-83.

[2] U. Linker, B. Kersten, T. Linker, *Tetrahedron* **1995**, *51*, 9917-9926; T. Linker, B. Kersten, U. Linker, K. Peters, E.-M. Peters, H. G. von Schnering, *Synlett* **1996**, 468-470; T. Linker,. *J. Prakt. Chem. Chem. Zt.* **1997**, *339*, 488-492.
Reviews: G. G. Melikyan, *Synthesis* **1993**, 833-850; J. Iqbal, B. Bhatia, N. K. Nayyar, *Chem. Rev.* **1994**, *94*, 519-564; P. I. Dalko, *Tetrahedron* **1995**, *51*, 7579-7653; B. B. Snider, *Chem. Rev.* **1996**, *96*, 339-363; G. G. Melikyan, *Org. React.* **1996**, *49*, 427-675.

[3] T. Linker, K. Hartmann, T. Sommermann, D. Scheutzow, E. Ruckdeschel, *Angew. Chem., Int. Ed. Engl.* **1996**, *35*, 1730-1732; T. Linker, *GIT Fachz. Lab.* **1996**, 1167-1169.

[4] Reviews: R. J. Ferrier, *Adv. Carbohydr. Chem. Biochem.* **1969**, *24*, 199-266; F. W. Lichtenthaler, In *Modern Synthetic Methods 1992*; Scheffold, R., Ed.; VCH: Weinheim, **1992**; pp 273-376; S. J. Danishefsky, M. T. Bilodeau, *Angew. Chem., Int. Ed. Engl.* **1996**, *35*, 1380-1419.

[5] R. J. Ferrier, N. Prasad, *J. Chem. Soc. C* **1969**, 581-586; B. Fraser-Reid, *Acc. Chem. Res.* **1996**, *29*, 57-66.

[6] T. Linker, T. Sommermann, F. Kahlenberg, *J. Am. Chem. Soc.* **1997**, *119*, 9377-9384.

Homogeneous Catalysis in Supercritical Carbon Dioxide: A Better Solution ?

Sabine Kainz, Daniel Koch and Walter Leitner*

Max-Planck-Institut für Kohlenforschung, Postfach 10 13 53, 45466 Mülheim an der Ruhr, Germany, Fax: +49-208-306 2993, E-Mail: leitner@mpi-muelheim.mpg.de

1 Supercritical Carbon Dioxide as a Solvent with Unique Properties

Supercritical carbon dioxide ($scCO_2$), i.e. carbon dioxide heated and compressed beyond its critical point ($T_c = 31.0°C$, $p_c = 73.8$ bar, $d_c = 0.467$ g mL^{-1}), is a technically well established solvent for extraction and purification processes like supercritical fluid chromatography (SFC) or supercritical fluid extraction (destraction, SFE) [1]. Destraction of green coffee beans using $scCO_2$ technology was developed in the Max-Planck-Institute for Coal Research in Mülheim in the late sixties [2] and is nowadays used worldwide for the production of approximately 100 000 t decaffeinated coffee per year. The unique combination of gas- and liquid-like properties featured by supercritical fluids has also stirred interest in their use as solvents for chemical syntheses and in particular $scCO_2$ has been suggested as an ecologically benign and economically feasible reaction medium for homogeneously metal catalysed reactions [3, 4].

Transition metal catalysed olefin metathesis [5] in $scCO_2$ was studied in conjunction with the group of A. Fürstner and provides an illustrative example for how the specific properties of $scCO_2$ can increase the scope of a well-established transformation in a remarkable way [6]. Compressed and especially supercritical CO_2 serves efficiently as a substitute solvent for all kinds of metathesis processes and in particular for ring closing metathesis (RCM). Remarkably simple isolation of products and recovery of catalysts in active form is possible exploiting the extractive properties of the supercritical fluid. Even more intriguingly, the product distribution during RCM of conformationally not pre-organized dienes like **1** can be controlled by variation of the density of the supercritical reaction medium (Scheme 1).

The macrocyclic compound **2** exhibits a strong musk-like odour on its own and can be hydrogenated to give the commercial perfume ingredient exaltolide®. It was produced via ring closing metathesis (RCM) in up to 88% yield at densities of the supercritical phase $d \geq 0.65$ g mL^{-1}. Oligomers were obtained almost exclusively by ADMET (acyclic diene metathesis) at lower densities. This density effect resembles grossly the dilution principle in solution and may well be of the same generality [6]. Furthermore, the use of $scCO_2$ as a reaction medium allows the use of carbene complexes $[(R_3P)_2Cl_2Ru=C(H)R')]$ as catalysts for the metathesis of substrates with unprotected N-H functionalities. The deactivation of the ruthenium carbenes, which is normally observed with these type of substrates, is probably suppressed by CO_2 acting as a protecting group through reversible formation of the corresponding carbamic acids [6].

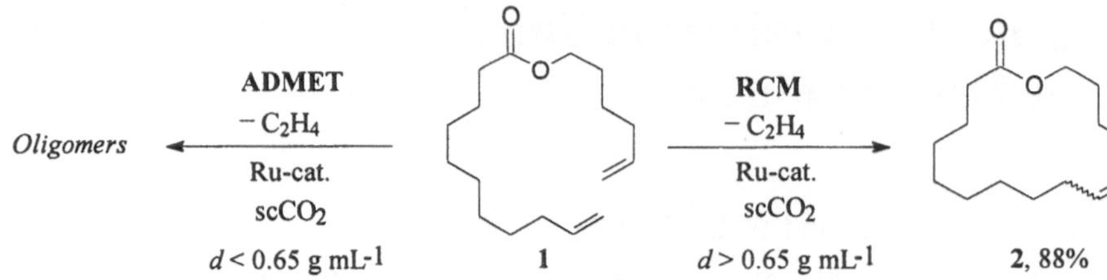

Scheme 1 Transition metal catalysed olefin metathesis of diene **1** in scCO$_2$ and density effect on product distribution (Ru-cat. = [(Cy$_3$P)$_2$Cl$_2$Ru=C(H)C(H)=CPh$_2$].

In general, chemical interactions of CO$_2$ with catalytically active intermediates may offer additional potential for activity and selectivity control [7] and the possibility to use scCO$_2$ as solvent and C1-building block provides further impetus for its use as an alternative to conventional solvents [8, 9]. The absence of a gas/liquid phase boundary is yet another potential technical advantage for the use of scCO$_2$ as reaction medium for catalytic reactions that involve gaseous reaction partners [3, 9a].

2 Tailor-Made Catalysts for the Use in Supercritical Carbon Dioxide

Obviously, scCO$_2$ has quite a remarkable potential for application as a reaction medium for homogeneous catalysis which is yet to be fully explored. The solvent properties of compressed CO$_2$ are, however, often not compatible with transition metal catalysts [10]. In particular complexes bearing aryl substituted phosphorous donor ligands, which have proven to bee extremely useful in homogeneous catalysis, are in general poorly soluble in scCO$_2$.

Although solubility in the supercritical phase is by no means always a necessary prerequisite for catalysis under these conditions [6, 11], high solubility is in most cases a decisive factor for the practical use of metal complexes as catalysts in scCO$_2$ [3, 9a, 12] and it is of course crucial for mechanistic investigations in this medium [3]. Our concept [13] for a generally applicable methodology to increase the solubility of arylphosphorous ligands and their metal complexes in scCO$_2$ is outlined in Figure 1 for complexes containing a bidentate chelating phosphane ligand: Appropriate solubilizers (waved lines) are fixed at the aryl moiety in a way that keeps structural and electronic changes at the active center (dashed oval) to a minimum compared to the unsubstituted parent complexes.

Long perfluoroalkyl chains of the formula R = (CH$_2$)$_x$(CF$_2$)$_y$F (x > 1, y > 3) were chosen as solubilizers, because the weak intermolecular van der Waals' forces between perfluorinated groups are similar to those found in compressed CO$_2$ [14]. Furthermore, two or more CH$_2$ groups are known to block the strong electron withdrawing effect of long fluorinated chains efficiently [15]. We have developed a synthesis that readily provides perfluoroalkyl substituted arylbromides like **3** [13, 16] in reasonable to good yields on a hundred gramm scale. They give access to the whole spectrum of arylphosphorous ligands with PAr$_n$-groups either directly or via the key intermediates Ar$_2$PCl (**6**). Representative examples are summarized in scheme 2 and include analogs of well known ligands like triphenylphosphine (tpp) or diphos (dppe) as well as the first example of a chiral perfluoroalkyl substituted phosphorous ligand. The variable substitution pattern of the bromides **3** makes this synthetic approach extremely flexible [17].

152

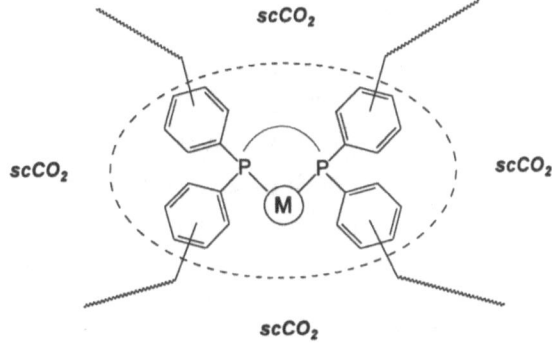

Figure 1 General concept for the development of scCO$_2$ soluble catalysts based on aryl phosphanes: The solubilisers (waved lines) in the periphery of the catalyst do not interfer with the catalytically active center (dashed oval).

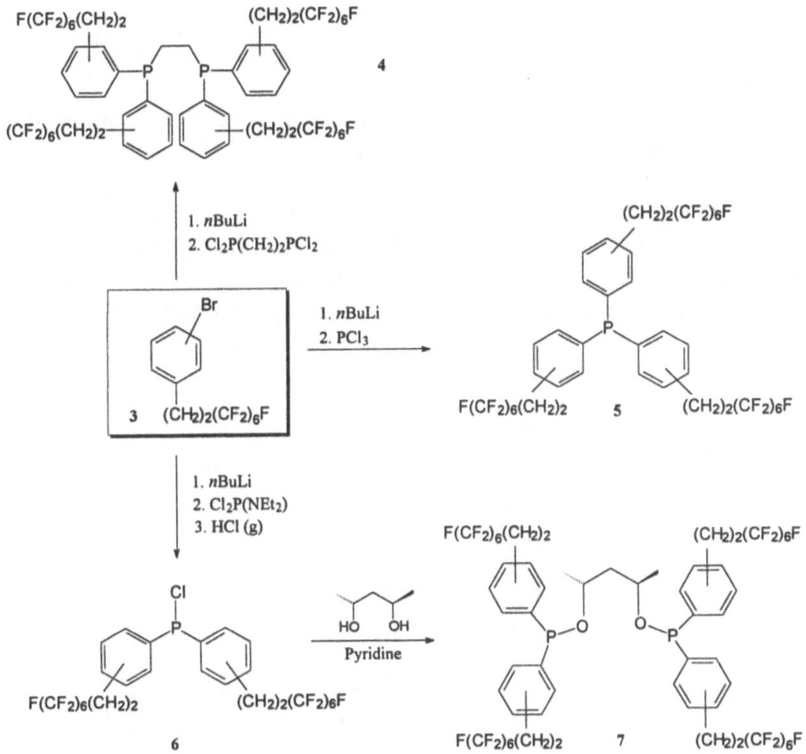

Scheme 2 Synthesis of selected perfluoroakyl substituted arylphosphorous ligands.

The spectroscopic data of the free ligands and of their metal complexes (Table 1) are almost identical to those of the unsubstituted parent compounds [13]. Comparison of the ^{103}Rh chemical shifts in **12a-c** and in the parent compound [(dppe)Rh(hfacac)] ($\delta = 438$, [18]) indicates, that the properties of the metal center are almost completely retained (**12b,c**) provided that steric interactions leading to distortions in the coordination geometry (**12a**) are avoided.

153

Table 1: Transition metal complexes with perfluoroalkyl substituted aryl phosphorous ligands and selected spectroscopic data [a].

trans-[(p-4)RuCl$_2$], 9 δ (^{31}P) = 43.9	[(p-4)Ir(cod)](BAr$_f$), 10 δ (^{31}P) = 47.1	[(m-5)$_3$RhCl], 11 δ (^{31}P) = 51.1, dt, J = 188/38 Hz δ (^{31}P) = 36.2, dd, J = 142/38 Hz δ (^{103}Rh) = –290
[(o-4)Rh(hfacac)], 12a δ (^{31}P) = 66.5, d, J = 197 Hz δ (^{103}Rh) = 705	[(m-4)Rh(hfacac)], 12b δ (^{31}P) = 71.8, d, J = 196 Hz δ (^{103}Rh) = 461	[(p-4)Rh(hfacac)], 12c δ (^{31}P) = 70.5, d, J = 196 Hz δ (^{103}Rh) = 459

[a] hfacac = hexafluoroacetylacetonate, cod = 1,5 cyclooctadiene, BAr$_f$ = B(3,5-(CF$_3$)$_2$C$_6$H$_3$)$_3$

The remarkable high solubility of the new complexes in compressed CO$_2$ is immediately apparent from visual inspection of the bright orange solutions of complexes **12** using window equipped high pressure reactors. The solubility can be quantified using UV/Vis-spectroscopy and was found to be at least two orders of magnitude higher than those reported previously for weakly soluble, yet already useful phosphine catalysts [12a, 13]. The novel ligands allow for the first time NMR spectroscopic investigations on transition metal phosphorous complexes in scCO$_2$ using standard high pressure sapphire NMR tubes (Figure 2) [19].

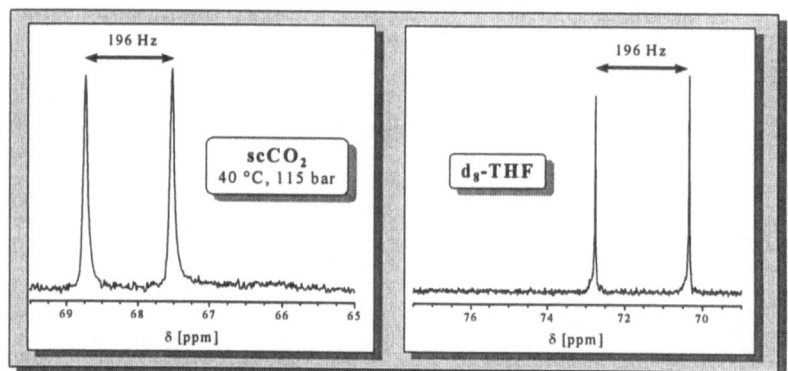

Figure 2 ^{31}P{^{1}H}-NMR spectra of complex **12b** in THF solution and in scCO$_2$.

3 Rhodium Catalysed Reactions in Supercritical Carbon Dioxide

The potential of the novel ligands for catalysis was briefly investigated for typical homogeneous rhodium catalyzed reactions. We found recently that the complex [(dppp)Rh(hfacac)] (dppp = Ph$_2$P(CH$_2$)$_3$PPh$_2$) is an efficient and highly selective catalyst for chemoselective hydrogenation of dienes to monoenes in a variety of organic solvents [20]. Isopren for example is quantitatively hydrogenated under 5 bar H_2 at 80°C in a dimethylsulfoxide/NEt_3 mixture to give the isomeric methyl substituted butenes. Only 3-methyl-1-butene, which is formed in amounts of less than 10%, is further reduced to isopentane upon prolonged reaction times. The same high selectivity is achieved in the hydrogenation of isopren in $scCO_2$ using **12b** as a catalyst, but the reaction occurs considerably more slowly under these conditions. At present, our data do not yet allow us to distinguish between the various possible explanations (concentration, solvent or ligand effect?) for the reduced rate.

Highly promising results which clearly demonstrate the applicability of our concept have been obtained in the rhodium catalyzed hydroformylation of 1-octene (**13**) to the isomeric aldehydes **14a-b** (Scheme 3). As expected, the catalytic system Rh/PPh_3, which is well established in organic solvents, proved not suitable for application in $scCO_2$, whereas a catalyst formed *in situ* from [(cod)Rh(hfacac)] and ligand *m*-**5** was highly active under otherwise identical conditions [13]. The intense yellow colour of the homogeneous supercritical phase confirmed the solubility of the catalytic active species containing *m*-**5** during the whole course of reaction. Up to 100 % conversion of **18** were obtained under typical conditions and side reactions such as hydrogenation or the formation of further isomeric aldehydes were not observed. The positive influence of the ratio *m*-**5**/Rh on the *n/iso* ratio **14a/14b** in $scCO_2$ parallels the trends typically encountered with rhodium catalysts containing monodentate phosphine ligands in homogeneous solution [21].

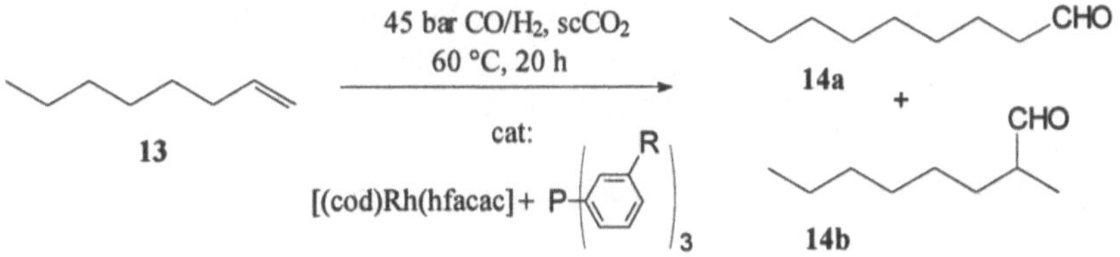

conditions	n/iso	conv.
0.25 mol% [Rh], R = H, P/Rh = 8.5	3.1	23 %
0.34 mol% [Rh], R = $(CH_2)_2(CF_2)_6F$, P/Rh = 3.8	3.7	100 %
0.42 mol% [Rh], R = $(CH_2)_2(CF_2)_6F$, P/Rh = 6.6	4.8	92 %

Scheme 3 Hydroformylation of 1-octene (**13**) in $scCO_2$ using rhodium phosphine catalysts.

4.   References and Notes

[1] a) M. A. McHugh, V. J. Krukonis, *Supercritical Fluid Extraction*, Butterworth, Boston, 2nd ed., **1994**; b) G. Brunner, *Gas Extraction*, Springer, New York, **1994**.

[2] K. Zosel, *Angew. Chem.* **1978**, *90*, 748; *Angew. Chem. Int. Ed. Engl.*, **1978**, *17*, 702.

[3] J. W. Rathke, R. J. Klingler, T. R. Krause, *Organometallics* **1991**, *10*, 1350.

[4] Recent reviews: a) M. Poliakoff, S. M. Howdle, S. G. Kazarian, *Angew. Chem.* **1995**, *107*, 1409; *Angew. Chem. Int. Ed. Engl* **1995**, *34*, 1275; b) P. G. Jessop, T. Ikariya, R. Noyori, *Science*

1995, *269*, 1065; c) D. A. Morgenstern, R. M. LeLacheur, D. K. Morita, S. L. Borkowsky, S. Feng, G. H. Brown, L. Luan, M. F. Gross, M. J. Burk, W. Tumas in *Green Chemistry* (Eds.: P. T. Anastas, T. C. Williamson), *ACS Symp. Ser. 626*, American Chemical Society, Washington DC, **1996**, pp. 132; d) E. Dinjus, R. Fornika, M. Scholz in *Chemistry under Extreme or Non-Classical Conditions* (Eds.: R. van Eldik, C. D. Hubbard), Wiley, New York, **1996**, pp. 219.

[5]   a) K. J. Ivin, J. C. Mol, *Olefin Metathesis and Metathesis Polymerization*, Academic Press, New York, **1997**; b) R. H. Grubbs, S. J. Miller, G. C. Fu, *Acc. Chem. Res.* **1995**, *28*, 446.

[6]   A. Fürstner, D. Koch, K. Langemann, W. Leitner, *Angew. Chem.* **1997**, *109*, in the press.

[7]   For the coordination chemistry of CO_2 and its relevance for catalysis see: W. Leitner, *Coord. Chem. Rev.* **1996**, *153*, 257.

[8]   For reviews on CO_2 as a raw material see: a) W. Leitner, *Angew. Chem.* **1995**, *107*, 2391; *Angew. Chem. Int. Ed. Engl.* **1995**, *34*, 2207 and ref.[7]; b) A. Behr, *Carbon Dioxide Activation by Metal Complexes*, VCH, Weinheim, **1988**.

[9]   a) P. G. Jessop, T. Ikariya, R. Noyori, *Nature* **1994**, *368*, 231; b) M. T. Reetz, W. Könen, T. Strack, *Chimia* **1993**, *47*, 493.

[10] M. G. Mason, J. A. Ibers, *J. Am. Chem. Soc.* **1982**, *104*, 5153.

[11] For example: a) O. Kröcher, R. A. Köppek, A. Baiker, *J. Chem. Soc. Chem. Commun.* **1996**, 1497; b) *ibid.* **1997**, 453; c) C. D. Mistele, H. H. Thorp, J. M. DeSimone, *J. M. S. Pure Appl. Chem.* **1996**, *A33*, 953; d) D. J. Darensbourg, N. White Stafford, T. Katsurao, *J. Mol. Catal. A: Chem.* **1995**, *104*, L1.

[12] For example a) M. J. Burk, S. Feng, M. F. Gross, W. Tumas, *J. Am. Chem. Soc.* **1995**, *117*, 8277; b) J. Xiao, S. C. A. Nefkens, P. G. Jessop, T. Ikariya, R. Noyori, *Tetrahedron Letters* **1996**, *37*, 2813; c) M. E. Super, E. Berluche, C. Costello, E. Beckman **1997**, *Macromolecules* **1997** *30*, 368.

[13] S. Kainz, D. Koch, W. Baumann, W. Leitner, *Angew. Chem.* **1997**, *109*, 1699; *Angew. Chem. Int. Ed. Engl.* **1997**, *36*, 1628.

[14] K. Harrison, J. Goveas, K. P. Johnston, E. A. O'Rear III, *Langmuir* **1994**, *10*, 3536.

[15] I. T. Horváth, J. Rábai, *Science*, **1994**, *266*, 72.

[16] S. Kainz, Diplomarbeit, MPI für Kohlenforschung/Universität Duisburg, **1996**. We thank Prof. G. Dyker for the possibility to conduct an external Diplomarbeit.

[17] In general, we use the following nomenclature: The prefix (z-H^xF^y) is placed in front of the established ligand acronym, whereby z gives the position of the perfluorinated alkyl chain in the aryl moiety relative to the carbon bound to phosphorus or oxygen. In this paper, the prefices *o-*, *m-*, *p-* are used in conjunction with the compound numbers of scheme 2 to designate the substitution pattern.

[18] K. Angermund, W. Baumann, E. Dinjus, R. Fornika, H. Görls, M. Kessler, C. Krüger, W. Leitner, F. Lutz *Chem. Eur. J.* **1997**, *3*, 755.

[19] a) D. C. Roe, *J. Magn. Res.* **1985**, *63*, 388; b) The tube used here was manufactured in the laboratories of C. J. Elsevier, J. van't Hoff Research Institute, University of Amsterdam.

[20] K. Wittmann, W. Leitner, to be published.

[21] C. D. Frohning, C. W. Kohlpaintner in *Applied Homogeneous Catalysis with Organometallic Compounds Vol. 1* (B. Cornils, W. A. Herrmann, Eds.), VCH, Weinheim **1996**, pp. 47

Stereoselective Hydroformylation of Acyclic Olefins with the Aid of a Catalyst Directing Group

Bernhard Breit

Fachbereich Chemie, Philipps-Universität Marburg, Hans-Meerweinstr., D-35043 Marburg, Germany

1   Introduction

Transition metal catalyzed C/C-bond forming reactions have evolved over the past two decades into powerful tools for the construction of carbon backbones in organic synthesis. These reactions allow also for the generation of a new stereocenter employing either a chiral catalyst (reagent control) or substrate-based asymmetric induction (substrate control). Although the latter approach seems attractive in regard to synthetic efficiency, it is not utilized frequently.

However, we recently found that a rational design of such processes becomes possible with the aid of a *Catalyst Directing Group* (CDG), a group that acts via a reversible catalyst coordination [1-4]. This enables the substrate to be attached in a chelating and therefore highly ordered manner to the catalytically active transition metal center within the stereochemistry defining step of the catalytic cycle, which we assumed to be mandatory to achieve high levels of stereoselectivity.

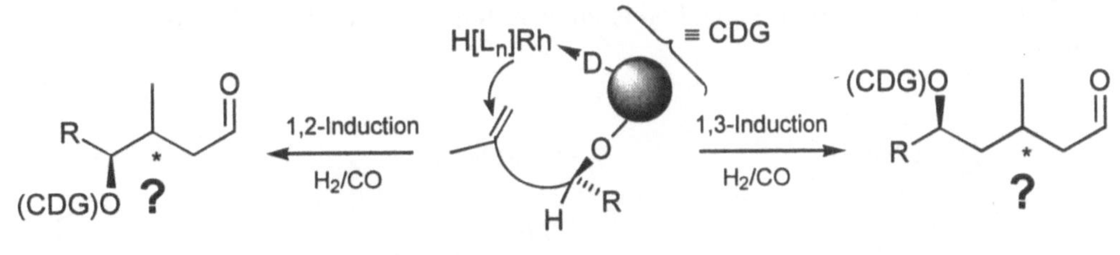

As a first test for this concept we chose the industrially important rhodium-catalyzed hydroformylation of acyclic olefins. We felt that this reaction could become of particular synthetic interest as an attractive C1 homologation starting out with readily available chiral olefins, employing the inexpensive synthesis gas as carbon source, thereby introducing the synthetically valuable aldehyde functionality under rather mild reaction conditions. In particular, the hydroformylation of acyclic methallylic alcohols and homomethallylic alcohols would generate structural building blocks bearing hydroxyl and methyl substituted stereo-centers in either a 1,2- or 1,3-relation. Both structural motifs are widespread in the polyketide class of natural products, which makes their stereoselective synthesis a rather attractive task. However, no efficient variants to control diastereoselectivity on hydroformy-lation of acyclic olefins such as methallylic and homomethallylic alcohols were known.

2 Stereoselective Hydroformylation of Acyclic Methallylic Al-cohols - 1,2-Asymmetric Induction

Applying the criteria for an effective *catalyst directing group*, which have been outlined by us previously [1-3], lead to the introduction of the *ortho*-diphenylphosphino benzoate group (*o*-DPPB group) as the presumably ideal catalyst directing group for this particular reaction.

Attachment of the *o*-DPPB group to the methallylic alcohol substrates was achieved employing the DCC/DMAP esterification protocol. Subsequent hydroformylation of the methallylic *o*-DPPB esters provided in mostly excellent yields the corresponding *syn*-aldehydes as the major diastereomers in diastereomer ratios of up to 96 : 4.

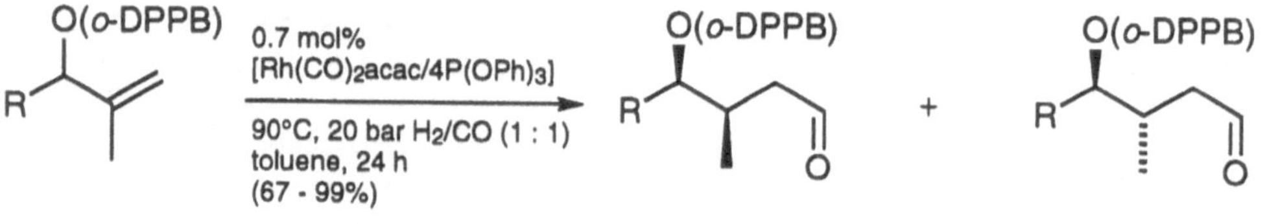

The reaction conditions were found to be compatible with a wide range of substrates. In particular those with branched substituents at the stereogenic center being attached either via an sp^3- or an sp^2 hybridized carbon atom were found to give the highest diastereoselec-tivities.

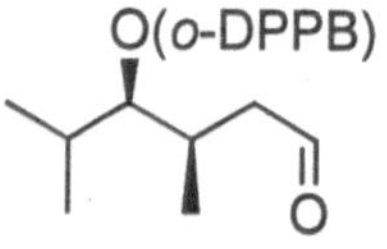

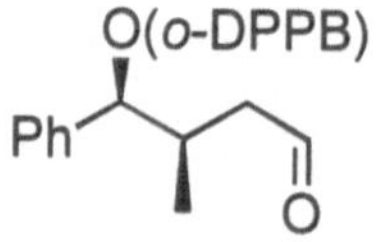

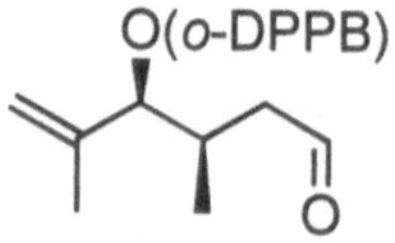

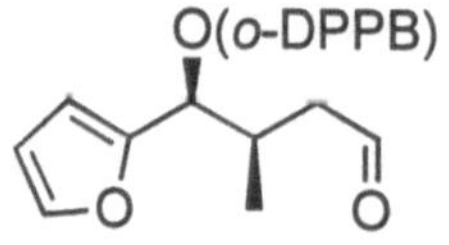

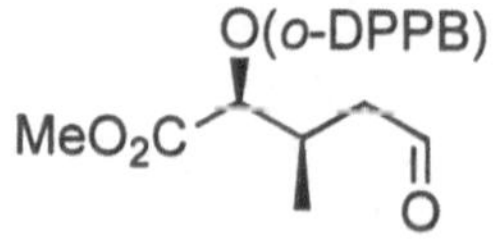

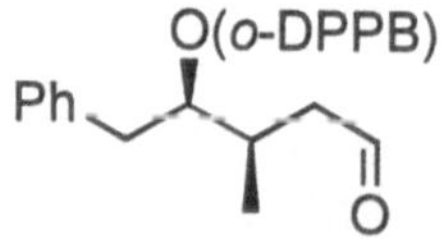

Furthermore, the reaction could be used for the efficient construction of stereotriades, structural building blocks with three consecutive stereocenters, which are considered as central structural building blocks of the polyketide class of natural products.

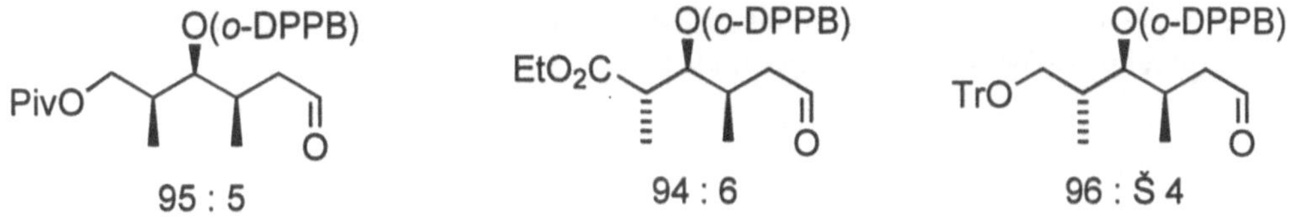

3 Stereoselective Hydroformylation of Acyclic Homomethallylic Alcohols - 1,3-Asymmetric Induction

Although the catalyst directing *o*-DPPB group has been developed originally for methallylic alcohol substrates to amplify 1,2-asymmetric induction, however, we recently found, that the same catalyst directing group allows also for high 1,3-asymmetric induction in the hydroformylation of homomethallylic alcohols [4]. Thus, homomethallylic *o*-DPPB esters **3** could be transformed into the corresponding aldehydes **4** in good to excellent yields and with *anti*-diastereoselectivities of up to 91%.

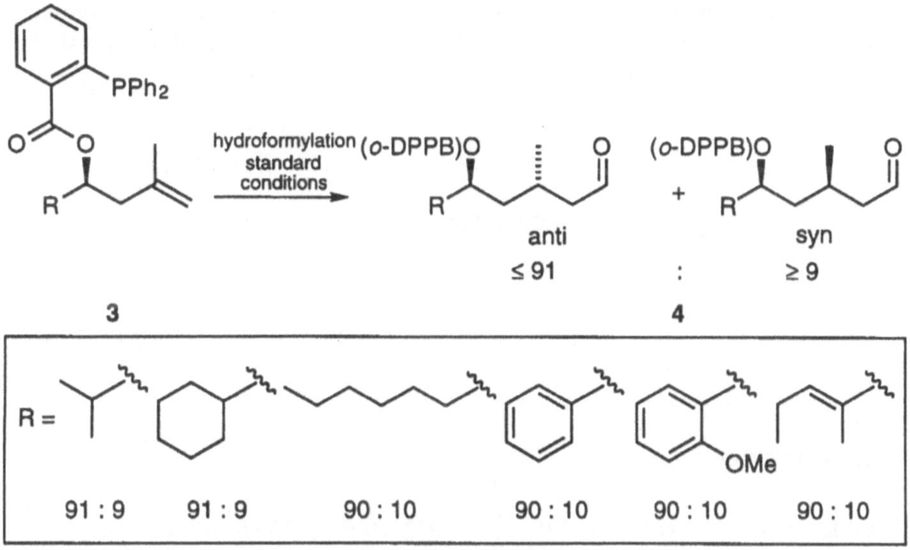

Interestingly, no significant influence of the diastereoselectivity on the nature of the substituent R could be detected. Thus, primary and secondary alkyl as well as aryl substituents were tolerated and gave diastereomer ratios independent on the nature of R of ca. 91:9.

Remarkable is also the chemo-, regio- and stereoselective hydroformylation of a ho-momethallylic alcohol system in the presence of a trisubstituted alkenyl side chain.

Furthermore, a significant dependence of the diastereoselectivity on the reaction temperature was observed. In agreement with these findings a conformational preference inherent to the homomethallylic alcohol system itself seemed to be the origin of the observed diastereoselectivities. Such a preferred substrate conformation could be detected experimentally employing 2D NOESY NMR spectra [4].

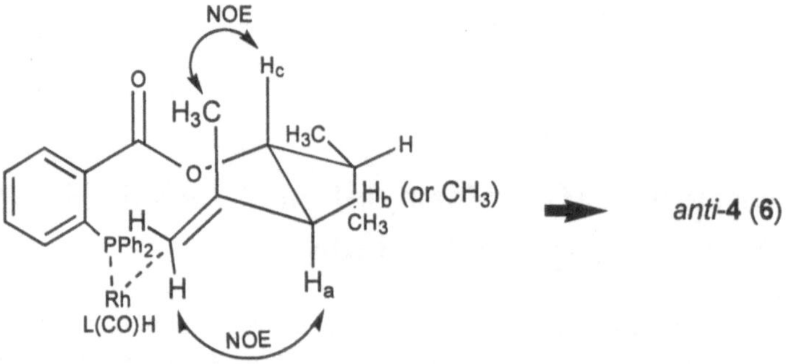

Consequently, we probed the influence of an additional methyl group bearing stereocenter in *anti* relation and in direct neighborhood to the directing oxygen bearing stereocenter (→**5**). Such a substituent should enhance the basic conformational preference of the ho-momethallylic alcohol system.

Interestingly, and in accord with this model a substantial increase in diastereoselectivity on hydroformylation of the corresponding *o*-DPPB ester **5** was detected, to furnish the aldehyde **6** in a diastereomer ratio of 96 : 4 [4].

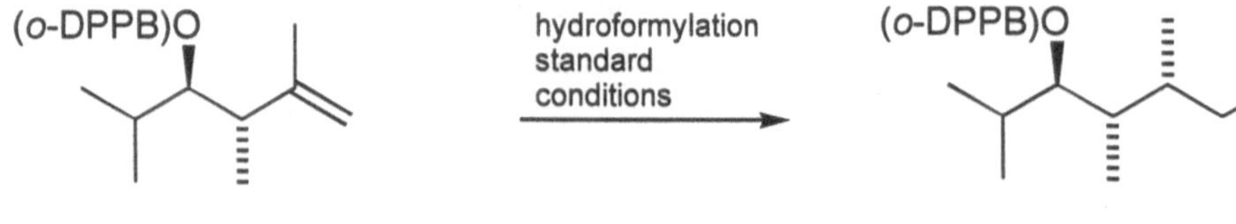

4 References

1. B. Breit, *Angew. Chem.* **1996**, *108*, 3021-3023; *Angew. Chem. Int. Ed. Engl.,* **1996**, *35*, 2835-2837.
2. B. Breit, *Liebigs Ann/Recueil.* **1997**, 1841-1851.
3. B. Breit in *Organic Synthesis via Organometallics*, G. Helmchen, J. Dibo, D. Flubacher, B. Wiese (Eds.), Vieweg, Braunschweig Wiesbaden **1997**, 139-146.
4. B. Breit, *J. Chem. Soc., Chem. Commun.* **1997**, 591-592.

Fluorenyl Complexes of Zirconium and Hafnium as Catalysts for the Olefin Polymerization

Helmut G. Alt, Syriac J. Palackal, Konstantinos Patsidis, Roland Zenk and Michael Schmid

Laboratorium für Anorganische Chemie der Universität Bayreuth, D-95440 Bayreuth, Germany

In the past years metallocene complexes of group 4 metals have attracted great attention because of their potential as homogeneous catalysts for olefin polymerization. In the meantime many parameters are known that determine the activity, stereospecifity and thermal stability of such metallocene catalysts. Our approach to this field deals with fluorenyl complexes of zirconium and hafnium: they can be bridged or unbridged metallocene complexes, containing one or two fluorenyl ligands.

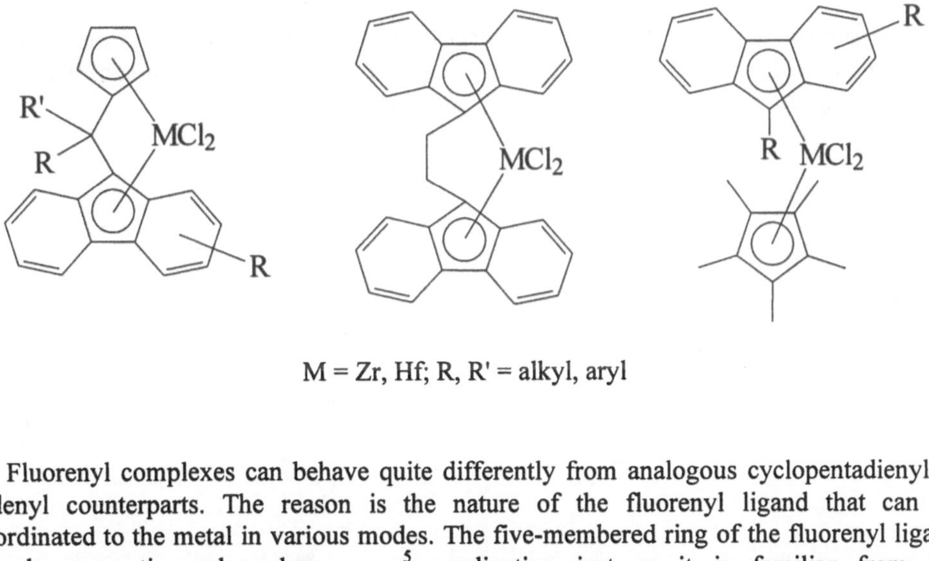

M = Zr, Hf; R, R' = alkyl, aryl

Fluorenyl complexes can behave quite differently from analogous cyclopentadienyl or indenyl counterparts. The reason is the nature of the fluorenyl ligand that can be coordinated to the metal in various modes. The five-membered ring of the fluorenyl ligand can be aromatic and undergo a η^5-coordination just as it is familiar from the cyclopentadienyl ligand. However, this is only possible when electrons from the two anellated six-membered rings contribute electrons. Because of this situation fluorenyl ligands can readily undergo ring-slippage reactions from $\eta^5 \rightarrow \eta^3 \rightarrow \eta^1$. For all these cases corresponding complexes have been characterized by X-ray structures. For the first time the X-ray structure of a η^1-bonded fluorenyl complex, $(C_5H_4Me)_2Zr(\eta^1\text{-}C_{13}H_9)Cl$, can be presented.

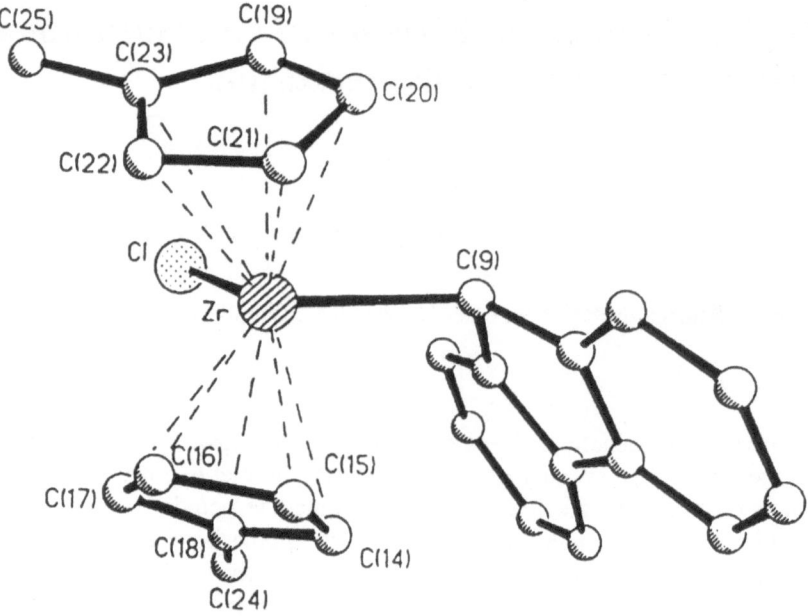

Various substituted ansa metallocene complexes containing cyclopentadienyl and fluorenyl units were tested for propylene polymerization and the influence of various substituents at various ring positions at the fluorenyl ligand was investigated.

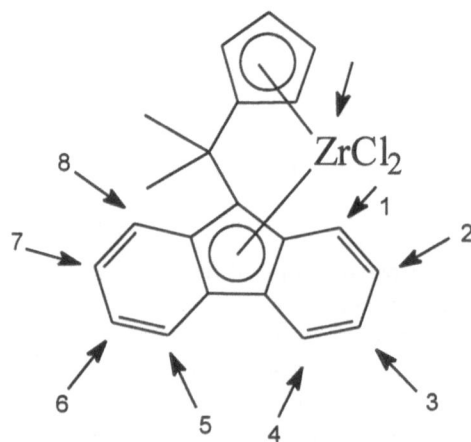

Table 1

Complex	Activity	T_i max	M_η	T_{m2}	r
	[kg PP/(mmol cat. h)]	[°C]	[kg/mol]	[°C]	[%]
H	33.5	59.5	82	135.0	94.6
2,7-Me$_2$	11.2	59.4	80	131.2	94.9
4-Me	35.5	61.9	63	111.5	
4,5-Me$_2$	9.3	62.0	29	107.9	77.4
2,7-Ph$_2$	17.6	58.9	65	132.5	92.7
2,7-Mes$_2$	77.1	61.0	150	132.7	n.b.
2,7-(tert-Bu)$_2$	54.8	60.4	74	142.0	93.5
2,7-(MeO)$_2$	0.3	58.7	20	96.3	n.b.
2,7-Cl$_2$	20.7	60.9	n.b.	n.b.	92.9
2,7-Br$_2$	26.8	58.9	60	131.0	90.5
3,4-Benzo	34.9	61.8	37.5	121.1	93.7
4,5-Benzo	10.5	60.0	154	n.b.	(72)

Though the substituents are comparatively far away from the center of olefin coordination they still have a strong influence on the activity of such catalysts and polymer parameters like molecular weight, tacticity and chain branching.

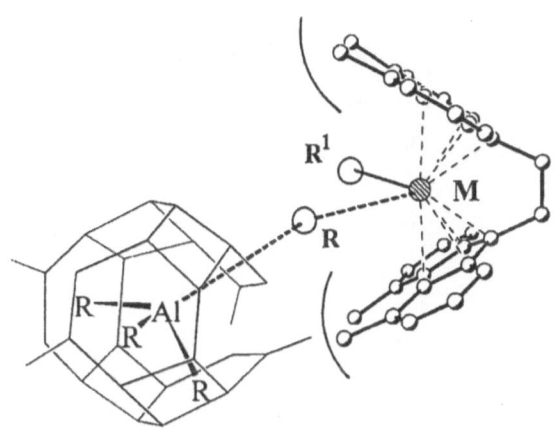

$M = Zr, Hf$ $R = Me$
$R^1 =$ polymer chain

This could be due to steric requirements for the orientation of the growing polymer chain during the polymerization process.

Another positive effect of fluorenyl ligands is their bulkiness. It is assumed that the actual metallocene catalyst for olefin polymerization consists of a metallocene monomethyl cation that is stabilized by a bulky counter anion such as a perfluoronated tetra phenyl borate or a methylalumoxane anion of still unknown composition. Because of their bulkiness fluorenyl ligands can prevent the approach of these bulky counter anions and as a result "naked" metallocene monomethyl cations with high activity become available. Two methyl substituents in the 4 and 5 positions of the fluorenyl ligand cause an increase in activity for ethylene polymerization by a factor of 6 compared with the unsubstituted complex (2.6 10^3 kg PE/mmol cat. h).

Another possibility to produce a naked metallocene monomethyl cation is the anellation of additional rings to the fluorenyl ligand.

Such a step can increase the activity for ethylene polymerization by a factor of 8.

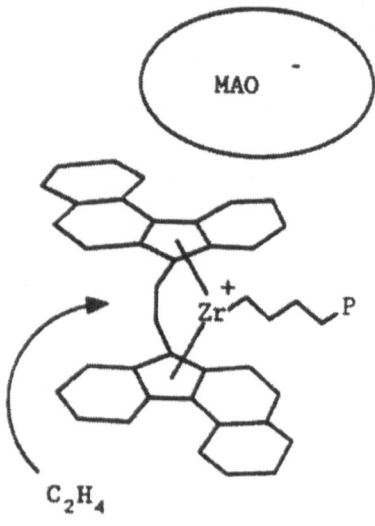

Orthometalated Rhodium(II) Compounds. Synthesis, Characterization and Catalytic Applications

Pascual Lahuerta

Departamento de Química Inorgánica, Universitat de Valencia, Burjasot-Valencia, Spain

Introduction

Intramolecular coordination compounds have received a great deal of interest[1]. Although many compounds with only one metal atom involved in the metalocycle (I) have been described, intramolecular coordination compounds containing metal-metal bonds (II) are relatively rare[2].

In 1985 F.A. Cotton and coworkers reported that the reaction of rhodium acetate and triphenylphosphine in pure acetic acid as solvent, yields $Rh_2(AcO)_2$ $[(C_6H_4)PPh_2]_2\cdot(AcOH)_2$ (1)[3]. No net redox process occurs during the reaction, and formally the displacement of two acetate ligands by two orthometalated phosphines entails merely the transfer of protons from the ortho positions of phenyl rings to the acetate anions. The phosphine ligands are mutually *cis* and bridge the two rhodium atoms addopting a head-to-tail (H-T) configuration.

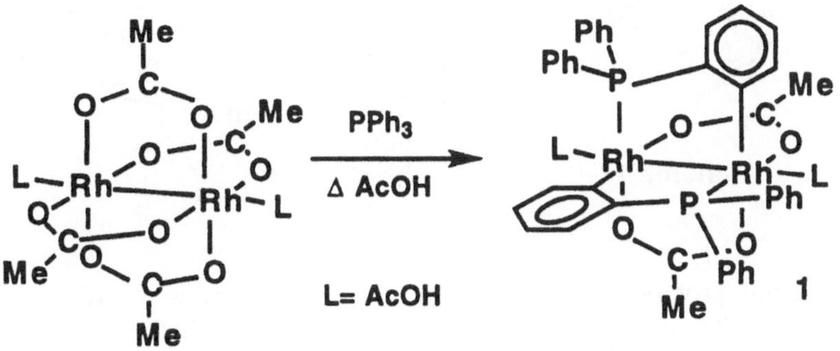

A different synthetic route yielded rhodium(II) compounds of general formula $Rh_2Br(\mu^2\text{-pz})_2[\mu^2\text{-}(C_6F_4)PPh_2](CO)L_2$ (**2**) (pz = pirazolyl and substituted pirazolyl groups; $L_2 = (CO)_2$ or $\eta^2\text{-}P(o\text{-}BrC_6F_4)Ph_2$) that also have a bridging arylphosphine anion. They were prepared in low or moderate yield by thermal reaction of $[Rh(\mu^2\text{-pz})(CO)_2]_2$ with PCBr [PCBr=$P(o\text{-}BrC_6F_4)Ph_2$] in refluxing toluene[4]. This metalation reaction involves 2c-2e oxidative addition with cleavage of the C-Br bond of the phosphine.

The reaction of rhodium acetate with phosphines was a more general approach to the synthesis of metalated rhodium compounds. Thus, a systematic study of the different variables was performed for this particular metalation reaction.

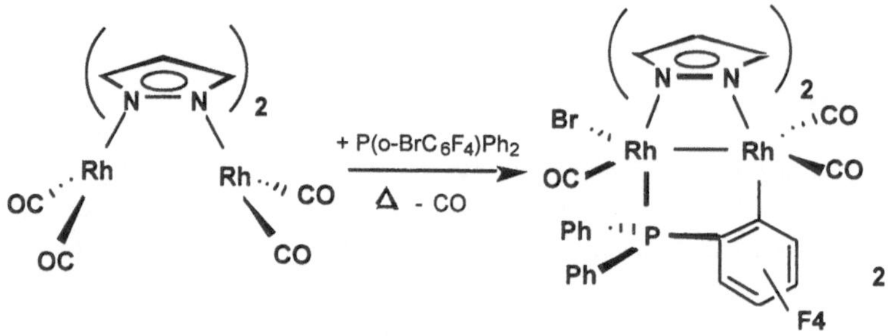

Based on the isolation of the intermediate compounds, a reaction pathway (Scheme 1) was proposed. Initially there was coordination of one phosphine with the formation of adduct I. This phosphine rearranges from axial to equatorial coordination with partial displacement of one acetate group. Proton transfer in this species II yields the mono-metalated compound III, that further reacts with one additional phosphine to form the doubly metalated compound Va, via an intermediate equatorial compound IVa. If more than equimolar amount of phosphine is present in solution, species I-III will be in equilibrium with the corresponding adducts having a second coordinated phosphine.

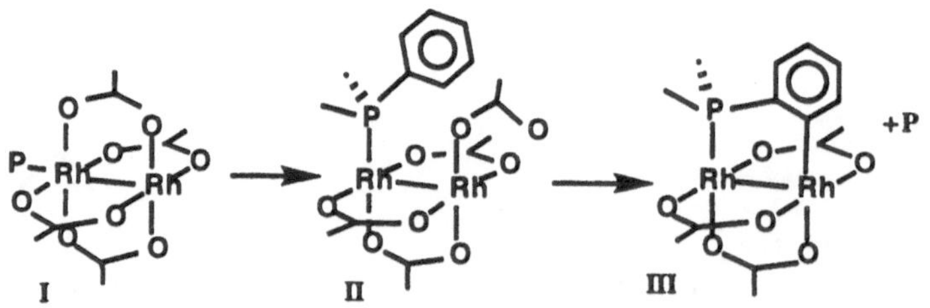

Scheme 1

These results prompted us to isolate reaction intermediates in order to study single reaction steps of this metalation process. A species of type I has been only characterized when P is $P(o\text{-}MeOC_6H_4)Ph_2$[5]. In the solid state two units of formula $Rh_2(O_2CCH_3)_4.P(o\text{-}$

166

MeOC$_6$H$_4$)Ph$_2$ (**3**) are associated forming a centrosymmetric dimer of dimers. One oxygen atom from one acetate in one Rh$_2$P unit is axially coordinating one Rh atom on another Rh$_2$P fragment. Osmometric measurements confirm that in CH$_2$Cl$_2$ solution, I is in the monomeric form.

If PPh$_3$ or other non-functionalized phosphines were used, the intermediate compound II in the reaction I→III was quite reactive and it was only detected in solution in low concentration or not detected at all. However, when ortho functionalized phosphine P(o-XC$_6$H$_4$)Ph$_2$ (X = MeO, Cl) was used, II is formed in measurable amounts. It was found that photochemical irradiation of **3** in CHCl$_3$ at room temperature affords the equatorial compound Rh$_2$(μ-O$_2$CCH$_3$)$_3$(η^1-O$_2$CCH$_3$)[η^2-(C$_6$H$_4$)P(o-MeOC$_6$H$_4$)Ph] (H$_2$O) (**4**) in high yield.[6]

An additional compound of type II which was structurally characterized[7] is Rh$_2$(μ-O$_2$CCF$_3$)$_3$(η^1-O$_2$CCF$_3$)[η^2-(C$_6$H$_4$)P(o-ClC$_6$H$_4$)Ph](H$_2$O) (**5**). The observation of high field signals (30-45 ppm) in the ^{31}P NMR spectrum, (38.2 ppm for **4** and 41.3 ppm for **5**), is characteristic of the presence of a non-metalated P ligand in equatorial coordination. In both structures one phenyl ring is oriented to the partially displaced carboxylate group in a situation that makes the transfer of one proton from the phenyl ring to the carboxylate very favorable. The quelating coordination of the phosphine seems to be responsible of the relative stability of these intermediates, that in solution undergo slow metalation reaction to form species of type III. Triphenylphosphine and other non-functionalyzed phosphines give very reactive equatorial intermediates that can only be detected in small concentration when the reaction progress is monitored by ^{31}P NMR spectroscopy.

Mono metalated species of type III have been obtained for several phosphines with yields in the range 60-90%.[8,14] The crystal structure has been determined for a selected number of compounds. There are two structural features that we should mention. The Rh-O bond *trans* to the carbon atom, 2.218(4)Å is longer than that trans to the P atom, 2.163(3)Å, and both bonds are longer than the other four involving acetate ligands (in the range 2.025(4)-2.073(4)Å). The OH groups of the axially coordinated acetic acid molecules are bound to the oxygen atoms of the acetate ligand *trans* to the phoshine, through intramolecular hydrogen bonds (O—O distances 2.586(6) and 2.705(5)Å).[8a]

167

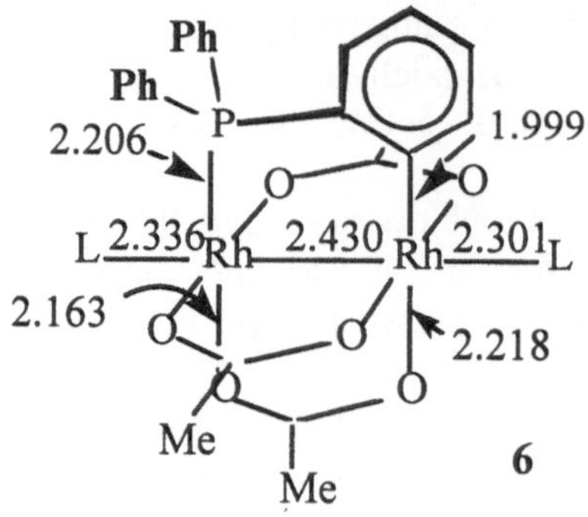

Equat.Rh-O
2.025-2.073 Å

L = CH$_3$CO$_2$H

6

Stepwise exchange reactions of CH$_3$COO$^-$ groups by CD$_3$COO$^-$ groups are observed for **6** in CDCl$_3$/CH$_3$COOD mixtures. The first step involves a fast exchange of the acetate group *trans* to the metalated phosphine as well as exchange of the two axial molecules of acetic acid. In a second and slower step the exchange of the other two acetate groups occurs.[9]

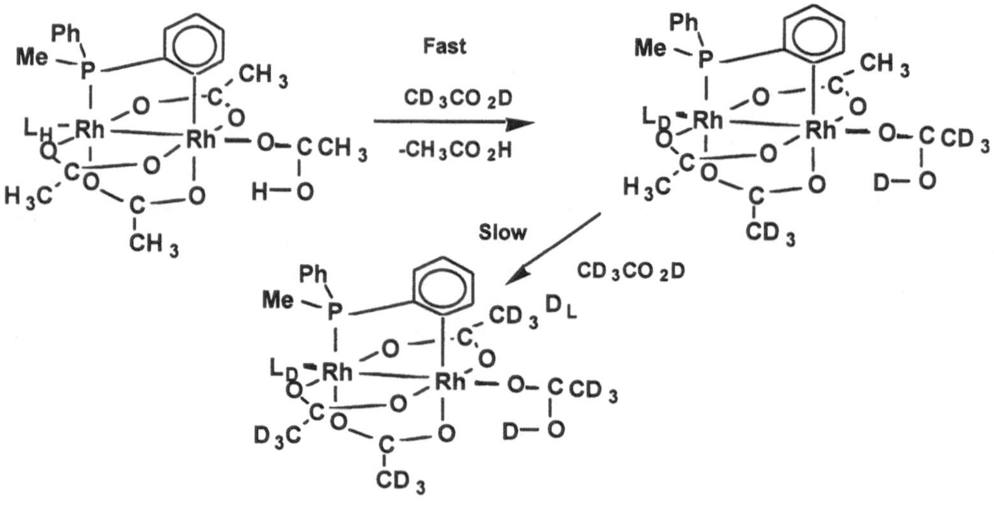

One additional observation is the reversibility of the metalation process. A clear evidence of this reversibility comes from the observation that Rh$_2$(O$_2$CCH$_3$)$_3$ [(C$_6$H$_4$) P(C$_6$H$_5$)$_2$].(HO$_2$CCH$_3$)$_2$ in CD$_3$COOD, undergoes H/D exchange at the ortho positions of the phenyl ring (90% exchange after 3 hours of reflux). This can only be explained by an electrophilic attack at the rhodium-carbon bond by the acetic acid, wich produces the protonation of the ortho aromatic carbon atom, followed by a cyclometalation at one of the ortho C-H bonds.[9]

168

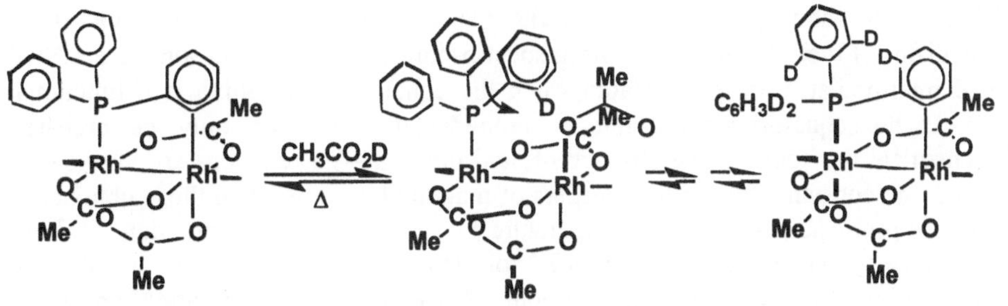

As it is shown in Scheme 2, the second metalation, starting from III, must take place following two pathways, as doubly metalated compounds with head-to-tail (pathway A) and head-to-head (pathway B) were obtained, in variable amounts, depending on the reaction conditions. In general, when stoichiometric amount of phosphine was used, compounds with head-to-tail structure are obtained. However if excess of phosphine was used the head-to-head compound was formed in some extend.Three doubly metalated compounds with head-to-head structure have been structurally characterized by X-ray methods.[10,11]

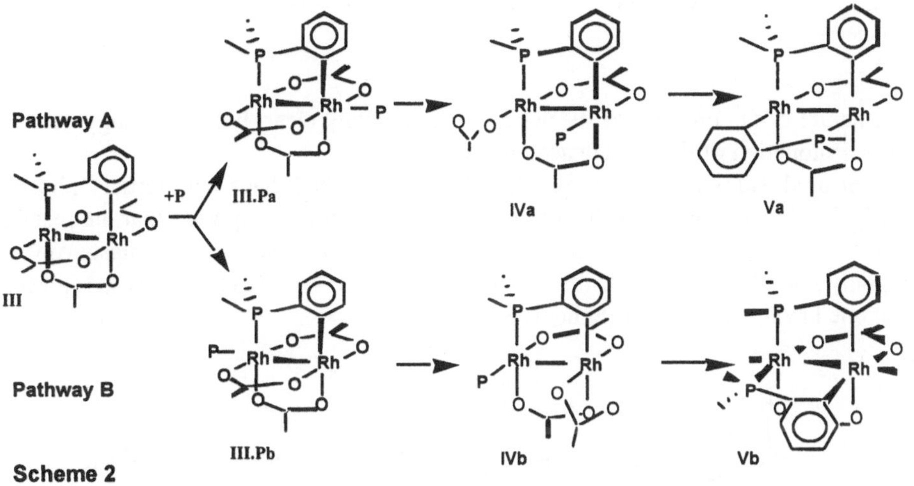

Scheme 2

In order to get some insight about the factors affecting these two pathways, a variable temperature ^{31}P NMR study of the reaction system was performed. It was confirmed that the reaction of the mono metalated compound $Rh_2(O_2CCH_3)_3 [(C_6H_4)PPh_2]\cdot(HO_2CCH_3)_2$ (6) with small amount of phosphine ([P]/[Rh$_2$] < 0.8) yields two adducts: 6.Pa having both P atoms (from the axial and the metalated ligands) attached to the same rhodium and 6.Pb that has each P bonded to a different rhodium atom.[12] These two adducts are in equilibrium and the [6.Pa]/[6.Pb] ratio increases as the temperature decreases from +25° to -60° C. If more phosphine is used, the bisadduct, 6.P$_2$, is also detected. Similar equilibrium occurs for other mono metalated compounds.[13]

169

These two pathways A and B occur via equatorial intemediates IVa and IVb. Intermediates of type IVb have only been spectroscopically detected in some cases.[14] These type of intermediates are quite stable when the equatorial phosphine is a functionalized ligand. If the equatorial phosphine is monodentate they can only be prepared by photochemical methods. In the absence of acid the IVa→Va reactions are very slow with observed rate constants two or three orders of magnitude smaller than those obtained in the presence of protic acid at the same temperature.[15] The enthalpies of activation $\Delta H^{\ddagger}$ are in a narrow range of values around 70 kJ mol^{-1} for the acid-catalyzed reaction and about 15-20 kJ mol^{-1} higher for the thermal reaction. The entropies of activation are negatives ($\Delta S^{\ddagger} = - 60/-100$ J K^{-1} mol^{-1}). The experimental data were fitted to a rate equation of the type

$$k_{obs} = k[H^+]/(Ke +[H^+])$$

where k is the limiting first order rate constant and Ke is the equilibrium constant for the acid-base equilibria previous to the rate-limiting step.

$$IVa + H^+ \underset{Ke}{\rightleftharpoons} [IVa.H^+] \xrightarrow{k} Va$$

The nature of the protonated species is not completely clear, but it was suggested that it involves protonation of the chelate acetate group.

Tocher et al. studied the reaction of rhodium tetracarboxylates with tertiary phosphines having both alkyl and aryl substituent groups.[16] These authors reported that the reaction of rhodium acetate with PMePh$_2$, in the presence of excese of pivalic acid, is diastereoselective, giving only one of the three possible isomers for Rh$_2${m^2-O$_2$C(CH$_3$)$_3$}$_2$ [(C$_6$H$_4$)PMePh]$_2$, (7) that one having structure II (Scheme 3, R= Me).

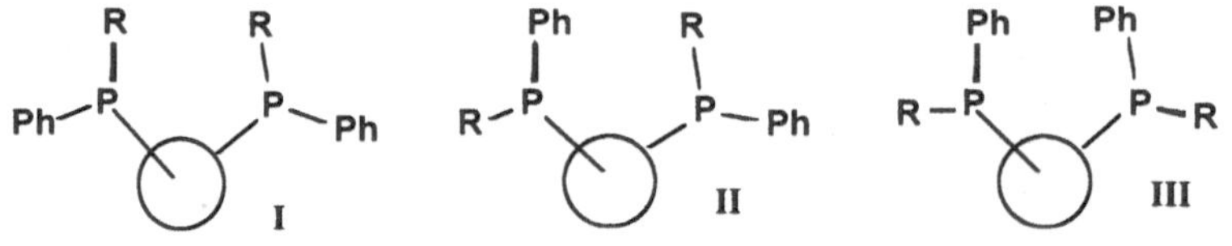

Scheme 3

R = Me, C$_6$F$_5$

Further studies confirmed that the same reaction in the absence of pivalic acid yields Rh$_2$(μ^2-O$_2$CCH$_3$)$_2$[(C$_6$H$_4$)PMePh]$_2$ (8). ^{1}H and ^{31}P NMR spectra of the crude reaction mixture indicated that only one isomer was formed. It was confirmed by X-ray crystalography that 8 has structure I.[17]

There is only one example reported of a rhodium (II) compound metalated at one alkyl group. This compound of formula Rh$_2$(μ^2-O$_2$CCH$_3$)$_2$[μ^2(CH$_2$)PPh$_2$] [μ^2(C$_6$H$_4$)

PPh$_2$]·PPh$_3$.(9) was obtained by serendipity from the reaction of Rh$_2$(μ^2-O$_2$CCH$_3$)$_3$ [(C$_6$H$_4$)PMePh] (10) with two moles of triphenylphosphine. Compound 9 is also obtained by reaction of 11 with one mol of triphenylphosphine. Small amounts of 9 are also detected in solution in the reaction 10→11.[18]

The reaction of rhodium acetate and P(C$_6$F$_5$)Ph$_2$ is considerably slow. Mixtures of mono and bi metalated compounds are always obtained even after long heating in mixtures of toluene/acetic acid. The three isomers are detected in solution by ^{31}P NMR spectroscopy. The isomer having different configuration at each P atom (Scheme 3, isomer II, R=C$_6$F$_5$) is the dominant product (70%). The other two isomers are in 15%, (endo-endo) and <5% (exo-exo). The two major isomers have been characterized by X-ray crystallography.[19].

The study of the reactivity of P(m-CH$_3$·C$_6$H$_4$)$_3$ showed that only one of the two different ortho positions available for the metalation is used. The crystal structure of the bi metalated compound confirmed that the metalation occurs at the carbon atom trans to the methyl group.[20]

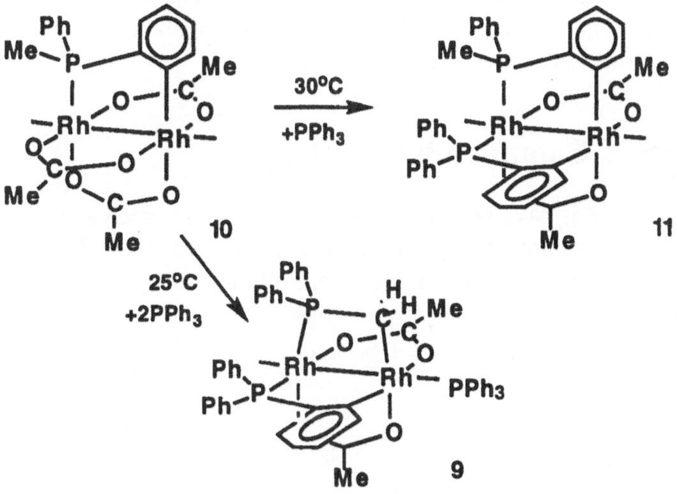

Chiral orthometalated compounds

The reaction of triphenlylphosphine with rhodium acetate under achiral conditions leads to a racemic mixture of products 1/1′. The resolution of this mixture has been achieved exchanging the bridging acetates by chiral carboxylates and further separation of the resulting diastereoisomers. The readily available and unexpensive tosylated proline has been used for this purpose.[22]

The resulting diastereoisomers 1-pro/1′-pro were separated by standard column chromatography on SiO$_2$ and treatment with trifluoroacetic acid gave 1-CF$_3$CO$_2$H and 1′-CF$_3$CO$_2$H. Their enantiomerical purity was confirmed by addition of chiral naphtylethyl-amine to NMR-samples of these enantiomers and to the racemic mixture.

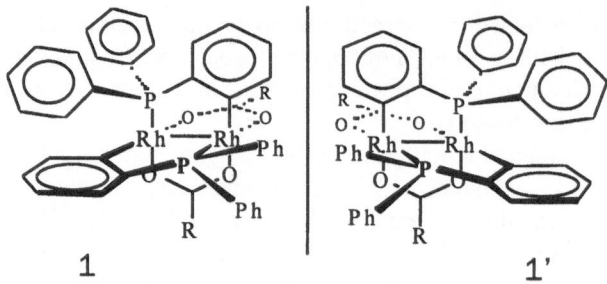

1   1'

The ^{31}P NMR spectra were recorded at different temperatures. Finally crystals were grown for **1-CF3CO2H** and **1'-CF3CO2H** with trifluoroacetic acid and pyridine in axial position and the structures determined by X-ray diffraction. By the same way the absolute configuration was assigned to the chiral centers.

Chiral metalated rhodium compounds have been also prepared using the chiral ligand PC*H =phenyl-(2S,5S)-2,5-bismethylphospholane. The reaction of rhodium acetate and PC*H under standard conditions allowed the isolation of mono and bimetalated compounds. The equatorial intermediates have been observed.[22]

Orthometalated Compounds in Catalysis

Rhodium (II) compounds have become the premier choice in catalytic transformation of α-diazo compounds to produce cyclopropanation, aliphatic carbon-hydrogen bond insertion, heteroatom-hydrogen bond insertion, aromatic substitution, and ylide formation.

It is accepted that these catalytic reactions occur via a metallocarbene intermediate.[23] The structure suggested for this species is shown above. The success of the rhodium (II) catalysts to produce intramolecular decomposition processes in an intermolecular or intramolecular way. Rhodium (II) acetate has shown very efficient to generate transient electrophilic metal carbenes from α-diazo compounds with capability to insert into unactivated C-H bonds.[24] Results have demonstrated that, in intramolecular reactions, the formation of the five-membered ring is a favored process for diazo compounds .[25] Studies have also indicated that the nature of the bridging ligands on the dirhodium (II) metal has a marked influence on the regio- and chemoselectivities of the reaction.[26]

The behavior of bimetalated rhodium (II) compounds with head-to-tail [H-T] configuration (type Va) in catalytic transformation of α-diazo compounds has been also studied. These compounds have some unique aspects that make them different from rhodium tetracarboxylates or tetraamidates more frequently used.[26] They contain very polarizable ligands, as aromatics rings attached to rhodium, and it is possible to modulate the electronic and steric properties of the metal centers by changing the carboxylate groups and the substituents on the metalated and non metalated rings of the phosphine. Results have confirmed[27] that changes in these ligands modify indeed the activity and selectivity of these rhodium (II) compouds to a grand extent.

Thus, in the competition between C-H insertion reaction and aromatic substitution[27] Rh (II) compounds with basic phosphines, as triphenyl phosphine, and acetate ligands exhibit poor chemoselectivity and reactivity. This value compares well with that reported for rhodium (II) caprolactamate. The selectivity is increased if the basicity of the phosphine is reduced introducing electron-withdrawing substituents in the non metalated phenyl ring as in the case of pentafluorophenyldiphenyl phosphine. The most effective way to increase both reactivities and chemoselectivities is found to be the use of a basic phosphine as triphenyl phosphine and high electrophilic carboxylic acid as perfluorobutyrate.

C-H Insertion vs Aromatic Substitution

We have also studied   competitive C-H insertion reactions mediated by this type of catalysts. In this type of reactions the complex with triphenyl phosphine and acetate demonstrated to be both reactive and very selective, yielding only the tertiary C-H insertion product. When used a catalyst with a less basic phosphine $P(C_6H_5)_2(C_6F_5)$, slighty decreased both reactivity and selectivity. More electrophilic catalysts with trifluoroacetates as carboxylates were less selective.

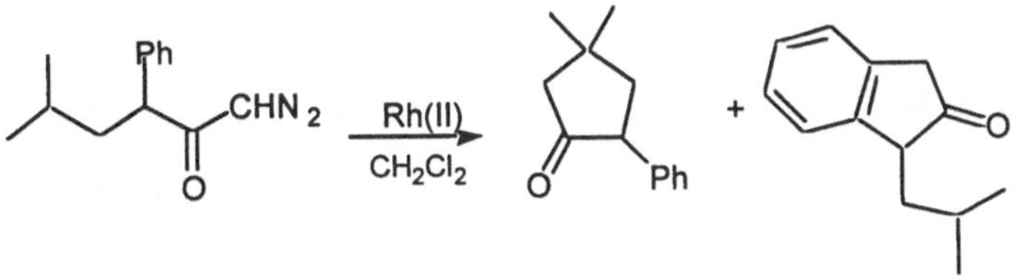

Catalyst	yield (%)	Relative ratio	
		C-H insertion	Aromatic substitution
Rh$_2$(O$_2$CCH$_3$)$_4$	96	65	35
Rh$_2$(O$_2$CC$_3$F$_7$)$_4$	96	0	100
Rh$_2$(cap)$_4$	64	70	30
Rh$_2$(O$_2$CCH$_3$)$_2$[C$_6$H$_4$P(C$_6$H$_5$)$_2$]$_2$	67	50	50
Rh$_2$(O$_2$CCH$_3$)$_2$[C$_6$H$_4$P(C$_6$F$_5$)(C$_6$H$_5$)]$_2$	82	80	20
Rh$_2$(O$_2$CC$_3$F$_7$)$_2$[C$_6$H$_4$P(C$_6$H$_5$)$_2$]$_2$	95	100	0

We also tested the reactivity and selectivity of mono metalated complexes. These catalysts though reactive were in general less selective than doubly metalated compounds with head-to-tail configuration. Finally, preliminary studies of reactivity of doubly metalated compounds with head-to-head configuration showed that this type of compounds are not promising catalysts for C-H insertion reactions.

Tertiary Aliphatic C-H Insertion vs Secondary aliphatic C-H Insertion

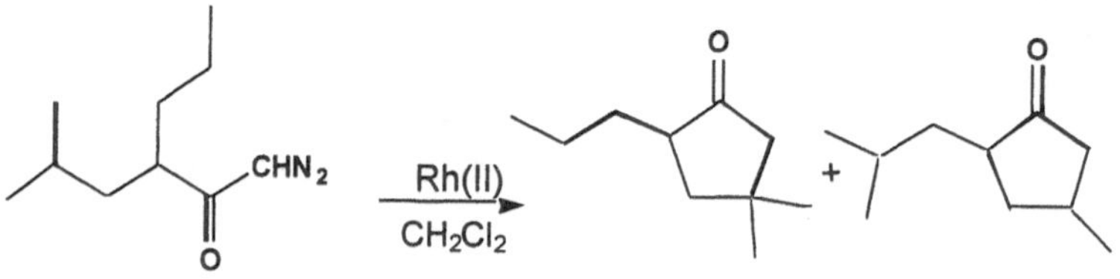

Catalyst	Yield (%)	Relative ratio	
		Tertiary C-H Insertion	Secondary C-H Insertion
Rh$_2$(O$_2$CCH$_3$)$_4$	99	79	21
Rh$_2$(O$_2$CCH$_3$)$_2$ [C$_6$H$_4$P(C$_6$H$_5$)$_2$]$_2$	99	100	0
Rh$_2$(O$_2$CCH$_3$)$_2$[C$_6$H$_4$P(C$_6$F$_5$)(C$_6$H$_5$)]$_2$	91	93	7
Rh$_2$(O$_2$CC$_3$F$_7$)$_2$[C$_6$H$_4$P(C$_6$H$_5$)$_2$]$_2$	99	70	30

Acknowledgements

The contribution of all the coworkers and students that have participated in this work is greatly acknowledged. Their dedication made possible this work. Financial support from CICYT and Generalitat Valenciana is also acknowledged.

References

1. Omae, *Coord. Chem. Rev.,* **1980,** *32,* 235.
2. (a) Barder, T.J.; Tetrick, S.M.; Walton, R.A.; Cotton, F.A.; Powell, G.L. *J. Am. Chem. Soc.* **1984,** *106,* 1323. (b) Bennett, M.A.; Bhargava, S. K.;Griffiths, K.D.; Robertson, G.B. *Angew.Chem., Int. Ed. Eng.* **1987,** *26,* 260. (c) Bennett, M.A.; Barghava, S.K.; Ditzel, E.J.; Robertson, G.B.; Willis, A.C. *J. Chem. Soc., Chem. Commun.* **1987,** 1613 (d) Anold, D.P.; Bennett, M.A.; Bilton, M.; Robertson, G.B.*J. Chem. Soc., Chem. Commun.* **1982,** 115 (e) Anold, D.P.; Bennett, M.A.; McLaughlin, G.M.; Robertson, G.B.; Whittaker, M.J. *J. Chem. Soc., Chem. Commun.* **1983,** 32 (f) Anold, D.P.; Bennett, M.A.; McLaughlin, G.M.; Robertson, G.B. *J. Chem. Soc., Chem. Commun.* **1983,** 34.
3. Chakravarty, A.R.; Cotton,F.A.; Tocher, D.A.; Tocher, J.H. *Organometallics,* **1985,** *4,* 8.
4. (a) Barceló, F.; Lahuerta, P.; Ubeda, M.A.; Foces-Foces, C.; Cano, F.H.; Martínez-Ripoll, M. *J.Chem. Soc., Chem. Commun.,* **1985,** 43. (b) Barceló, F.; Lahuerta, P.; Ubeda, M.A.; Foces-Foces, C.; Cano, F.H.; Martínez-Ripoll, M. *Organometallics,* **1988,** *7,* 584
5. Alarcón, C.J.; Lahuerta. P.; Peris, E.; Ubeda, M.A.; Aguirre, A.; Garcia-Granda, S.; Gómez-Beltrán, F. *Inorg. Chim. Acta,* **1997,** *254,* 177.
6. Alarcón, C.J.; Estevan, F.; Lahuerta, P.; Ubeda, M.A.; Gonzalez, G.; Martinez, M. *Inorg. Chim. Acta,* in press.
7. Lahuerta, P.; Peris,E.; Ubeda, M.A.; García-Granda, S.; Diaz, M.R. *J. Organomet. Chem.,* **1993,** *455* , C10.
8. (a) Lahuerta, P.; Payá, J.; Peris, E.; Pellinghelli, M.A.; A.Tiripicchio, A. *J. Organomet. Chem.* **1989,** *373,* C5. (b) Lahuerta, P.; Payá, J.; Garcia-Granda, S.; Gómez-Beltrán, F.; Anillo, A. *J. Organomet. Chem.* **1993,** *443,* C14. (c) Lahuerta, P.; Latorre, J.; Peris, E.; Sanaú, M.; Garcia-Granda, S. *J. Organomet. Chem.* **1993,** *456,* 279. (d) Garcia-Granda, S.; Lahuerta, P.; Latorre, J.; Martinez, M.;Peris, E.; Sanaú, M.; Ubeda, M.A. *J. Chem. Soc.Dalton Trans.* **1994,** 539.
9. (a) Lahuerta, P.; Peris, E.; *Inorg. Chem.,* **1992,** *31* , 4547.
10. (a) Cotton,F.A.; Barceló, F.; Lahuerta, P.; Llusar, R.; Payá,J.; Ubeda, M.A. *Inorg. Chem.,* **1988,** *27,* 1010.
11. Lahuerta, P.; Payá, J.; Pellinghelli, M.A.; A.Tiripicchio, A. *Inorg. Chem.,* **1992,** *31,* 1224
12. Lahuerta, P.; Estevan, F. unpublished results.
13. Borrachero, M.V.; Estevan, F.; Lahuerta, P.; Paya, J.; Peris, E. *Polyhedron,* **1993,** *12,* 1715
14. Lahuerta, P.; Payá, J.;Solans, X.; Ubeda, M.A. *Inorg. Chem.,* **1992,** *31,* 385.
15. Gonzalez, G.; Lahuerta, P.; Martinez, M.; Peris, E.; Sanaú, M. *J. Chem. Soc. Dalton Trans.* **1994,** 545.,
16. Morrison, E.C.; Tocher, D.A. *Inorg. Chim. Acta,* **1989,***157* , 139.
17. Lahuerta, P.; Estevan, F. unpublished results.
18. Borrachero,M.V.; Estevan, F.; García-Granda,S.; Lahuerta, P.; Latorre, J.; Peris, E.; Sanaú, M. *J. Chem. Soc. Chem. Commun.,* **1993,** 1864.

19. Estevan, F.; Lahuerta, P.; Pérez-Prieto, J.; Stiriba, S. E. *Organometallics*, submitted.
20. Lahuerta, P.; Martinez-Mañez, R.; Payá, J.;Peris, E.; Diaz, W. *Inorg. Chim. Acta*, **1990**,*173* , 99.
21. Lahuerta, P.; Bieger, K.; Sanaú, M. *J.Chem. Soc. Chem. Commun.*, submitted.
22. Estevan, F.; Krüger, P.; Lahuerta, P.; Werner, H. unpublished results.
23. Doyle, M. P. *Chem. Rev.* **1986**, 86.
24. For reviews: (a) Padwa, A.; Austin, D. *Angew. Chem. Int. Ed. Engl.* **1994**, 33, 1797. (b) McKervey, M. A.; Ye, T. *Chem. Rev.* **1994**, *94*, 1091. (c) McKervey, M. A.; Doyle, M. P.; *J. Chem. Soc. Chem. Commun.* **1997**, 983-1072. (d) Doyle, M. P. In *Homogeneous Transition Metal Catalysts in Organic Synthesis*; Moser, W. R., Slocum, D. W., Eds.; ACS Advanced Chemistry Seies 230; American Chemical Society: Washington, DC, **1992** ; Chapter 30. (e) Taber, D. F. *Comprehensive Organic Synthesis*, Eds.; Pergamon Press: Oxford, **1991** ; Vol.3 pp 1045. (f) Taber, D. F. *Methods of Organic Chemistry*; Houben-Weyl, Helmchen, G.; Hoffman, W.; Mulzer, J.; Schaumann, E., Eds.; George Thieme Verlag: Stuttgart, **1995**, pp 1127. (g) Wulff, W. D. *Comprehensive Organic Synthesis*; Trost, B. M., Fleming, I., Eds.; Pergamon Press; New York, **1990**; Vol.5.
25. For relevant examples in Rh (II) mediated C-H insertion, see: (a) Doyle, M. P.; Dyatkin, A. B.; Roos, G. H. P.; Cañas, F.; Pierson, D. A.; Van Basten, A.; Muller, P.; Polleux, P. *J. Am. Chem. Soc.* **1994**, *116*, 4507. (b) Wang, P.; Adams, J. *J. Am. Chem. Soc.* **1994**, *116*, 3296. (c) Doyle, M. P.; Westrum, L. J.; Wolthuis W., N. E.; See, M. M.; Boone, W. P.; Bagheri, V.; Person, M. M. *J. Am. Chem. Soc.* **1993**, *115*, 958. (d) Taber, D. F.; Amedio, Jr., J. C.; Raman, K. *J. Org. Chem.* **1988**, *53*, 2984.

Monomeric Organo-Copper(I) Compounds as Organyl Transfer Reagents

Heinrich Lang[*] and Wolfgang Frosch

Anorganische Chemie, Technische Universität Chemnitz-Zwickau, Straße der Nationen 62, D-09107 Chemnitz

Organo-copper compounds can extensively be used as transfer reagents for the regiospecific and stereo- as well as chemoselective introduction of organic fragments in organometallic and organic synthesis [1]. However, there is only little known about the nature of the reactive species in copper-mediated reactions [1, 2]. Out of this reason mononuclear and therefore well-defined organo-copper(I) compounds are of potential interest. Recently, we could show that the chelating effect of the organometallic 1,4-diynes of structural type **A** are well suitable for generating those compounds (Type **B** molecule) [3].

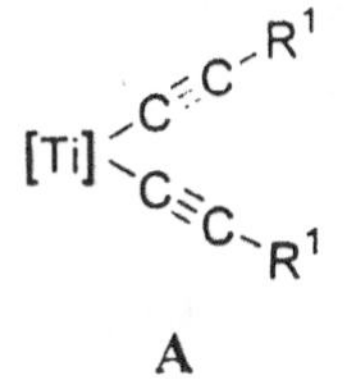

$[Ti] = (\eta^5\text{-}C_5H_4SiMe_3)_2Ti$

R^1, R^2 = singly bonded organic ligand

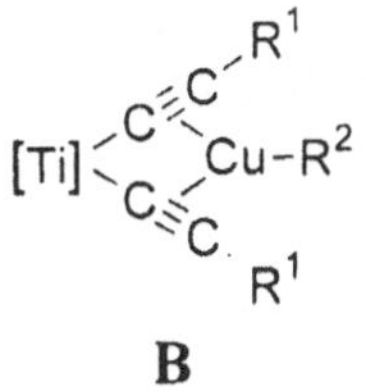

A	B

In type **B** molecules the $R^1C\equiv C\text{-}[Ti]\text{-}C\equiv CR^1$ building block acts as a host, while the copper(I)-R^2 moiety acts as a guest. The copper atom possesses a trigonal-planar environment, comprising η^2-coordination of both alkynyl ligands of the bis(alkynyl)-titanocene entity and η^1-bonding of the organic group R^2; the $(\eta^2\text{-alkyne})_2CuR^2$ moiety thereby represents a 16 valence electron complex fragment.

This article focuses on selected reactions of monomeric organo-copper(I) compounds (Type **B** molecules).

Reaction Chemistry

1 Metathesis Reactions

Monomeric copper(I) methyl (**1**) spontaneously reacts with acidic substrates by elimination of methane to produce the heterobimetallic complexes **2 - 6** [2: R^1 = SiMe$_3$, ^{t}Bu; X = C≡CR^3 with R^3 = SiMe$_3$, ^{t}Bu, Ph, CO$_2$Me; **3**: R^1 = SiMe$_3$; X = N(SiMe$_3$)$_2$; **4**: R^1 = ^{t}Bu; X = PPh$_2$; **5**: R^1 = SiMe$_3$, ^{t}Bu; X = SPh, SCH$_2$Ph, S^{i}Pr; **6**: R^1 = SiMe$_3$, ^{t}Bu; X = O^{t}Bu, OPh]. The driving forces for these reactions are the differences between the appropriate pK_a-values as well as the volatility of the generated methane.

Whereas complexes **2**, **4** and **5** are remarkably stable, compounds **3** and **6** slowly start to decompose in solution: Cleavage of the carbon-silicon bond of one of the TiC≡C-SiMe$_3$ units affords - on elimination of N(SiMe$_3$)$_3$ (compound **3**) or Me$_3$SiOR (compound **6**) - quantitatively the dimeric copper-titanium acetylide **7**. Dimeric **7** contains two heterobimetallic TiC≡CCu building blocks in which the C≡C entity within the TiC≡CCu unit is η^2-coordinated to a second copper atom, thus forming an alkynyl bridged dimer [3].

Additionally, **7** can be obtained by starting out from {[Ti](C≡CSiMe$_3$)$_2$}CuEt (**8a**) and {[Ti](C≡CSiMe$_3$)$_2$}Cu^{n}Bu (**9a**). In solution these compounds slowly start to eliminate ethene (**8a**) or 2-butene (**9a**) along with HSiMe$_3$ [3].

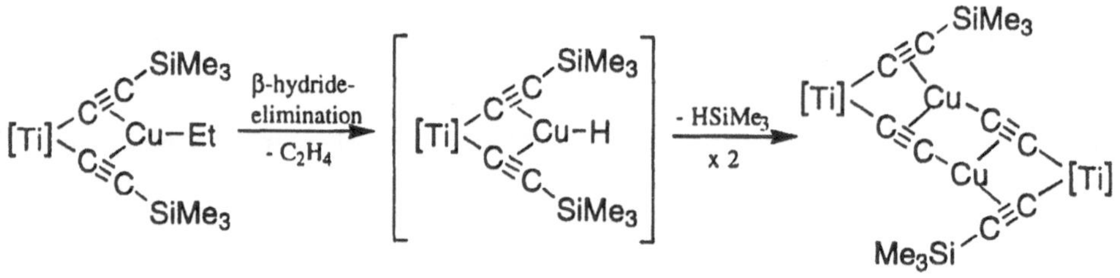

Scheme 1 Formation of dimeric **7**.

Scheme 1 shows that the decomposition route of **8a** is the β-hydride elimination pathway. Accordingly, the copper(I) hydride complex **10a** is formed as intermediate first [4].

To avoid the cleavage of the carbon-silicon bond, observed for C≡CSiMe$_3$ ligands, the SiMe$_3$ groups are replaced by ^{t}Bu units. However, it is found that {[Ti](C≡C^{t}Bu)$_2$}CuEt (**8b**) also initially undergoes β-hydride elimination to form the corresponding copper(I) hydride, $\{[Ti](C\equiv C^tBu)_2\}CuH$ (**10b**), which undergoes cleavage of the titanium-carbon σ-bond of the Ti-$C\equiv C^tBu$ unit, yielding $[CuC\equiv C^tBu]_n$, " $[Ti]H_2$ " and $\{[Ti](C\equiv C^tBu)_2\}CuC\equiv C^tBu$ [4].

2 Carbon-Carbon Coupling Reactions

Thermolysis of monomeric copper(I) acetylides, $\{[Ti](C\equiv CSiMe_3)_2\}CuC\equiv CR^3$ (**2**: $R^3 = SiMe_3$, ^tBu, Ph), produces $[Ti](C\equiv CSiMe_3)_2$, $R^3C\equiv CC\equiv CR^3$ and copper(0). The latter compounds are formed by homolytical cleavage of the copper-carbon σ-bond followed by C-C coupling of two $C\equiv CR^3$ units [3]. Oxidative coupling of two terminal alkynes $HC\equiv CR^3$ by copper(I) ions to produce symmetrical diynes represents the *Glaser reaction* [5].

Moreover, a way to synthesize unsymmetrical diynes $R^3C\equiv CC\equiv CR^4$ is given by the reaction of $\{[Ti](C\equiv CR^1)_2\}CuC\equiv CR^3$ (**2**) with equimolar amounts of $BrC\equiv CR^4$. This finding corresponds to the well-known *Cadiot-Chodkiewicz coupling* [6].

C-C bond formation can also be achieved by the reaction of monomeric organo-copper(I) compounds with organo halides R^5-Hal (Hal = Br, I) as presented in Scheme 2.

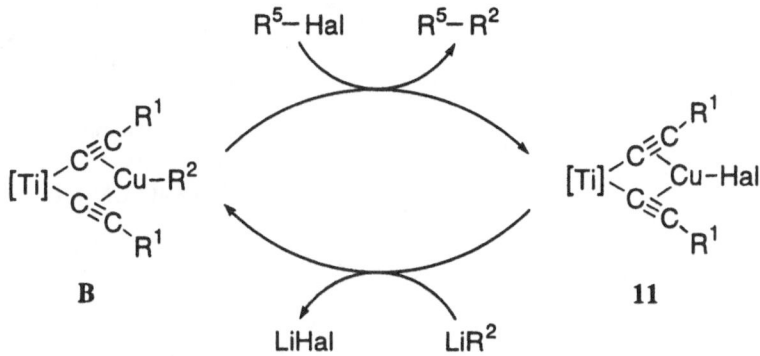

Scheme 2 Copper(I)-mediated carbon-carbon bond formation.

The starting materials $\{[Ti](C\equiv CR^1)_2\}CuR^2$ (a type **B** molecule) can be regained by reacting the monomeric copper(I) halides **11** with *e.g.* lithium organyles.

A further type of C-C bond formation is the reaction of organo-copper(I) compounds with acyl chlorides, a *Gilman type reaction* [2], yielding unsymmetrical ketones.

3 Halogenation Reactions

Treatment of $\{[Ti](C\equiv CR^1)_2\}CuR^2$ (type **B** molecule) with one equivalent of Hal_2 (Hal = Br, I) affords R^2-Br and $\{[Ti](C\equiv CR^1)_2\}CuHal$ (**11**). The latter compound can easily be transferred to the appropriate organo-copper(I) starting materials by treatment with *e.g.* lithium organyles.

4 Decarboxylation Reactions

The monomeric copper(I) carboxylates $\{[Ti](C\equiv CR^1)_2\}CuO_2CR^2$ (**12**: R^2 = Me, Ph) can be utilized as model compounds for the *Hunsdiecker reaction* [7]: a way for decreasing the length of carbon chains by a C_1 unit. This reaction corresponds to the replacement of a CO_2 group in mercury and silver carboxylates by an halogen atom. Careful heating of **12** causes these molecules to eliminate CO_2 to form the corresponding $\{[Ti](C\equiv CR^1)_2\}$ CuR^2 species first (Scheme 3). Treatment of the latter molecules with one equivalent of Br_2 results in the formation of R^2-Br and $\{[Ti](C\equiv CR^1)_2\}CuBr$ (**11a**), which can be transferred to the starting copper(I) carboxylates by its reaction with $^1/_n\,[AgO_2CR^2]_n$ (Scheme 3).

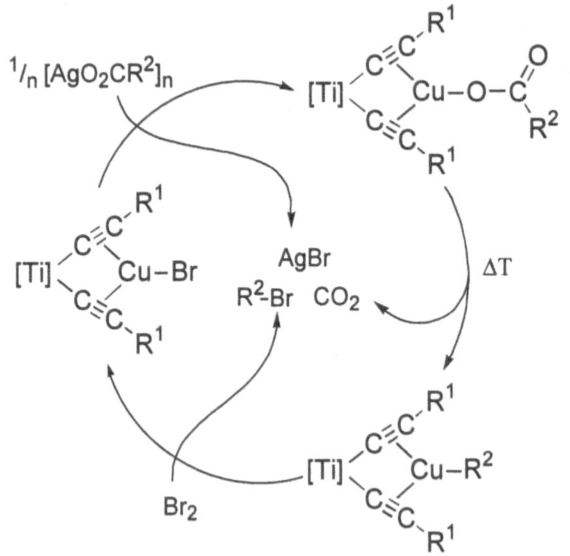

Scheme 3 Decarboxylative bromination of copper(I) carboxylates [3].

References and Notes

[1] a) Carruthers, W. *Comprehensive Organometallic Chemistry*, Eds. Wilkinson, G.; Stone, F.G.A.; Abel, E.W., Vol. 7, 685 - 729, Pergamon Press, **1982**. b) Bähr, G; Burba, P. in Houben-Weyl *Metallorganische Verbindungen* 1970, XIII/1, 735 - 761. c) Hegedus, L.S. *J. Organomet. Chem.* **1994**, *477*, 269 - 362.

[2] Gilman, H.; Straley, J.M. *Recl. Trav. Chim. Pays-Bas* **1936**, *55*, 821.

[3] a) Lang, H.; Weinmann, M. *Synlett* **1996**, 1 - 10 and lit. cited therein. b) Lang, H.; Köhler, K.; Blau, S. *Coord. Chem. Rev.* **1995**, *143*, 113.

[4] Köhler, K.; Pritzkow, H.; Lang, H. *J. Organomet. Chem.*, in press.

[5] Glaser, C. *Chem. Ber.* **1869**, *2*, 422.

[6] Chodkiewicz, W. *Ann. Chim. (Paris)* **1958**, *2*, 819.

[7] Hunsdiecker, Cl.; Hunsdiecker, H.; Vogt, E. *Chem. Zentralblatt* **1937**, *I*, 2258.

Unsymmetrical Chelating Ligands: The First Peralkylated Phosphino(stibino)- and Phosphino(arsino)methanes and Their Organometallic Rhodium Complexes

H. Werner*, D. Stalke, J. Wolf, M. Manger, U. Schmidt, O. Gevert, M. Laubender, and M. Teichert

Institut für Anorganische Chemie, Universität Würzburg, D-97074 Würzburg, Germany

Summary

Ligands of the general composition $R_2PCH_2ER'_2$ (E = Sb, As) containing bulky alkyl or cycloalkyl groups R, R' have been prepared following a two-step procedure. Using $[RhCl(C_8H_{12})]_2$ as the starting organometallic material, neutral as well as cationic complexes, which contain the new sterically hindered ligands either in a mono- or bidentate coordination mode, are available.

Synthesis of the ligands $R_2PCH_2ER'_2$ (E = Sb, As)

Using ICH_2SnPh_3 as the starting material, metalation by nBuLi generates $LiCH_2SnPh_3$ which, in the presence of TMEDA (tetramethylethylenediamine), reacts with $ClPR_2$ to yield the stannylated phosphanes $R_2PCH_2SnPh_3$. Transmetalation of the latter with PhLi generates in situ the corresponding lithiomethylphosphane R_2PCH_2Li, which reacts with a chloro (or bromo) stibane or arsane in the presence of TMEDA to give the final products $R_2PCH_2ER'_2$ in high yield [1].

CH_2SnPh_3 → (1) nBuLi / 2) TMEDA / 3) $ClPR_2$) → $R_2PCH_2SnPh_3$

$R_2PCH_2SnPh_3$ → (1) PhLi, $- SnPh_4$ / 2) TMEDA / 3) XER'_2) → $R_2PCH_2ER'_2$

E = Sb	R	R'
	iPr	iPr
	iPr	tBu
	Cy	tBu

E = As	R	R'
	iPr	iPr
	iPr	tBu
	iPr	Cy
	Cy	Cy

Synthesis and reactivity of rhodium(I) complexes containing ligands of the type $R_2PCH_2ER'_2$ (E = Sb, As)

Treatment of $[RhCl(C_8H_{12})]_2$ with $R_2PCH_2ER'_2$ affords either the mononuclear compounds $RhCl(\eta^4-C_8H_{12})(\kappa-P-R_2PCH_2ER'_2)$ (1) or (for E = As) the novel dinuclear chelate complex

181

$[\{Rh(\eta^4\text{-}C_8H_{12})\}\{Rh(\kappa^2\text{-}As,P\text{-}tBu_2AsCH_2PiPr_2)\}(\mu\text{-}Cl)_2]$ **(2)**. Cationic derivatives of the general composition $[Rh(\eta^4\text{-}C_8H_{12})(\kappa^2\text{-}P,E\text{-}R_2PCH_2ER'_2)]X$ **(3)** ($X = PF_6$, BPh_4) are obtained from **1** and $AgPF_6$ or $NaBPh_4$, respectively.

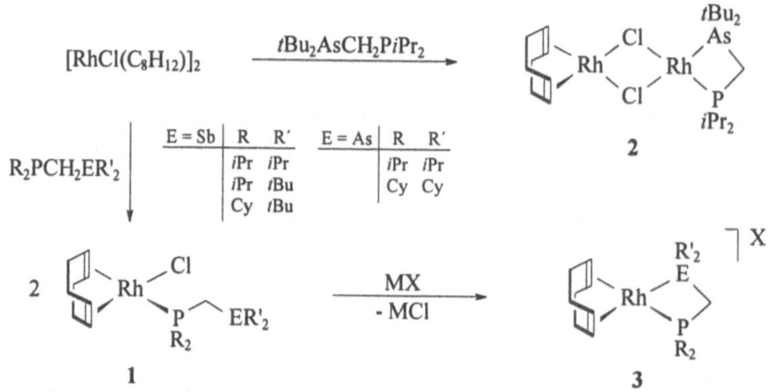

The cationic complexes **3** are versatile starting materials for the synthesis of a variety of organometallic rhodium compounds. The reaction with diazomethane proceeds by insertion of the methylene unit into the Rh-E bond and the corresponding complexes $[Rh(\eta^4\text{-}C_8H_{12})\{\kappa^2\text{-}C,P\text{-}CH_2E(R')_2CH_2PR_2\}]X$ **(4)** were obtained. For E = Sb, they are the first examples of transition metal compounds with Sb-ylides as ligands.

The cyclooctadiene unit of the chelate complexes **3** can easily be replaced by treatment with H_2. The reaction of the resulting intermediate depends on the type of the corresponding anion X^- and the reaction conditions in particular. In the presence of CF_3CO_2H, the hexafluorophosphate derivatives of **3** are converted unexpectedly to the novel dinuclear hydrido-bridged rhodium(III) complexes $[\{RhH(\kappa^2\text{-}P,E\text{-}R_2PCH_2ER'_2)\}_2(\mu\text{-}H)(\mu\text{-}O_2CCF_3)_2]PF_6$ **(5)** [2]. Moreover, from **3** with $X = BPh_4$ halfsandwich-type compounds of the general composition $(\eta^6\text{-}C_6H_5BPh_3)Rh(\kappa^2\text{-}P,E\text{-}R_2PCH_2ER'_2)$ **(6)** were obtained.

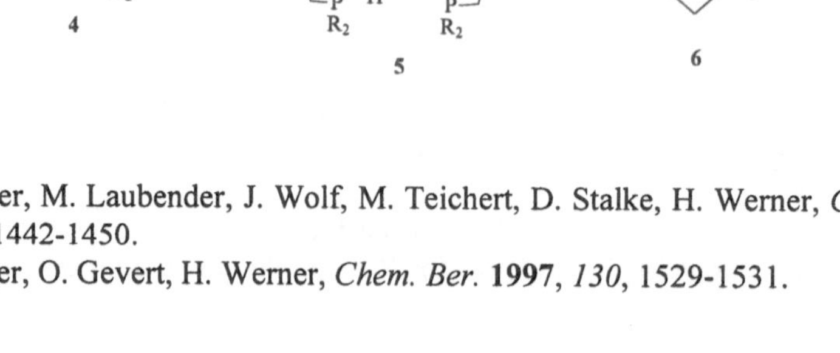

References

[1] M. Manger, M. Laubender, J. Wolf, M. Teichert, D. Stalke, H. Werner, *Chem. Eur. J.* **1997**, *3*, 1442-1450.

[2] M. Manger, O. Gevert, H. Werner, *Chem. Ber.* **1997**, *130*, 1529-1531.

C-C, C-P and C-N Coupling Reactions Mediated by Square-Planar Cumulenylidene Rhodium Complexes

H. Werner*, J. Wolf, M. Laubender, I. Kovacik, and O. Gevert

Institut für Anorganische Chemie, Universität Würzburg, D-97074 Würzburg, Germany

Summary

In square-planar allenylidenerhodium(I) complexes $trans$-[RhX(=C=C=CRR')(PiPr$_3$)$_2$] the cumulated ligands show versatile usability in C-C, C-P and C-N coupling reactions to generate new species which are hardly accessible by other synthetic routes. The highly reactive metallacumulene $trans$-[RhCl(=C=C=C=C=CPh$_2$)(PiPr$_3$)$_2$] was prepared which, when reacted with CH$_2$N$_2$, affords the hitherto unknown hexapentaene ligand H$_2$C=C=C=C=C=CPh$_2$.

Introduction

Since the first isolation of a transition metal complex containing an allenylidene ligand in 1976 [1] a variety of metallabutatrienes of the elements Cr, Mn, Fe and Ru have been prepared and studied both theoretically and experimentally [2]. Following the Selegue method for the preparation of [C$_5$H$_5$Ru(=C=C=CPh$_2$)(PMe$_3$)$_2$]PF$_6$ [3] we recently reported an efficient synthetic approach to electronically as well as coordinatively unsaturated allenylidenerhodium(I) complexes $trans$-[RhCl(=C=C=CRR')(PiPr$_3$)$_2$] [4] and investigated their reactivity.

High Selectivity in C-C Bond Formations

The hydroxo allenylidene derivatives in the scheme shown below, which were prepared from the chloro complexes and KOtBu [5], are useful precursors for the introduction of anionic ligands R'O⁻ derived from Brønsted acids R'OH as well as for C-C coupling reactions to afford special alkynes by nucleophilic attack at the γ-carbon atom of the Rh=C=C=C unit.

In the reaction of chloro substituted allenylidene complexes with diazomethane the α-carbon atom of the :C=C=CRR' ligand is connected with :CH$_2$ leading to terminally π-bonded buta-trienes which can be converted to thermodynamically more stable products with internally bonded cumulenes. Interestingly, for R = Ph and p-C$_6$H$_4$OMe these C-C coupling products are also available from the reaction of the allenylidenerhodium compounds with methyl-iodide including the formal cleavage of CH$_3$I into HI and :CH$_2$ [6].

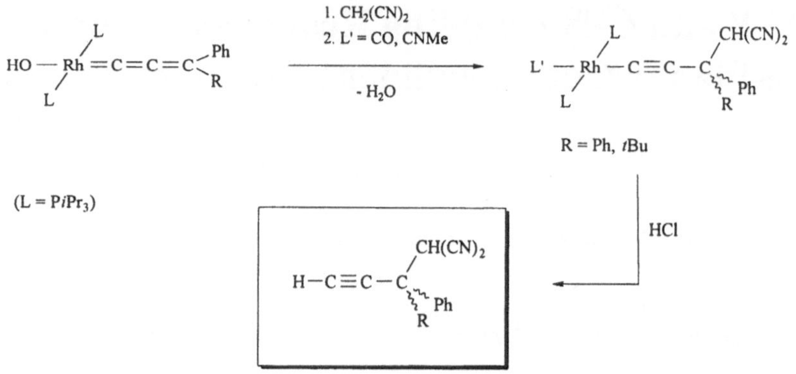

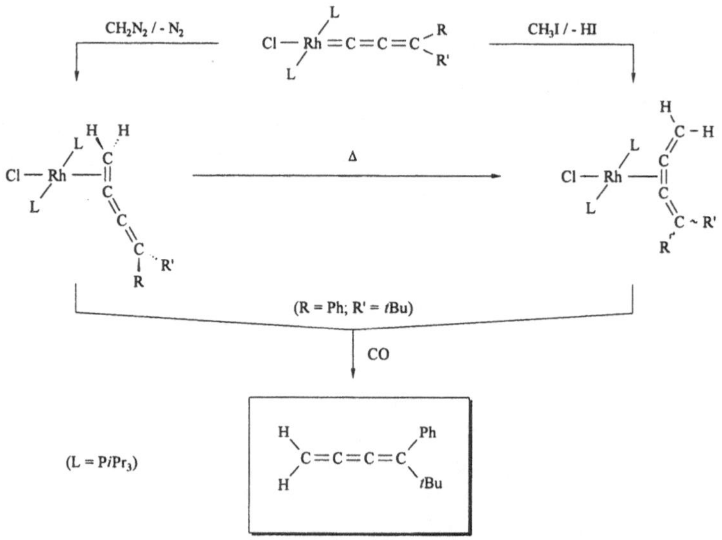

1. CH$_2$(CN)$_2$
2. L' = CO, CNMe

$-$ H$_2$O

HO $-$ Rh $=$C$=$C$=$C$\big\langle$ $^{Ph}_{R}$ (L above and below)

L' $-$ Rh $-$ C$\equiv$C$-$C$\big\langle$ CH(CN)$_2$, Ph, R

R = Ph, tBu

(L = PiPr$_3$)

H$-$C$\equiv$C$-$C $-$ CH(CN)$_2$, Ph, R

HCl

CH$_2$N$_2$ / $-$ N$_2$

Cl $-$ Rh $=$C$=$C$=$C$\big\langle$ $^{R}_{R'}$ (L above and below)

CH$_3$I / $-$ HI

Cl $-$ Rh (with H, H, C=C=C, R', R cluster)

Δ

Cl $-$ Rh (with C$-$H, C, C$\sim$R', R cluster)

(R = Ph; R' = tBu)

CO

(L = PiPr$_3$)

H, H $\;$ C$=$C$=$C$=$C $\big\langle$ $^{Ph}_{tBu}$

Generation of a iPr$_3$P=C=C=CPh$_2$ Ligand via C-P Coupling

On treatment of the allenylidene precursor trans-[RhCl(=C=C=CPh$_2$)(PiPr$_3$)$_2$] with chlorine an η^1-bonded phosphacumulene-ylide is formed by a [1,2]-shift of one PiPr$_3$ ligand from the metal to the α-carbon atom of the [Rh]=C=C=CPh$_2$ unit [7].

C-N Coupling Reactions in Azido(allenylidene)rhodium Complexes

In the presence of CO the azido ligand in trans-[Rh(N$_3$)(=C=C=CRR')(PiPr$_3$)$_2$] can be transferred to the γ-carbon atom of the allenylidene unit. Spontaneous loss of N$_2$ and isomerization affords a σ-bonded vinylic moiety [8] which can be cleaved by acids to give acrylonitriles.

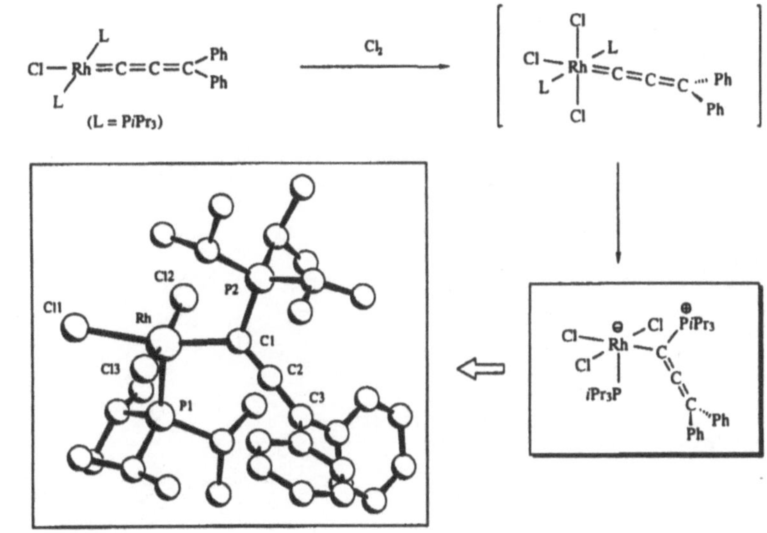

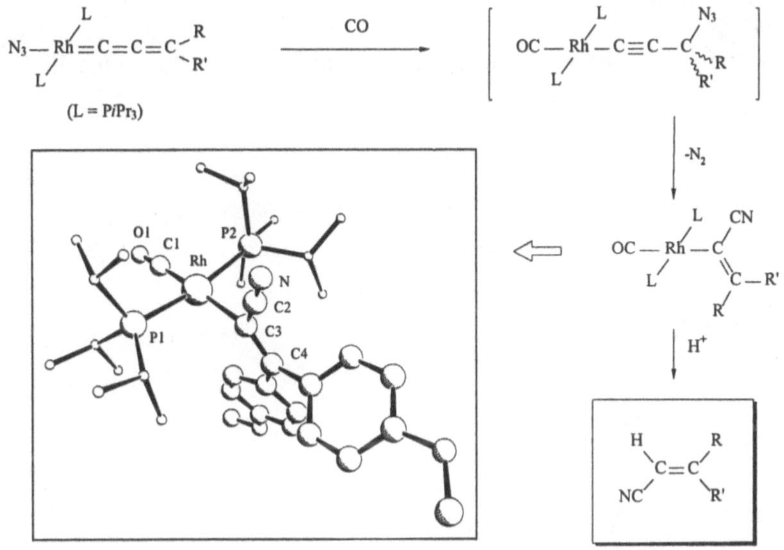

Synthesis of a Square-Planar Pentatetraenylidene Complex

In the coordination sphere of rhodium the silylated diynol HC≡CC≡CC(OSiMe$_3$)Ph$_2$ can be converted into an alkinyl-substituted vinylidene which by formal abstraction of HOSiMe$_3$ provides a synthetic route to an unprecedented, highly reactive square-planar rhodium(I) complex with a cumulated C$_5$ chain [9].

185

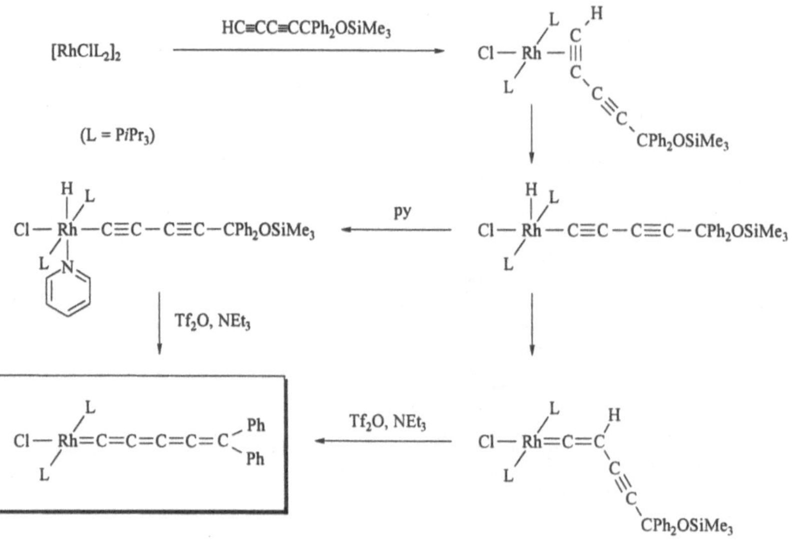

Treatment of the pentatetraenylidene complex with diazomethane affords a mixture of the isomers **A** and **B** of which the thermodynamically more stable isomer **B** can be separated by fractional crystallization. Its structure has been determined crystallographically [9].

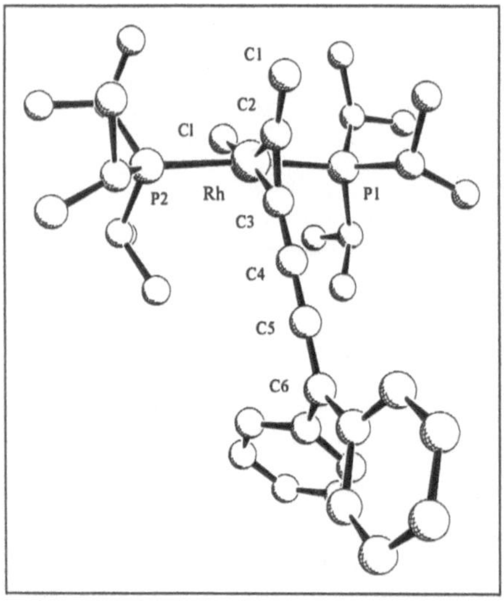

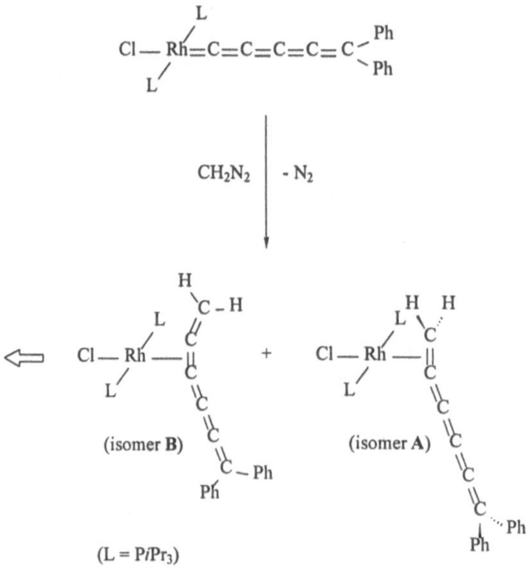

References

[1] E.O. Fischer, H.J. Kalder, A. Frank, F.H. Köhler, G. Huttner, *Angew. Chem.* **1976**, *88*, 683-684; *Angew. Chem. Int. Ed. Engl.* **1976**, *15*, 623-624.

[2] a) M.I. Bruce, *Chem. Rev.* **1991**, *91*, 197-257; b) B.E.R. Schilling, R. Hoffmann, D.L. Lichtenberger, *J. Am. Chem. Soc.* **1979**, *101*, 585-591; c) N.M. Kostic, R.F. Fenske, *Organometallics* **1982**, *1*, 974-982.

[3] J.P. Selegue, *Organometallics* **1982**, *1*, 217-218.

[4] a) H. Werner, T. Rappert, *Chem. Ber.* **1993**, *126*, 669-678; b) H. Werner, T. Rappert, R. Wiedemann, J. Wolf, N. Mahr, *Organometallics* **1994**, *13*, 2721-2727.

[5] H. Werner, R. Wiedemann, M. Laubender, J. Wolf, B. Windmüller, *J. Chem. Soc.,Chem. Commun.* **1996**, 1413-1414.

[6] H. Werner, M. Laubender, R. Wiedemann, B. Windmüller, *Angew. Chem.* **1996**, *108*, 1330-1332; *Angew. Chem. Int. Ed. Engl.* **1996**, *35*, 1237-1239.

[7] H. Werner, *J. Chem. Soc.,Chem. Commun.* **1997**, 903-910.

[8] M. Laubender, H. Werner, *Angew. Chem.* **1997**, in press.

[9] I. Kovacik, M. Laubender, H. Werner, submitted for publication.

Structure and Reactivity of 2,6-Bis(diphenylphosphinomethyl)pyridine Complexes of Rhodium(I) and Palladium(II)

Christine Hahn[a], Rudolf Taube[b]*

[a] Universität Würzburg, Institut für Anorganische Chemie, Am Hubland, D-97074 Würzburg, Germany
[b] Technische Universität München, Anorganisch-Chemisches Institut, Lichtenbergstraße 4, D-85747 Garching, Germany

In order to clarify the reaction steps considered crucial for the rhodium complex catalyzed hydroamination [1] of ethylene the structure and the reactivity of rhodium(I) and palladium(II) complexes of the tridentate ligand 2,6-bis(diphenylphosphinomethyl)pyridine (PNP) [2] as model compounds for the catalyst complex $[Rh(C_2H_4)(PPh_3)_2(Me_2CO)]PF_6$ [3] were investigated.

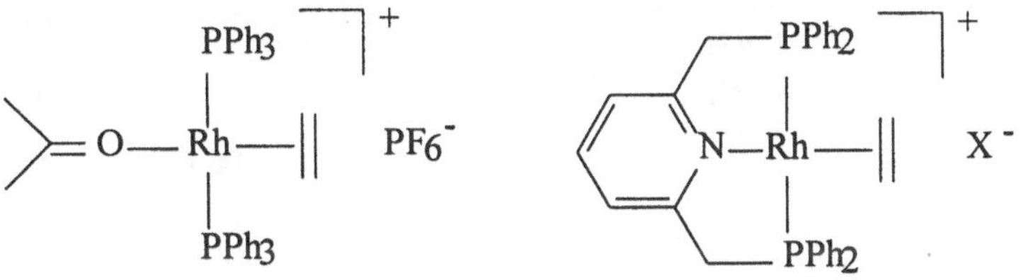

Catalyst complex [3]   Model complex

Scheme 1

Starting from the known bis(ethylene)-complexes $[Rh(C_2H_4)_2(solv)_2]X$ [4] the cationic ethylene-PNP-Rh(I)-complexes $[Rh(PNP)(C_2H_4)]X$ (X = BF_4 **1a**, PF_6 **1b**, CF_3SO_3 **1c**) were prepared by addition of one equivalent of the PNP-ligand. The structurally analogous cationic styrene complex $[Rh(PNP)(CH_2CHPh)]BF_4$ **2** could be obtained by substitution of the ethylene from the complex **1a** with an excess of styrene. The results of the IR and NMR spectroscopic characterization as well as the X-ray crystal structure analysis show the analogy to the structure of the catalyst complex [5].

The cationic olefin-Rh(I) complexes **1** and **2** react with secondary amines such as piperidine, $HNMe_2$, and $HNEt_2$ by complete substitution of the olefin to the corresponding amine complexes $[Rh(PNP)(HNR_2)]X$ **3-5** without any indication of a nucleophilic attack.

189

The reversible substitution of the amine by an excess of the olefin suggests the presence of a thermodynamic equilibrium between the olefin and the amine complexes (cf. eq. 1), which is supported by NMR spectroscopy.

$$[Rh(PNP)(CH_2CHR)]X + HNR_2 \rightleftharpoons [Rh(PNP)(HNR_2)]X + CH_2CHR \qquad (1)$$

R = H **1a-c**, Ph **2**

HNR_2 = piperidine; X = BF_4 **3a**, PF_6 **3b**, SO_3CF_3 **3c**
$HNMe_2$; X = BF_4 **4**
$HNEt_2$; X = BF_4 **5**

The organo-PNP-Rh(I)-complexes [Rh(PNP)R] (R = Me **6**, Ph **7**) which could be synthesized from the ethylene complex **1a** and 3 equivalents of LiR in THF (cf.eq. 2) and were characterized by NMR spectroscopy and mass spectrometry were used for the investigation of protolytic stability and reactivity of the Rh-C σ-bond which play an essential role in the catalytic cycle.

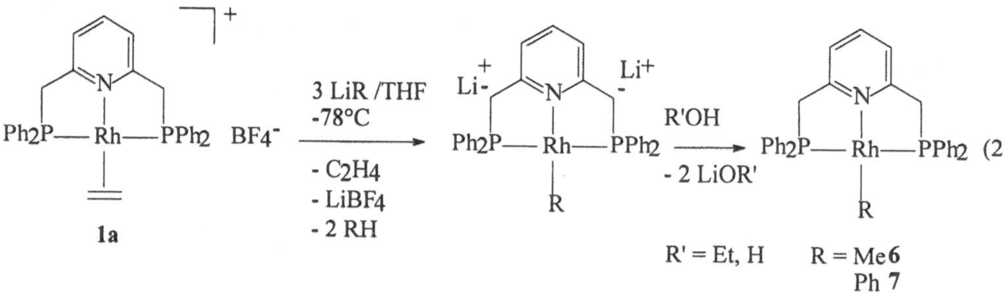

The reaction of the organo-Rh(I)-complexes with HSO_3CF_3 led immediately to the release of the corresponding hydrocarbon. Upon protolysis the remaining $[Rh(PNP)]^+$-fragment can be trapped with olefins (ethylene or COD) by the formation of the corresponding cationic olefin complexes **1b** and **8** (cf. eqs. 3 and 4).

$$Rh(PNP)R] + HSO_3CF_3 \xrightarrow{\ THF\ /\ C_2H_4\ } [Rh(PNP)(C_2H_4)]SO_3CF_3 + HR \qquad (3)$$

6, 7 $\qquad\qquad\qquad\qquad\qquad\qquad\qquad\qquad$ **1b**

When using $HNMe_3Cl$ as protic acid in THF the clevage of the Rh-C σ-bond in **6** and **7** succeeds also immediately while the formed chloro complex [Rh(PNP)Cl] precipitates from the reaction solution. That means that the reductive elimination step from the proposed hydrido organo-PNP-rhodium(III) intermediate should be irreversible and without any kinetic inhibition because it is independently if a non-coordinating or an coordinating anion is present.

$$2\ [Rh(PNP)R]\ +\ 2\ HSO_3CF_3 \xrightarrow[-\ 2\ C_6H_6]{THF\ /\ COD}$$

7

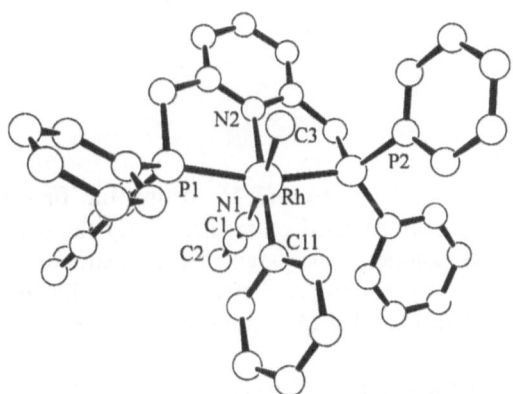

$$2\ SO_3CF_3^-\quad (4)$$

8

The stable organo-PNP-Rh(III)-complexes [Rh(PNP)RI$_2$] (R = Me **9**, Ph **10**) and the bis(organo)-Rh(III)-complexes [Rh(PNP)R(Me)I] (R = Me **11**, Ph **12**) can easily be obtained by oxidative addition of I$_2$ and CH$_3$I. It was found that the complex [Rh(PNP)(Ph)(Me)I] **12** reacts with TlBF$_4$ in acetone by reductive elimination of toluene, whereas in acetonitrile the stable cationic bis(organo)-Rh(III)-complex [Rh(PNP)(Ph)(Me)(CH$_3$CN)]BF$_4$ **13** is formed which was characterized by X-ray structure analysis (cf. fig 1).

Figure 1 Molecule structure of [Rh(PNP)(Ph)(Me)(CH$_3$CN)]$^+$. Selected bond length [Å] and angles [°]: Rh-N(1) 2.152(4), P(1)-Rh-P(2) 164.36(4), Rh-N(2) 2.157(5), N(2)- Rh-C(11) 176.71(2),Rh-C(3) 2.088(5), C(1)-N(1)-Rh 172.1(4),Rh-C(11) 2.054(5), N(1)-C(1)-C(2) 178.0(6),Rh-P(1) 2.2912(14), C(3)-Rh-C(11) 91.6(2), Rh-P(2) 2.3151(14), N(1)-Rh-C(3) 176.71(2).

The formation of π-complexes of olefins, the protolytic claevage of the Rh-C σ-bond and the change of the oxidation state (Rh^I→Rh^III) can be carried out at the [Rh(PNP)]^+ fragment supporting the reaction mechanism in essential steps and thus the suitability of the rhodium(I) for the catalysis of the hydroamination of olefins.

Furthermore the analogous dicationic olefin-PNP-Pd(II)-complexes [Pd(PNP)-(CH$_2$CHR)](BF$_4$)$_2$ (R = H **14**, Ph **15**) could be synthesized starting from the complex Pd(PNP)Cl$_2$ using the silver salt method in the presence of an excess of the olefin (cf. eq. 5). The Pd(II)-complexes **14** and **15** were characterized by NMR spectroscopy and a X-ray crystal structure analysis of the styrene complex **15** could be performed (cf. fig. 2).

$$\text{Pd(PNP)Cl}_2 \ + \ 2 \ \text{AgBF}_4 \ \xrightarrow[\text{- 2 AgCl}]{\text{CH}_2=\text{CHR/CH}_2\text{Cl}_2} \ \text{[structure]}^{2+} \ 2 \ \text{BF}_4^- \quad (5$$

R = H **14**
Ph **15**

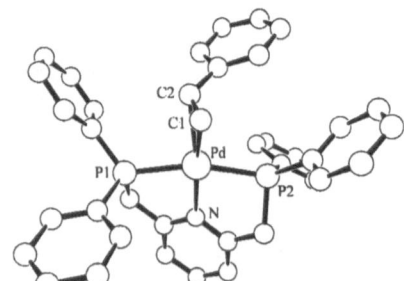

Figure 2 Molecule structure of [Pd(PNP)(CH$_2$CHPh)]$^{2+}$. Selected bond length [Å] and angles [°]:Pd-N 2.058(4), P(1)-Pd-P(2) 162.4(1), Pd-P(1) 2.314(1), P(1)-Pd-N 81.2(2), Pd-P(2) 2.330(2), P(2)-Pd-N 81.2(2), Pd-C(1) 2.165(7), C(1)-Pd-C(2) 33.7(4), Pd-C(2) 2.273(7), C(1)-C(2) 1.292(10).

The olefin-PNP-Pd(II)-complexes **14** and **15** exhibit the first examples of dicationic monoolefin-Pd(II) complexes. In comparison to their Rh(I) analogoues they are less stable in solid state as well as in solution. Compared to the isostructural cationic styrene-Rh(I) complex **2** (C=C: 1.383(7) Å) the distance of the C-C double bond of the coordinated styrene in the dicationic Pd(II) complex **15** is significantly shorter and no longer than in free styrene.

In contrast to the cationic olefin-PNP-Rh(I) complexes **1** and **2** the isostructural dicationic olefin-PNP-Pd(II) complexes **14** and **15** are susceptible to nucleophilic attack by secondary amines such as piperidine and dimethylamine forming corresponding β-aminoethyl-Pd(II) complexes since the coordination at the [Pd(PNP)]$^{2+}$ fragment strongly increases the electrophilic character of the olefin.

By the reaction of the styrene complex **15** with HNMe$_2$ the *anti*-Markownikow product was formed exclusively. Since in addition the corresponding ammonium salt

192

$(PhCH_2CH_2)_2NMe_2^+$ was also detected upon reductive degradation, a multiple addition can be suggested (cf. scheme 2). Unfortunately, as long as target is a catalytic hydroamination, the palladium(II) species seem unfavoured, since the Pd-C σ-bond in the β-aminoethyl complexes is too stable to be cleaved protolytically under the reaction conditions and can only undergo reductive cleavage with irreversible destruction of the complex.

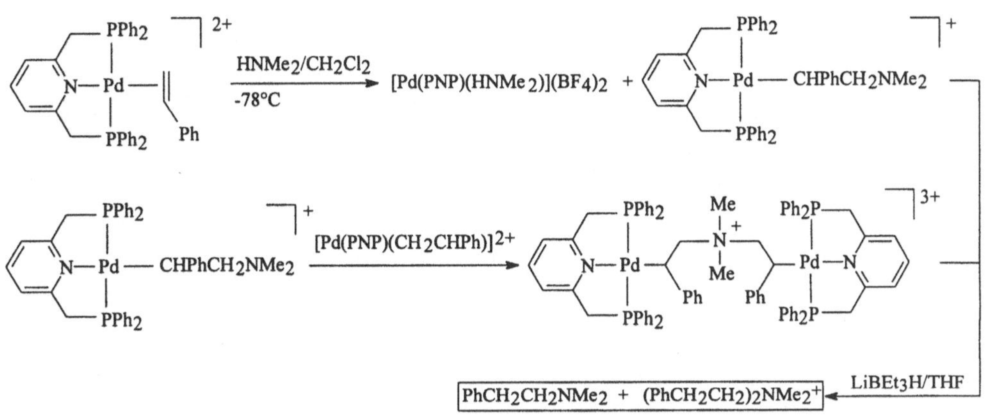

Scheme 2

Acknowledgements

We thank the Deutsche Forschungsgemeinschaft (Sonderforschungsbereich 347 of the University of Würzburg) for financial support. An ERASMUS visiting scholarship for C. H. at the University of Napoli, May - July 1996, and the Italian National Research Concil (CNR) is acknowledged. For X-ray structure analyses we thank Prof. Sieler from the Univerity of Leipzig, Prof. F. Giordano from the University of Napoli and Dr. E. Herdtweck and M. Spiegler from the TU München.

References

1. R. Taube in *Applied Homogeneous Catalysis with Organometallic Compounds* (Eds.: B. Cornils / W. A. Herrmann), Verlag Chemie, Weinheim (1996), vol. 1, pp. 507.
2. W.V. Dahlhoff, S. M. Nelson, J. Chem. Soc. (A) **1971**, 2184-2190.
3. E. Krukowka, R. Taube, D. Steinborn, DD 296 909 A5 (10.11. 88) 1988
4. R. R. Schrock, J. A. Osporn, *J. Am. Chem. Soc.* **1971**, *93*, 3089-3090.
5. (a) C. Hahn, R. Taube in Stereoselective Reactions of Metal-Activated Molecules (Eds.: H. Werner, J. Sundermeyer), Vieweg, Braunschweig, **1994**, S.123.
   (b) C. Hahn, J. Sieler, R. Taube, *Chem. Ber.* **1997**, *130*, 939-945.

C-E and E-E Bonds
(E = Heteroatom)

The Unusual Chemistry of Bimetallic Silyl Complexes

Pierre Braunstein, Michael Knorr, Thomas Stährfeldt and Christine Stern

Laboratoire de Chimie de Coordination (URA CNRS 416), Institut Le Bel, Université Louis Pasteur, 4 rue Blaise Pascal, F-67070 Strasbourg Cedex, France
E-mail: braunst@chimie.u-strasbg.fr

The chemistry of mononuclear organometallic complexes that contain metal-silicon bond has been under investigation for many years principally because of its relevance to a number of industrially important catalytic reactions [1]. In comparison, relatively little is known about the chemistry of silyl and related ligands in bimetallic analogoues. We felt that it was of interest to synthesize heterometallic complexes containing at least one metal-silicon bond in order to assess how reactivity may be influenced by the presence of both, metal-metal and metal-silicon bonds. The aims were to (i) take advantage of synergistic effects which could result from the presence of a heterometallic bond and (ii) explore the reactivity of the metal-silicon bond in a bimetallic environment. Because of the unique structural features revealed by our earlier work, we have focused principally on alkoxysilyl ($-Si(OR)_3$) (see below) and siloxyl ($-Si(OSiR_3)_3$) [2] complexes. Molecules containing such ligands, which have been very little studied so far in the bimetallic context, could provide interesting models for the study of interactions between metals and silica surfaces [3a] and also behave as new molecular precursors of interesting materials obtained *e.g.* by sol-gel condensation routes [3b-d].

Silicon-containing heterobimetallic complexes of the type $[(R'_3P)(OC)_3(R_3Si)Fe-ML_n]$ (M = main group element or transition metal, R = alkyl, aryl, alkoxy or Cl; R' = alkyl, aryl or alkoxy) can be conveniently prepared by reaction of the corresponding silylated metalates, *trans*-$[Fe(SiR_3)(CO)_3(PR'_3)]^-$, with metal halide complexes $[MXL_n]$ [4-9]. An alternative synthetic route consists of the oxidative addition of the Fe-H bond of the parent hydrido-silyl complexes *mer*-$[HFe(SiR_3)(CO)_3(PR'_3)]$ to a low-valent metal centre [7,10-14]. We have successfully applied both methods in the presence of an assembling ligand, e.g. $Ph_2PCH_2PPh_2$ (dppm), $Ph_2PNHPPh_2$, $Ph_2PC_5H_4N$ or PR_2H, which is originally bonded to the functional metalate or its corresponding hydride.

1 Occurrence and Properties of μ_2-η^2-Si-O Bridges Between Different Metals.

The pendant phosphorus donor of the Fe-bound dppm ligand in $K[Fe\{Si(OR)_3\}(CO)_3(dppm\text{-}P)]$ has been used to assist the formation of Fe-M bonds and five-membered ring structures of the type Fe(μ-dppm)M with a number of different metals. For M = Ag, Zn, In, Pd, Pt or Rh, the unprecedented the formation of a μ_2-η^2-Si-O bridge of type **A** between the metals was observed [7-10,14-17] :

197

$$\begin{array}{c}
\text{Fe} \text{---} \text{M} \\
\text{RO}^{\cdots}\overset{|}{\underset{RO}{\text{Si}}}\text{---}\overset{O}{\underset{R}{}} \\
\text{RO} \quad \textbf{(A)}
\end{array}$$

This new type of bridging system has been more rarely observed with weakly or unsupported metal-metal bonded Fe-Zn, Fe-Cd, Fe-In, Re_2 and Os_3 complexes [15, 17-20] . The labile character of the dative oxygen → metal interaction can be detected by variable temperature ^{1}H NMR spectroscopy since two different resonances for the R groups were observed below the coalescence temperature. The dynamic exchange between the OR groups may be explained by rapid rotation of the silyl ligand about the Fe-Si bond. The labile oxygen → metal donor interaction provides the metal M with a masked, potentially vacant coordination site and this could be exploited in chemical reactivity studies that focus on the attachment and activation of small molecules. We have therefore explored such reactions with CO, isocyanides or olefins with the aim of forming carbon-carbon bonds around a bimetallic template [21a].

2 Migratory Insertion Reactions of CO and Olefins

The migratory insertion of small molecules such as CO, organic isocyanides [21b] and olefins into metal carbon bonds has been recognized as an elementary step of considerable importance in organometallic chemistry and catalysis [22-24]. Several mononuclear Pd(II) catalytic systems have been described recently in which stepwise successive insertion reactions lead *e.g.* to polyketone chain growth [25-29]. Relatively little is known about insertion reactions of related *heterometallic* alkyls, perhaps owing to the limited number of such complexes and/or their instability [30-35].

Carbonylation of the methyl complexes [(OC)₃Fe{μ-Si(OMe)₂ (OMe)}(μ-dppm)MMe] (**1** M = Pd, **2** M = Pt) afforded the corresponding acyl derivatives [(OC)₃Fe{μ-Si(OMe)₂ (OMe)}(μ-dppm)MC(O) Me] (**3** M = Pd, **4** M = Pt) (equation 1) [14,21a].

Olefin insertion into the Pd-acyl bond of **3** occurred quantitatively after 3 h in the presence of norbornene (nbe) to give **5** whose ^{1}H NMR spectroscopic data indicate that *cis* addition of Pd-C(O)Me across the *exo* face of norbornene has occurred (equation 2) [21a] .

$$\text{(1)}$$

1 M = Pd; 2 M = Pt   3 M = Pd; 4 M = Pt

The IR spectroscopic data confirmed the expected co-ordination of the acyl group to palladium, resulting in a five-membered chelate rather than in a Fe-Si-O→Pd four-membered ring.

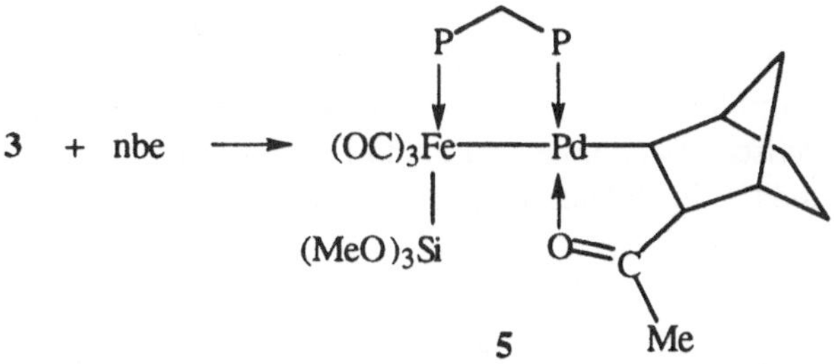

$$3 \ + \ nbe \ \longrightarrow \ \text{(structure 5)} \tag{2}$$

A similar result was obtained with the more reactive norbornadiene (nbd) substrate. Further CO-insertion into its Pd-C bond took place under a CO atmosphere and 7 was isolated as a stable product that was fully characterized. Olefin insertion into the Pd-acyl bond is more facile than CO insertion into the Pd-alkyl bond, whereas CO insertion is only possible into the Pd-alkyl bond and not into the Pd-acyl bond. This accounts for the perfect alternation of CO and nbe units in the oligomeric chain of 8 but the exact length of the polyketone is not yet known (equation 3).

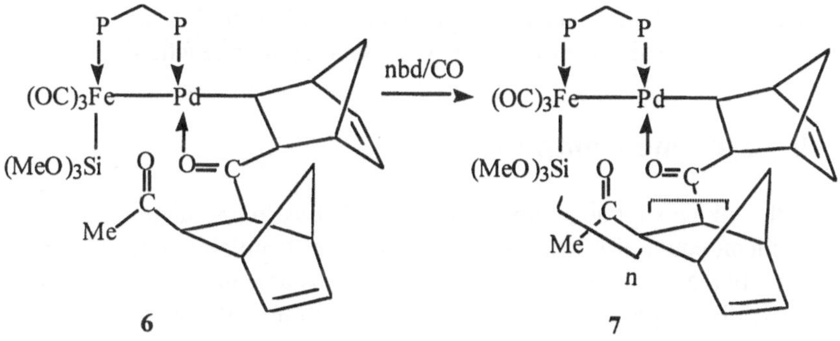

$$\text{(structure 6)} \ \xrightarrow{\ nbd/CO\ } \ \text{(structure 7)} \tag{3}$$

Note that mononuclear complexes related to 8 with $n = 0$-3 have been recently characterized [27,36-38].

The mechanism of all these insertion reactions follows a sequence which contains the elementary steps involved in CO migratory insertion: opening of the labile SiO→Pd bond, coordination of the substrate, cis-migration, isomerization and closing of the Fe-Si-O→Pd ring. Although five-coordinated intermediates have not been detected and need not be invoked in this mechanism, their involvement in the course of the displacement of the O→M bond cannot be excluded.

3 Intramolecular Silyl Migration from One Metal to Another

In order to evaluate the role of the assembling ligand, we wished to examine the reactivity of iron-platinum complexes closely related to those described above, but in which a three electron μ_2-diaryl(or dialkyl)phosphido bridging ligand, replaces of the four-electron donor μ-dppm ligand (equation 4). This modification led to a completely different chemistry: when CO was bubbled through a solution of **8**, selective substitution of the ligand L (L = PPh$_3$) trans to the phosphido-bridge occurred leading to **9**. This was followed by an unexpected, quantitative isomerization in solution to **10** in which the Si atom is now bonded to Pt, whilst the Pt-bound CO ligand has migrated to the Fe centre [11].

$$
\begin{array}{ccc}
\underset{\mathbf{8}}{
\begin{array}{c}
\mathrm{Ph_2} \\
\mathrm{P} \\
(OC)_3Fe \!-\!\!-\! Pt \\
(MeO)_3Si \qquad L
\end{array}
}
& \overset{+\,CO}{\underset{-\,L}{\longrightarrow}} &
\underset{\mathbf{9}}{
\begin{array}{c}
\mathrm{Ph_2} \\
\mathrm{P} \\
(OC)_3Fe \!-\!\!-\! Pt \\
(MeO)_3Si \qquad C \\
O
\end{array}
}
\end{array}
\qquad \longrightarrow \qquad
\underset{\mathbf{10}}{
\begin{array}{c}
\mathrm{Ph_2} \\
\mathrm{P} \\
(OC)_4Fe \!-\!\!-\! Pt \\
\qquad L
\end{array}
\;Si(OMe)_3}
\qquad (4)
$$

The result of this dyotropic-type rearrangement was established by the X-ray structure analysis of **10**. The rate of this unprecedented silyl shift appears to depend mainly upon the steric requirements of the silyl ligand. Cross-over experiments indicated that this silyl transfer reaction is an intramolecular isomerization. Further studies are in progress to determine the parameters which control this new silicon migration reaction.

4 Bimetallic Silylene Complexes

Our studies on the alkoxysilyl and siloxyl complexes led us to speculate on whether other substituents at silicon could also lead to the occurrence of hemilabile interactions with the adjacent metal. Only the -SiF$_3$ ligand had been found to be able to bridge between iron and palladium [39]. It also appears that the reactivity of tris(amino)silanes toward transition metals has not yet been examined, although tris(amino)silyl complexes have been obtained indirectly [40]. The amino substituents would be expected to have a considerable influence on the properties of the iron-silicon bond.

In a pioneering work, G. Schmid reported the synthesis of the first base-stabilized silylene complex [Fe(CO)$_4$\{SiMe$_2$(NEt$_2$H)\}] [41]. We prepared the amine-stabilized bis(dimethylamino)silylene complex [Fe(CO)$_4$\{Si(NMe$_2$)$_2$(NMe$_2$H)\}] (**11**) by oxidative-addition of the Si-H bond of HSi(NMe$_2$)$_3$ to the photochemically generated Fe(CO)$_4$ fragment [42]. A hydrido, tris(dimethylamino)silyl complex is most likely to be the primary product but is unstable owing to proton migration to the more basic nitrogen atom (equation 5).

$$Fe(CO)_5 \ + \ HSi(NMe_2)_3 \ \xrightarrow[-\ CO]{h\nu\,/\,hexane}$$

$$(OC)_4Fe\!-\!\underset{\underset{NMe_2}{|}}{\overset{\overset{\displaystyle H}{|}}{Si}}\!-\!NMe_2 \quad\longrightarrow\quad (OC)_4Fe\!=\!\underset{\underset{NMe_2}{|}}{\overset{\overset{\displaystyle NHMe_2}{\nearrow}}{Si}}\!-\!NMe_2 \qquad (5)$$
$$\mathbf{11}$$

The reactions of **11** with [Pt(C$_2$H$_4$)(PR$_3$)$_2$] (R = Ph, *p*-tolyl) yielded new complexes to which we assign structures **12** on the basis of analytical and spectroscopic data.

$$(OC)_4Fe$$
$$\parallel \longrightarrow Pt \overset{\nearrow PR_3}{\underset{\searrow PR_3}{}}$$
$$\underset{Me_2N}{Si}\!\cdots NMe_2$$
$$\mathbf{12}$$

The dimethylamine ligand of **12** and the Pt-bound ethylene ligands have been displaced leading to new bimetallic silylene complexes. It wass surprising that the amine ligand in **11** was so readily displaced by a metal fragment. The resulting silylene complex is now stabilized by an *electron-acceptor* fragment (*i.e.* PtL$_2$) instead of an *electron-donor* group, as observed so far [43].

In order to evaluate the influence of increased electron density at the iron centre on the behaviour of such complexes, we investigated the photochemically induced oxidative addition reaction of [Fe(CO)$_4${P(OMe)$_3$}] with HSi(NMe$_2$)$_3$ (Scheme 2). To our great surprise, this led to the amine-stabilized iron-silylene complex **13** in which two methoxy groups on phosphorus have exchanged with two amino groups on silicon [44]. Since **13** was only obtained in *ca.* 30-40% yield (based on ^{31}P{^{1}H} NMR, other products not yet characterized), we attempted an alternative synthesis based on the reaction of [Fe(CO)$_4${P(NMe$_2$)$_3$}] with HSi(OMe)$_3$. It also led to **13** but in *ca.* 80% yield. These cross-experiments indicate that the subtle balance between the oxophilicity of the phosphorus and silicon centres may lead to highly selective substituent exchange reactions. A remarkable rearrangement of this type was also observed with OEt/NMe$_2$ groups and was shown to be metal-promoted [47].

Deprotonation of **13** with excess KH in THF afforded the metalate complex K[Fe(CO)$_3${P(OMe)(NMe$_2$)$_2$}{Si(OMe)$_2$(NMe$_2$)}] which on reaction with [CuCl(PPh$_3$)] at low temperature in a 1:1 Fe/Cu stoichiometry yielded [(OC)$_3${(Me$_2$N)$_2$(MeO)P}Fe{μ-Si(OMe)$_2$(NMe)$_2$}Cu(PPh$_3$)] (**14**), without any further rearrangement of the ligands (Scheme 2).

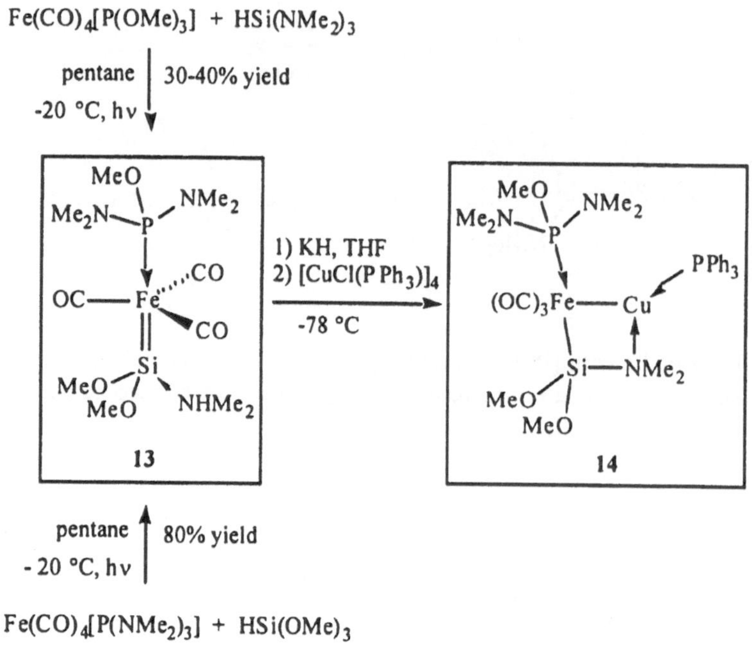

Fe(CO)4[P(OMe)3] + HSi(NMe2)3
pentane
-20 °C, hv
30-40% yield
MeO
Me2N
P
NMe2
CO
OC
Fe
CO
Si
MeO
MeO
NHMe2
13
1) KH, THF
2) [CuCl(PPh3)]4
-78 °C
MeO
Me2N
P
NMe2
PPh3
(OC)3Fe
Cu
Si
NMe2
MeO
MeO
14
pentane
- 20 °C, hv
80% yield
Fe(CO)4[P(NMe2)3] + HSi(OMe)3

Scheme 2

Complex **14** provides the first example of a bridging aminosilyl ligand and this significantly extends the scope of the studies performed with the trisalkoxysilyl ligands. That the amino group rather than either of the alkoxy groups is involved in the bridge formation may be explained by its higher basicity.

5 Conclusions

We have seen that bimetallic complexes containing silyl ligands may display unique structures and reactivity patterns that are directly related to a subtle interplay between the metals and the ligands. The hemilabile behaviour of the bridging -$Si(OR)_3$ ligand in Fe-Pd complexes allowed controlled insertion reactions to be performed under mild conditions into a preformed Pd-carbon bond. By altering the nature of the assembling ligand (μ-PR_2 vs. μ-dppm) but keeping the metals and the silyl ligand unchanged, we found a completely different chemistry and discovered the first example of intramolecular silyl migration from one metal to another. Finally, the use of aminosilyl ligands led to the formation of new silylene complexes, to unprecedented examples of metal-mediated substituent exchange reactions between phosphorus and silicon, and to the characterization of the first complexes containing a bridging aminosilyl ligand. There is no doubt that the rich chemistry associated with the presence of silicon ligands in bi- or polymetallic complexes will lead to more exiting results.

Acknowledgements

We are very grateful to the coworkers who participated in this research, to the Centre National de la Recherche Scientifique for support, the Deutsche Forschungsgemeinschaft for a Habilitation grant to M.K., the Ministère des Affaires Etrangères (Paris) and the Deutscher Akademischer Austauschdienst (Bonn) (Procope 93035), the Commission of the European Communities and Johnson Matthey PLC for a generous loan of $PdCl_2$ and $PtCl_2$.

References

[1]   See for example: (a) Braunstein, P.; Knorr, M. *J. Organomet. Chem.* **1996**, *500*, 21. (b) *The Chemistry of Organic Silicon Compounds*; Patai, S.; Rappoport, Z. (Eds.); Wiley: Chichester, U.K., 1989. (c) Aylett, B. J. *Adv. Inorg. Chem. Radiochem.* **1982**, *25*, 1. (d) Seyferth, D. in *Organosilicon Chemistry, from Molecules to Materials,* Auner, N., Weis, J. (Eds.); VCH Publishers, Weinheim (Germany), 1994, p 269. (e) Schubert, U. *Transition Met. Chem.* **1991**, *16*, 136. (f) Hofmann, P. in *Organosilicon Chemistry, from Molecules to Materials,* Auner, N., Weis, J. (Eds.); VCH Publishers, Weinheim (Germany), 1994, p 231. (g) Speier, J. L. *Adv. Organomet. Chem.* **1979**, *17*, 407. (h) Curtis, M. D.; Epstein, P. S. *Adv. Organomet. Chem.* **1981**, *19*, 213. (i) Tilley, T. D. *Acc. Chem. Res.* **1993**, *26*, 22. (j) Chalk, A. J.; Harrod, J. F. *J. Am. Chem. Soc.* **1965**, *87*, 1133. (k) Chalk, A. J.; Harrod, J. F. *J. Am. Chem. Soc.* **1965**, *87*, 16. (l) Brookhart, M.; Grant, B. E. *J. Am. Chem. Soc.* **1993**, *115*, 2151. (m) Seitz, F.; Wrighton, M. S. *Angew. Chem. Int. Ed. Engl.* **1988**, *27*, 289. (n) Sharma H. K., Pannell K. H. *Chem. Rev.* **1995**, *95*, 1351. (o) Marciniec B. *New J. Chem.* **1997**, *21*, 815. (p) *Comprehensive Handbook on Hydrosilylation*, Marciniec B., (Ed.), Pergamon, Oxford, **1992**, Chapter 2. (q) Marciniec B.; Gulinski, J. *J. Organomet. Chem.* **1993**, *446*, 15. (r) Harrod, J. F.; Yun, S. S. *Organometallics* **1987**, *6*, 1381. (s) Laine, R. M.; Rahn, J. A.; Youngdahl, K. A.; Babonneau, F.; Hoppe, M. L.; Zhang, Z. F.; Harrod, J. F. *Chem. Mat.* **1990**, *2*, 464. (t) Liu, H. Q.; Harrod, J. F. *Organometallics* **1992**, *11*, 822. (u) Hengge, E.; Weinberger, M. *J. Organomet. Chem.* **1992**, *433*, 21. (v) Chang, L. S.; Corey, J. Y. *Organometallics* **1989**, *8*, 1885.

[2]   (a) Knorr, M.; Braunstein, P. *Bull. Soc. Chim. Fr.* **1992**, *129*, 663. (b) Braunstein, P.; Knorr, M. in *Metal-Ligand Interactions - Structure and Reactivity*, Russo, N.; Salahub, D. R. (Eds.); Kluwer Academic Publishers, Dordrecht (The Netherlands) - NATO ASI Ser., Ser. C, **1996**, *474*, 49-83. (c) Knorr, M.; Braunstein, P.; Tiripicchio, A.; Ugozzoli, F. *Organometalliccs* **1995**, *14*, 4910.

[3]   (a) Braunstein, P. in *Perspectives in Coordination Chemistry*, Williams, A. F.; Floriani, C.; Merbach, A. E. (Eds.); Verlag Helvetica Chimica Acta/VCH: CH-4010 Bâle, **1992**, p. 67-107. (b) *Proceedings of the First European Workshop on Hybrid Organic-Inorganic Materials*, Sanchez, C.; Ribot, F. (Eds.); *New J. Chem.* **1994**, *18*, 989. (c) Braunstein, P.; Cauzzi, D.; Predieri, G.; Tiripicchio, A. *J. Chem. Soc., Chem. Commun.* **1995**, 229. (d) Bachert, I.; Braunstein, P.; Hasselbring, R. *New J. Chem.* **1996**, *20*, 993.

[4]   Jetz, W.; Graham, W. A. G. *Inorg. Chem.* **1971**, *10*, 1647.

[5]   (a) Schubert, U.; Kunz, E.; Knorr, M.; Müller, J. *Chem. Ber.* **1987**, *120*, 1085. (b) Kunz, E.; Schubert, U. *Chem. Ber.* **1989**, *122*, 231.

[6]   Schubert, U. in *Organosilicon Chemistry, from Molecules to Materials,* Auner, N.; Weis, J. (Eds.); VCH Publishers, Weinheim (Germany), 1994, p 205.

[7]   Braunstein, P.; Knorr, M.; Tiripicchio, A.; Tiripicchio-Camellini, M. *Angew. Chem., Int. Ed. Engl.* **1989**, *28*, 1361.

[8]   Braunstein, P.; Knorr, M.; Schubert, U.; Lanfranchi, M.; Tiripicchio, A. *J. Chem. Soc. Dalton Trans.* **1991**, 1507.

[9]   Braunstein, P.; Faure, T.; Knorr, M.; Balegroune, F.; Grandjean, D. *J. Organomet. Chem.* **1993**, *462*, 271.

[10]   Braunstein, P.; Knorr, M.; Villarroya, B. E.; DeCian, A.; Fischer, J. *Organometallics* **1991**, *10*, 3714.

[11]   Braunstein, P.; Knorr, M.; Hirle, B.; Reinhard, G.; Schubert, U. *Angew. Chem. Int. Ed. Engl.* **1992**, *31*, 1583.

[12]   Reinhard, G.; Knorr, M.; Braunstein, P.; Schubert, U.; Khan, S.; Strouse, C. E.; Kaesz, H. D.; Zinn, A. *Chem. Ber.* **1993**, *126*, 17.

[13]   Knorr, M.; Stährfeldt, T.; Braunstein, P.; Reinhard, G.; Hauenstein,   P.;   Mayer,   B.; Schubert, U.; Khan, S.; Kaesz, H. D. *Chem. Ber.* **1994**, *127*, 295.

[14]   Braunstein, P.; Faure, T.; Knorr, M.; Stährfeldt, T.; DeCian, A.; Fischer, J. *Gazz. Chim. Ital.* **1995**, *125*, 35.

[15]   Balegroune, F.; Braunstein, P.; Douce, L.; Dusausoy, Y.; Grandjean, D.; Knorr, M.; Strampfer, M. *J. Cluster Sc.* **1992**, *3*, 275.

[16]   Braunstein, P.; Knorr, M.; Piana, H.; Schubert, U. *Organometallics* **1991**, *10*, 828.

[17]   Braunstein, P.; Knorr, M.; Strampfer, M.; DeCian, A.; Fischer, J. *J. Chem. Soc. Dalton Trans.* **1994**, 117.

[18]   Braunstein, P.; Douce, L.; Knorr, M.; Strampfer, M.; Lanfranchi, M.; Tiripicchio, A. *J. Chem. Soc. Dalton Trans.* **1992**, 331.

[19]   Adams, R. D.; Cortopassi, J. E.; Pompeo, M. P. *Inorg. Chem.* **1992**, *31*, 2563.

[20]   Adams, R. D.; Cortopassi, J. E.; Yamamoto, J. H. *Organometallics* **1993**, *12*, 3036.

[21]   (a) Braunstein, P.; Knorr, M.; Stährfeldt, T. *J. Chem. Soc., Chem. Commun.* **1994**, 1913. (b) Yamamoto, Y.; Yamazaki, H. *Coord. Chem. Rev.* **1972**, *8*, 225.

[22]   Anderson, G. K.; Cross, R. J. *Acc. Chem. Res.* **1984**, *17*, 67.

[23]   Yamamoto, A. *Organotransition Metal Chemistry*; Wiley: New York, 1986.

[24]   (a) Garrou, P. E.; Heck, R. F. *J. Am. Chem. Soc.* **1976**, *98*, 4115. (b) Calderazzo, F. *Angew. Chem.* **1977**, *89*, 305; *Angew. Chem. Int. Ed. Engl.* **1977**, *16*, 209. (c) DeShong, P.; Slough, G. A. *Organometallics* **1984**, *3*, 636.

[25]   Lai, T. W.; Sen, A. *Organometallics* **1984**, *3*, 866. (b) Sen. A. *Acc. Chem. Res* **1993**, *26*, 303.

[26]   (a) Brookhard, M.; Rix, F. C.; DeSimone, J. M. *J. Am. Chem. Soc.* **1992**, *114*, 5894. (b) Rix, F. C.; Brookhard, M.; White, P. S. *J. Am. Chem. Soc.* **1996**, *118*, 4764.

[27]   (a) van Asselt, R.; Gielens, E. E. C. G.; Rulke, R. E.; Vrieze, K.; Elsevier, C. J. *J. Am. Chem. Soc.* **1994**, *116*, 977. (b) Markies, B. A.; Kruis, D.; Rietveld, M. H. P.; Verkerk, K. A. N.; Boersma, J.; Kooijman, H.; Lakin, M. T.; Spek, A. L.; van Koten, G. *J. Am.Chem. Soc.* **1995**, *117, 5263*.

[28]   (a) Drent, E.; van Broekhoven, J. A. M.; Doyle, M. J. *J. Organomet. Chem.* **1992**, *417*, 235. (b) Drent, E.; Budzelaar, P. H. M. *Chem. Rev.* **1996**, *96*, 663.

[29]   (a) Milani, B.; Alessio, E.; Mestroni, G.; Zangrando, E.; Randaccio, L.; Consiglio, G. *J. Chem. Soc. Dalton Trans.* **1996**, 1021. (b) Milani, B.; Vicentini, L.; Sommazzi, A.; Garbassi, F.; Chiarparin, E.; Zangrando, E.; Mestroni, G. *J. Chem. Soc. Dalton Trans.* **1996**, 3139. (c) Milani, B.; Anzilutti, A.; Vicentini, L.; Sessanta o Santi, A.; Zangrando, E.; Geremia, S.;

Mestroni, G. *Organometallics*, **1997**, in press. (d) Ozawa, F.; Hayashi, T.; Koide, H.; Yamamoto, A. *J. Chem. Soc., Chem. Commun.* **1991**, 1469.

[30]   Longato, B.; Norton, J. R.; Hufman, J. C.; Marsella, J. A.; Caulton, K. G. *J. Am. Chem. Soc.* **1981**, *103,* 209.

[31]   (a) Antonelli, D. M.; Cowie, M. *Organometallics* **1991**, *10,* 2550. (b) Antwi-Nsiah, F.; Cowie, M. *Organometallics* **1992**, *11,* 3157. (c) Antwi-Nsiah, F.; Oke, O.; Cowie, M. *Organometallics* **1996**, *15,* 1042.

[32]   Jacobson, G. B.; Shaw, B. L.; Thornton-Pett, M. *J. Chem. Soc. Dalton Trans.* **1987**, 3079.

[33]   Ferrer, M.; Rossell, O.; Seco, M.; Braunstein, P. *J. Chem. Soc. Dalton Trans.* **1989**, 379.

[34]   (a) Arndt, L. W.; Bancroft, B. T.; Darensbourg, M. Y.; Janzen, C. P.; Kim, C. M.; Reibenspies, J.; Varner, K. E.; Youngdahl, K. A. *Organometallics* **1988**, *7,* 1302. (b) Finke, R. G.; Gaughan, G.; Pierpont, C.; Noordik, J. H. *Organometallics* **1983**, *2,* 1481. (c) Ozawa, F.; Park, J. W.; Mackenzie, P. B.; Schaefer, W. P.; Henling, L. M.; Grubbs, R. H. *J. Am . Chem. Soc.* **1989**, *111,* 1319.

[35]   Fukuoka, A.; Sadashima, T.; Endo, I.; Ohashi, N.; Kambara, Y.; Sugiura, T.; Miki, K.; Kasai, N.; Komiya, S. *Organometallics* **1994**, *13,* 4033. (b) Fukuoka, A.; Sadashima, T.; Sugiura, T.; Wu, X.; Mizuho, Y.; Komiya, S. *J. Organomet. Chem* **1994**, *473* 139.

[36]   Markies, B. A.; Verkerk, K. A. N.; Rietveld, M. H. P.; Boersma, J.; Kooijman, H.; Spek, A. L.; van Koten, G. *J. Chem. Soc.; Chem. Commun.* **1993**, 1317.

[37]   Brumbaugh, J. S.; Whittle, R. R.; Parvez, M.; Sen, A. *Organometallics* **1990**, *9*, 1735.

[38]   Markies, B. A.; Rietveld, M. H. P.; Boersma, J.; Spek, A. L.; van Koten, G. *J. Organomet. Chem.* **1992**, *424,* C12.

[39]   Braunstein, P.; Colomer, E.; Knorr, M.; Tiripicchio, A.; Tiripicchio-Camellini, M. *J. Chem. Soc., Dalton Trans.* **1992**, 903.

[40]   (a) Thum, G.; Malisch, W. *J. Organomet. Chem.* **1984**, *264,* C5. (b) Malisch, W.; Schmitzer, S.; Kaupp, G.; Hindahl, K.; Käb, H.; Wachtler, U. in *Organosilicon Chemistry*, Auner, M.; Weis, J. (Eds.), VCH, Weinheim, **1994**, p. 185.

[41]   Schmid, G.; Welz, E. *Angew. Chem. Int. Ed. Engl.* **1977**, *16*, 785.

[42]   Bodensieck, U.; Braunstein, P.; Deck, W.; Faure, T.; Knorr, M.; Stern, C. *Angew. Chem., Int. Ed. Engl.* **1994**, *33*, 2440.

[43]   (a) Zybill, C. *Topics in Current Chem.* **1991**, *160*, 1. (b) Leis, C.; Zybill, C.; Lachmann, J.; Müller, G. *Polyhedron* **1991**, *10*, 1163. (c) Corriu, R.; Lanneau, G.; Priou, C. *Angew. Chem. Int. Ed. Engl.* **1991**, *30*, 1130. (d) Don Tilley, T. in *The Silicon-Heteroatom Bond*, Patai, S.; Rappoport, Z. (Eds.), John Wiley, New York, **1991**, 245 and 309. (e) Leis, C.; Wilkinson, D. L.; Handwerker, H.; Zybill, C. *Organometallics* **1992**, *11*, 514. (f) Handwerker, H.; Leis, C.; Gamper, S.; Zybill, C. *Inorg. Chim. Acta* **1992**, *198-200,* 763. (g) Corriu, R. J. P.; Lanneau, G. F.; Chauhan, B. P. S. *Organometallics* **1993**, *12*, 2001 and references cited therein.

[44]   Braunstein, P.; Stern, C.; Strohmann, C.; Tong, N. *Chem. Commun.* **1996**, 2237.

Organotransition-Metal Dendrimers

Christine Valério, Ester Alonso, Jaime Ruiz, Jean-Luc Fillaut, Françoise Moulines, and Didier Astruc*

Laboratoire de Chimie Organique et Organométallique, UMR CNRS N° 5802, Université de Bordeaux I, 33405 Talence Cédex, France

A remarkable area of applications of organotransition-metal chemistry is the modification of the reactivity of molecules by temporary complexation onto a transition-metal center (1). Repetition of this principle after removal of the transformed ligand from the metal center has led to the powerful principle of catalysis. However, repetition without decoordination has not been exploited or even much underlined. Some years ago, we have disclosed the CpFe$^+$ induced polyfunctionalization of polymethylaromatics. We have now used this series of selective, high-yield reactions for the syntheses of dentritic cores. Indeed, dendrimers are monodisperse macromolecules that are promising in material science (2-5) and biology (5) and this article summarizes our first results in this area including synthetic and molecular-recognition aspects.

1 Syntheses of ferrocene dendrimers

The reaction of CpFe(mesitylene)$^+$PF$_6^-$ with excess KOH or t-BuOK in DME or THF (respectively) and allylbromide at ambient temperature for two days gives the nona-allylation complex resulting from the replacement of the nine benzylic protons by the nine allyl substituents (6) (Scheme 1, top). This is the result of nine deprotonation-allylation sequences proceeding in one pot. The pK$_a$ of the CpM(polymethylbenzene)$^+$ complexes (M = Fe or Ru) have been measured by the direct method using t-BuOK by ^{1}H-NMR and were found to be around 28 in DMSO (7). Thus, the deprotonation by these bases is reversible, but shifted towards the products by the following electrophilic attack which irreversibly leads to the formation of carbon-carbon bonds (6). The nona-allyl complex is subsequently photolyzed using visible light and regiospecific hydroboration by 9-BBN followed by oxidation using H$_2$O$_2$ (10) yields a nonol (6). A first nona-iron dendrimer was synthesized by reaction of this nonol with [FeCp(p-F-C$_6$H$_4$)][PF$_6$] (6) (Scheme 1). For the metal-dendrimer, a single reversible wave was obtained in cyclic voltammetry at -30°C in DMF corresponding to the cathodic reduction of d^{6}Fe(II) or d^{7}Fe(I). From the intensity of the current, the number of electrons involved in the process was found to be 8±1 (6). However, this wave is not reversible at room temperature and we have subsequently been seeking more robust redox systems for applications. Thus, a nona-amine was synthesized from the nonol by Michael reaction of acrylonitrile (11) followed by reduction of the nona-nitrile (11,12) (Scheme 2).

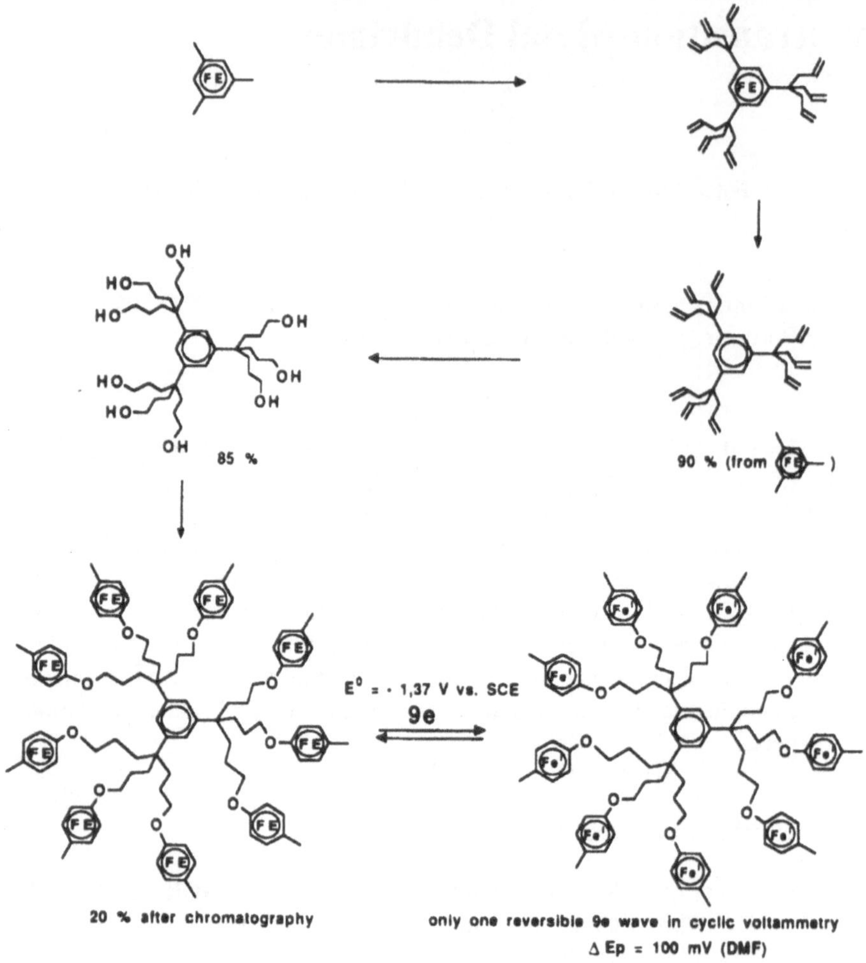

Scheme 1

The polyamines were synthesized with the aim to obtain polyamido-metallocene dendrimers by reactions with metallocenylcarbonylchlorides (8,9). The first obvious target was ferrocene units since the famous ferrocene/ferricinium redox couple had found considerable use as a redox sensor. Reactions between the polyamines and chlorocarbonylferrocene were performed at room temperature for 1-3 days in CH_2Cl_2 in the presence of NEt_3 (8,9):

$$\text{Dendri-NH2} + \text{FcCOCl} + \text{NEt}_3 \rightarrow \text{Dendri-NHCOFc} + \text{FcCOCl} + \text{NEt}_3\text{H}^+\text{Cl}^-$$

Only the 9- and 18-amine dendrimers gave soluble ferrocene dendrimers (Chart 1) whereas the expected 36- and 72-amido-ferrocene dendrimers were totally insoluble in all the solvents.

208

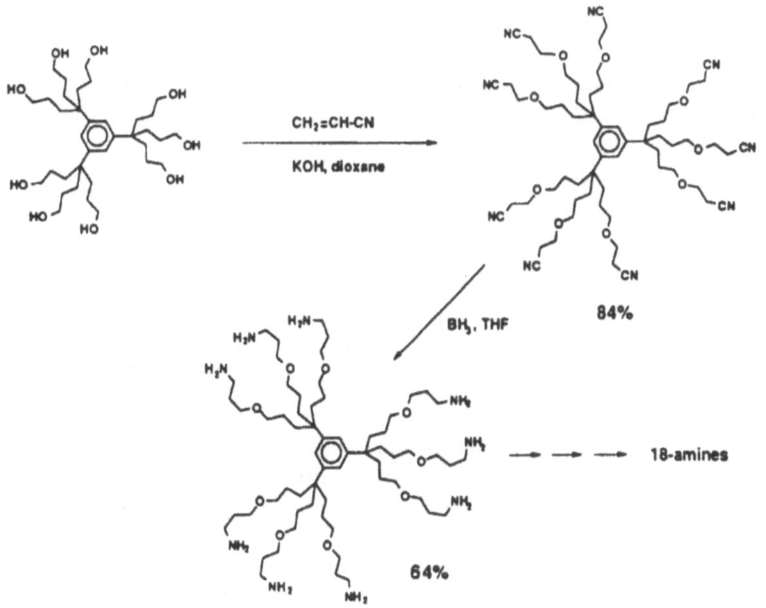

Scheme 2

The insolubility reached for the 36-Fc dendrimer is a sign of steric saturation at the surface which prevents the solvents from penetrating inside the dendrimer. This also means that at or above the 36-Fc generation the number of ferrocene units becomes lower than expected in a default-free dendrimer which is confirmed by the elemental analysis and molecular modelization. On the other hand, the 9-Fc and 18-Fc dendrimers were characterized by ^{1}H and ^{13}C-NMR and infrared spectra, correct elemental analysis and the molecular peaks of the MALDI-TOF mass spectra (MH$^+$: m/z = 3066 for 9-Fc and MNa$^+$: m/z = 6024 for 18-Fc) (8). These condensation reactions of the polyamines were also carried out with cationic transition-metal-sandwich complexes, namely [Co(CpCOCl)Cp][PF$_6$] or analogues and [Fe(CpCOCl)(arene)][PF$_6$] (13). The nona-metal dendrimers with these sandwiches could be synthesized and characterized by standard spectroscopic techniques and elemental analysis. Electrospray mass spectra recorded by E. Leize and A. van Dorsselaer from Strasbourg University showed the molecular peaks for the nona-cobalticinium- and nona-CpFe(toluene)$^+$ dendrimers. However, their solubilities are weaker than those of the ferrocene dendrimers and decrease as the bulk of the sandwich moiety increases (i.e. when the number of methyl groups increases on the arene ligand of the iron complex). The 18-Co dendrimer could also be made but its solubility in MeCN is very weak which makes NMR characterization less unambiguous and elemental analysis suggested the inclusion of water molecules. Thus, its purity is highly uncertain. Very recently, the synthesis of ferrocene dendrimers of closely related structure has also been communicated by the group of Moran and Cuadrado (14) and publication of metallodendrimers is very prolific (for beautiful examples, see references 14-23).

2 Ferrocene dendrimer as sensors

The area of anion recognition pioneered by Lehn (24-27) is of particular importance for its biological implications. Various types of sensors are known including redox sensors with macrocycles and tripods (24-39). The anion receptors designed so far are endo-receptors (24-31). On the other hand, dendrimers with redox sensors at the extremities of the branches could function as exo-receptors especially if the surface covered with redox sensors is not too far from steric saturation. At this point, it could mimic the surface of micro-organisms such as viruses. The ferrocene unit has long been used as a redox sensor since both Fe(II) and Fe(III) forms are stable enough for electrochemical scanning without loss of reversibility (38). The principle is that the redox potential of the Fe(II/III) redox system of the ferrocene unit is not the same in the presence and absence of substrate whose recognition is looked for.

In the meantime, the binding constant of the substrate with the host bearing the ferrocene unit close to the receptor is not the same in the neutral Fe(II) redox form of ferrocene and in its Fe(III) cationic form (39). These thermodynamic values are related by the thermodynamic cycle of Scheme 3.

The amidoferrocene fragment also has the benefit of the acidic amide hydrogen atom which can form a hydrogen bond with an oxygen atom of oxo-anions. Amidoferrocenes have indeed been used as redox sensors in tripodal units (37,38).

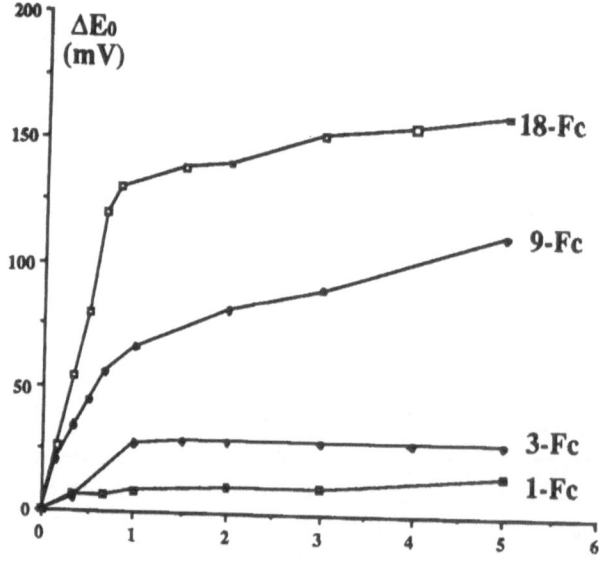

Number of equivalents of n-Bu$_4$N$^+$HSO$_4^-$ versus ferrocenic function

Figure 1: Titration of 1-Fc (1-Fc = [Fe(C$_5$H$_5$)(C$_5$H$_4$CONHCH$_2$CH$_2$OPh)]), 3-Fc, 9-Fc and 18-Fc (see Chart 1) by n-Bu$_4$N$^+$ HSO$_4^-$ monitored by CV. Concentrations in FcDs were 0.001 M, CH$_2$Cl$_2$, n-Bu$_4$N$^+$BF$_4^-$ (0.1 M), 20°C, reference electrode: SCE, auxiliary and working electrodes: Pt, scan rate: 100 mV.s^{-1}.

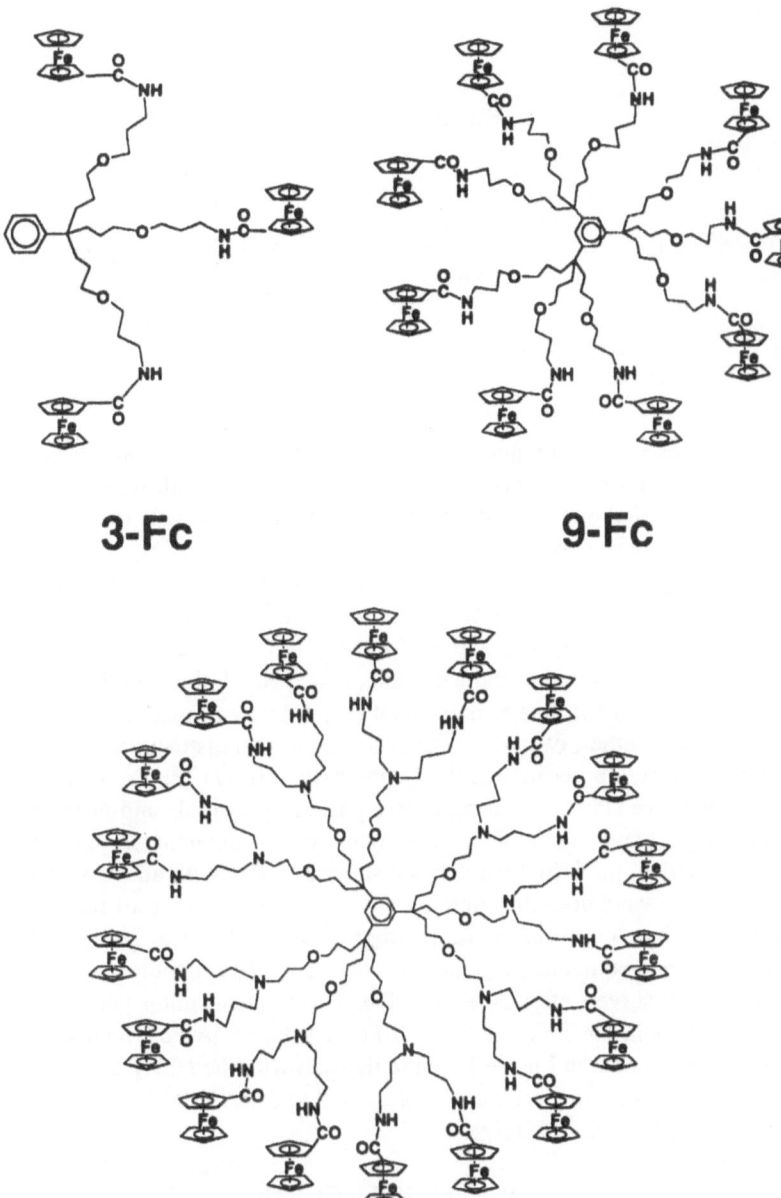

3-Fc

9-Fc

18-Fc

Chart 1

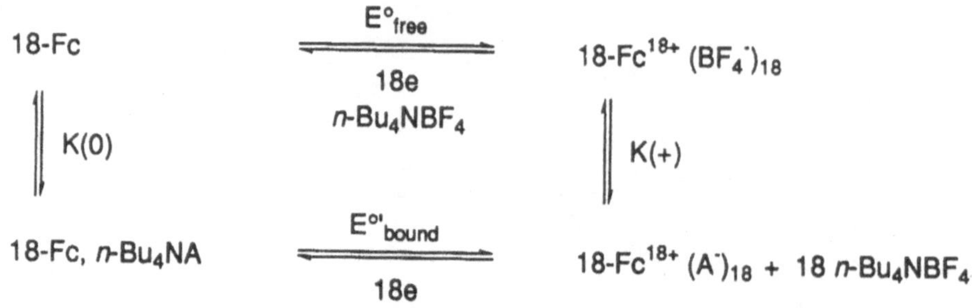

$$18\text{-Fc} \underset{\underset{n\text{-Bu}_4\text{NBF}_4}{18e}}{\overset{E^{\circ}_{free}}{\rightleftharpoons}} 18\text{-Fc}^{18+} (BF_4^-)_{18}$$

K(0)   K(+)

$$18\text{-Fc, } n\text{-Bu}_4\text{NA} \underset{18e}{\overset{E^{\circ\prime}_{bound}}{\rightleftharpoons}} 18\text{-Fc}^{18+} (A^-)_{18} + 18\ n\text{-Bu}_4\text{NBF}_4$$

Scheme 3

We have compared the 9-Fc and 18-Fc dendrimers with mono- and tripodal amidoferrocenes of closely related structure in order to investigate dendritic effects. Recognition studies have been carried out by cyclic voltametry and by ^{1}H NMR. In each case, titrations of the ferrocene dendrimers were effected by n-Bu$_4$N$^+$ salts of H$_2$PO$_4^-$, HSO$_4^-$, Cl$^-$ and NO$_3^-$. By far, the most informative results were obtained by cyclic voltametry (41-42), by scanning the Fe(II/III) wave. Before any titration, the cyclic voltamograms or the 9-Fc and 18-Fc dendrimers show a unique wave at 0.59V vs SCE in CH$_2$Cl$_2$ corresponding to the oxidation of the 9 redox centers which indicates that, as expected, the 9 or 18 redox centers are electrochemically equivalent thus, independent (when two equivalent redox centers are not so far away from each other, two waves are observed at two distinct potentials, even if there is no electronic connection because of the electrostatic effect) (37). For instance, in the case of H$_2$PO$_4^-$, a new wave starts appearing at less positive potentials and correlatively, the intensity of the initial wave starts decreasing. When one equivalent of anion per dendrimer branch has been added, the initial wave has disappeared and with addition of the anion, the intensity of the new wave does not increase any longer. In the case of the other anions, no new wave appears but the initial wave is progressively shifted to less positive potentials upon titration until one equivalent of anion has been added per dendrimer branch. The shifts ΔE° of potentials observed after addition of one equivalent anion per dendrimer branch considerably increase in the series: 1-Fc $\rightarrow$ 3-Fc $\rightarrow$ 9-Fc $\rightarrow$ 18-Fc which shows a dramatic dendritic effect represented in Figure 1 for the titration with the HSO$_4^-$ anion. The tripod and dedrimers involved in Figure 1 are those represented in Chart 1. The magnitude of interaction with the anion increased as follows:

$$H_2PO_4^- > HSO_4^- > Cl^- > NO_3^-$$

Both situations upon titration - appearance of a new wave and shift of the initial wave - have already been analyzed from the thermodynamic standpoint (39).

The ^{1}H-NMR monitoring of the titrations (40) is not as useful in the case of other anions as in the case of H$_2$PO$_4^-$ because as indicated above the interaction is weak. Indeed, equivalent points are very variable and very far from corresponding to one equivalent anion per branch, whereas they do so for the ferricinium form which more strongly binds the different anions.

212

In general, the ferricinium form of the tripod or dendrimer binds the anions relatively strong because of the synergy of the electrostatic attraction with the intermolecular hydrogen bond formed between the acidic amide H atom and the anionic substrate through an oxygen atom of an oxo-anion or the halogen anion. Both factors are important and if one of them is absent, the interaction gets loose and cannot be used for sensing (except in the case of $H_2PO_4^-$ for the dendrimers). This effect has previously been recognized and used (37,38).

Of special interest here is the dramatic dendritic effect observed for all the anions. Even if the synergy between the electrostatic and H-bonding is fulfilled, the $\Delta E°$ value is unobservable or small if the amido-ferrocene used is monometallic (1-Fc) or trimetallic (3-Fc). The shape selectivity designed in the dendrimer is crucial and its effect is much more marked for 18-Fc than for 9-Fc as the ferrocene termini are closer to each other when the dendritic generation increases. This dendritic effect is thus maximum for the generation (18-Fc) which precedes steric saturation by ferrocene groups on the dendrimer surface (36-Fc). It can be understood in the course of the dendritic synthesis as the insolubility of sterically saturated ferrocene dendrimers is complete in all the solvents. In the amido-ferrocene dendrimers, the amide H atom is located on the branch behind the ferrocene unit which provides the surface bulk. Thus, the anion has to reach the inside of the microcavity formed by the amido-ferrocene units at the surface of the dendrimer. These conditions become optimal for redox sensing and recognition by the close ferrocene units at the 18-Fc generation, since the channels allowing the entry of the anions into the surface microcavity to reach the amide H atom are as narrow as possible.

In conclusion, after having synthesized organometallic dendrimers, we have now been able to demonstrate a dendritic effect in molecular recognition. Other effects are expected with polycationic organometallic dendrimers and molecular recognition studies are presently in progress in our laboratories along this line.

Acknowledgement

We thank E. Leize and A. van Doresselaer (Université Louis Pasteur, Strasbourg) and J.-C. Blais and J. Guittard (Université Paris VI) for electrospray and MALDI-TOF mass spectra, respectively M.J. Hynes for providing and discussing his EQ NMR program (ref. 40), the Institut Universitaire de France (D.A.), the Université Bordeaux I, the CNRS, the Région Aquitaine and Rhône-Poulenc for financial support, the Ministère de la Recherche et de la Technologie for a thesis grant to C.V., and the Ministerio de Education y Ciencia (Spain) for a post-doctoral grant to E.A.

References

1. L.S. Hegedus *Transition Metals in the Synthesis of Complex Organic Molecules* University Science Books Mill Valley, Ca, **1994.**
2. D.A. Tomalia, H. Dupont Durst in *Topics Curr. Chem.* Vol 165, *Supramolecular Chemistry: Directed Synthesis and Molecular Recognition* Weber E Ed, Springer Verlag, Berlin, **1993**, 193.
3. G.R. Newcome, C.N. Moorefield, F. Vögtle *Dendritic Macromolecules: Concepts, Syntheses and Perspectives* VCH, Weinheim, **1996.**
4. J.M.J Fréchet *Science* **1994**, *263*, 1710.

5. Comprehensive review on dendrimers: N. Ardoin, D. Astruc *Bull. Soc. Chim. Fr.* **1995**, *132*, 875; review on dendrimers in biology: D. Astruc *C. R. Acad. Sci.* Paris, ser. II *b*, **1996**, *322*, 757.

6. F. Moulines, L. Djakovitch, R. Boese, B. Gloaguen, W. Thiel, J.-L. Fillaut, M.-H. Delville, D. Astruc *Angew. Che. Int. Ed. Engl.* **1993**, *105*, 1132; for review, see: D. Astruc *Top. Curr. Chem.* **1991**, *160*, 47.

7. H. Trujillo, C.M. Casado, D. Astruc *J. Chem. Commun.* **1995**, 7; M. C. Casado, T. Wagner, D. Astruc *J. Organomet. Chem.* **1995**, *502*, 143.

8. C. Valério, J.-L. Fillaut, J. Ruiz, J. Guittard, J.- C. Blais, D. Astruc *J. Am. Chem. Soc.* **1997**, *119*, 2588.

9. D. Astruc L'Actalité Chimique, **1996**, *7*, 69; D. Astruc, C. Valério, J.-L. Fillaut, J. Ruiz, J.-R. Hamon, F. Varret In *Magenetism, a Supramolecular Function* O. Kahn Ed. NATO ASAI Series, Kluwer, Dordrecht, **1996**, 107.

10. H. C. Brown, Y.M. Choi, S. Narasimham *Synthesis* **1981**, 605 and *J. Org. Chem.* **1982**, *47*, 3153. G.R.

11. Newkome, X. Lin, J.K. Young *Synlett* **1992**, 53.

12. E.M.M. de Brabander-van den Berg, E.W. Meijer *Angew. Chem. Int. Ed. Engl.* **1993**, *32*, 1308; C. Wörner, R. Müllhaupt *Angew. Chem. Int. Ed. Engl.* **1993**, *32*, 1306.

13 E. Buhlein, W. Wehner, F. Vögtle *Synthesis* (1978) 155; F. Vögtle, E. Weber *Angew. Chem. Int. Ed. Engl.* **1979** *18*, 753; R. Moors, F. Vögtle *Chem. Ber.* **1993**, *126*, 2133.

14. C. Valério, E. Alonso, J. Ruiz, D. Astruc, *New J. Chem.* **1997**, **21**, 000.

15. I. Cuadrado, C. M. Casado, B. Alonso, M. Moran, J. Losada, V. Belsky *J. Am. Chem. Soc.* **1997**, 119, 7613.

16. S. Campagna, G. Denti, S. Serroni, A. Juris, M. Venturi, R. Ricevuto, V. Balzani *Chem. Eur. J.* **1995**, *1*, 211 and references cited therein.

17. M. Slany, M. Bardaji, M.-J. Casanove, A.-M. Caminade, J. P. Majoral, B. Chaudret *J. Am. Chem. Soc.* **1995**, *117*, 9764.

18. J.W.J. Knapen, A.W. van der Made, J.C. de Wilde, P.W.N.M. van Leeuwen, P. Wijkens, D.M. Grove, G. van Koten *Nature* **1994**, **372**, 659.

19. S. Achar, R. J. Puddephatt *Angew. Chem. Int. Ed. Engl.* **1994**, *33*, 847; *J. Chem. Soc., Chem. Commun.* **1994**, 1895.

20. Y.-H. Liao, J.R. Moss *J. Chem. Soc. Chem. Commun.* **1993**, 1774; *Organometallics* **1995**, *14*, 2130.

21. E.C. Constable, P. Harverson, M. Oberholzer *Chem. Commun.* **1996**, 33 and 1821.

22. A. Miedaner, C.J. Curtis, R.M. Barkley, D.L. Dubois *Inorg. Chem.* **1994**, *33*, 5482; A.M. Herring, B.D. Steffey, A. Miedaner, S.A. Wander, D.L. Dubois *Inorg. Chem.* **1994**, *33*, 5482.

23. M. F. Ottaviana, S. Bossmann, J. Turro, D. A. Tomalia *J. Am. Chem. Soc.* **1994**, 116, 661.

24. C. Moucheron, A. Kirch-De Mesmaeker, A. Dupont-Gervais, E. Leize, A. van Dorsselaer *J. Am. Chem. Soc.* **1996**, *118*, 128.

25. J.-M. Lehn *Supramolecular Chemistry: Concepts and Perspectives*, VCH, Weinheim, 1995.

26. E. Graf, J.-M. Lehn *J. Am. Chem. Soc.* **1976**, *98*, 6403.

27. M.W. Hosseini, J.-M. Lehn *Helv. Chim. Acta* **1986**, *69*, 587.

28. B. Hasenknopf, J.-M. Lehn, B. O. Kneisel, G. Baum, D. Fenske *Angew. Chem. Int. Ed. Engl.* **1996**, *35*, 1838.

29. F.P. Schmidschen *Angew. Chem. Int. Ed. Engl.* **1977**, *16*, 720; *Chem. Ber.* **1981**, *114*, 597.

30. D. M. Rudkevitch, W. Verboom, Z. Brzozka, M.J. Palys, W.P.R.G.- Stauthamer, G. J. van Hummel, S. M. Franken, S. Harkema, J. F. J. Engbersen, D. N. Reinhoudt, *J. Am. Chem. Soc.* **1994**, *116*, 4341.

31. S. Valiyaveettil, J. F. J. Engbersen, W. Verboom, D. N. Reinhoudt *Angew. Chem. Int. Ed. Engl.* **1993**, *32*, 900.

32. F. Vögtle *Supramolecular Chemistry*, 2nd Ed, Wiley, Chichester, 1993.

33. A. W. Czarnik *Acc. Chem. Res.* **1994**, *27*, 302.

34. J. L. Atwood, K. T. Holman, J. W. Steed *Chem. Commun.* **1996**, 1401.

35. T. J. James, S. Sandanayake, S. Shinkai *Angew. Chem. Int. Ed. Engl.* **1996**, *325*, 1910.

36. T. Saiji, I. Kinoshita *J. Chem. Soc. Commun.* **1986**, 716.

37. T. E. Edmonds In *Chemical Sensors*, T. E. Edmonds Ed., Blackie, Glasgow, **1988**, p. 193.

38. A.E. Kaifer, S. Mendoza In *Comprehensive Supramolecular Chemistry* Vol. 1; G. W. Gokel Ed., Pergamon, Oxford, **1996**, chapter 19, p. 701.

39. P.D. Beer, *Chem. Commun.* **1996**, 689; *Advan. Inorg. Chem.* **1992**, *39*, 79.

40. S.R. Miller, D.A. Gustowski, Z.-h. Chen, G.W. Gokel, L. Echegoyen, A.E. Kaifer *Anal. Chem.* **1988**, *60*, 2021.

41. M.J. Hynes *J. Chem. Soc. Dalton Trans.* **1993**, 311.

42. J. B. Flanagan, S. Margel, A.J. Bard, F. C. Anson *J. Am. Chem. Soc.* **1978**, *100*, 4268.

43. D. Astruc, "*Electron Transfer and Radical Processes in Transition-Metal Chemistry*" VCH, New York, **1995**, Chapter 2.

Towards Metalloreceptors on the Basis of Square Pyramidal Coordination Caps

C. Dietz, S. Schmidt, F. W. Heinemann, and A. Grohmann*

Institut für Anorganische Chemie der Universität Erlangen-Nürnberg, Egerlandstr. 1, D-91058 Erlangen, Germany

Introduction

In only a few cases the Werner approach (generation of transition metal complexes by forming dative bonds) and the supramolecular aspect (utilization of secondary bonding interactions such as hydrogen bonding, electrostatic or charge transfer interactions) have been integrated in the design of the coordination sphere of a metal. The objective is to create mononuclear *functional metal complexes* which will modulate the reactivity of substrates [1]. Examples are:

- Iron complexes of derivatised porphyrins (capped, picket-fence porphyrins, etc.) as possible models for hemoglobin, catalase, peroxidase or cytochrome P-450. In these complexes, shielding of the apical coordination site is intended to reduce non-specific reactions of the substrate [2].
- Metal receptors derived from crown ethers [1]: these involve organometallic complexes (M = Rh, Pd, Pt) in which a crown ether ring is bonded directly to the metal centre through N, P or S atoms which are part of the ring. The crown ether oxygen atoms form hydrogen bonds to a neutral ligand (for example, NH_3, H_2O) which is covalently bound to the metal, thereby locking the ligand into position. Metal coordination results in ligand activation as it confers greater acidity on the N-H or O-H bonds [3]. Recently, a related calixarene derivative has been described where an organopalladium unit bridges the upper rim of a calix[4]arene in such a manner as to allow an aromatic amine which is coordinated to the metal centre to be "fixed" through additional[1] stacking interactions with the calixarene cavity [4].
- Metal receptors derived from tris(2-aminoethyl)amine ligands (tren): coordination of a suitably derivatised ligand to Zn^{2+} ion gives a complex of trigonal bipyramidal geometry in which the remaining apical site is surrounded by the functional groups of the derivatised tren ligand (3-hydroxyphenyl). The complex catalyses the hydrolysis of a 4-pyridine carboxylic acid ester in basic medium. Apparently this substrate coordinates to zinc *via* the pyridine nitrogen atom, and the phenoxide substituents on the ligand periphery accelerate the hydrolysis of the optimally positioned ester functionality through intramolecular attack [5].

217

Square Pyramidal Coordination Caps as a Novel Structural Motif in Coordination Chemistry

We are interested in simple pentadentate ligands which provide octahedrally coordinating transition metal ions with a square pyramidal coordination cap of C_{2v} symmetry. The ligands are thus required to have a donor atom with a lone pair orientated towards the centre of the octahedron in the apical position of the pyramid *and* four identical donor arms at the base. A square pyramidal coordination cap having these structural features would be a prototypical setup for studying the reactivity of *free* coordination sites in octahedrally coordinated complexes. Additionally, it should provide an attractive platform for the construction of *active ligand peripheries* through suitable derivatization of the equatorial positions, using achiral or chiral building blocks (Figure 1.1a). A further attraction of such a coordination cap lies in its possible incorporation into suitably substituted calix[4]arenes.

Specifically, we wish to determine the degree to which substituents S that have been anchored to the pentadentate ligand can favour the coordination of a sixth ligand at the apical site and modulate its reactivity. We expect activation of substrates L (Figure 1.1a) such as H_2O, O_2, N_2, RCN, $CO(NH_2)_2$ or carboxylic acid esters, as a consequence of being anchored to the metal receptor through covalent *and* non-covalent (or secondary) interactions. We also aim to create dinucleating ligands containing square-pyramidally coordinating head groups, whose complexes should allow the incorporation of substrates between the two metal centres.

In order to access square pyramidal coordination caps of the above specification we first synthesised a pentaamine ligand (**1**, Figure 1.1b), which is a pyridine derivative with four pendent primary amino groups. The ligand is obtained from 2,6-diethylpyridine by fourfold hydroxymethylation to give a tetraalcohol precursor (**2**), which is subsequently transformed into the pentaamine *via* tosylation (**3**), tosylate-azide exchange (**4**), and reduction [6]. The selectivity of the hydroxymethylation step may be increased by adding an equivalent amount of $MgSO_4$, which points to the Mg^{2+} ion acting as a template [7]. A complication of the ligand synthesis derives from the fact that all steps require the uniform derivatization of 4 functional groups at the same time. These reactions do not proceed quantitatively, and the

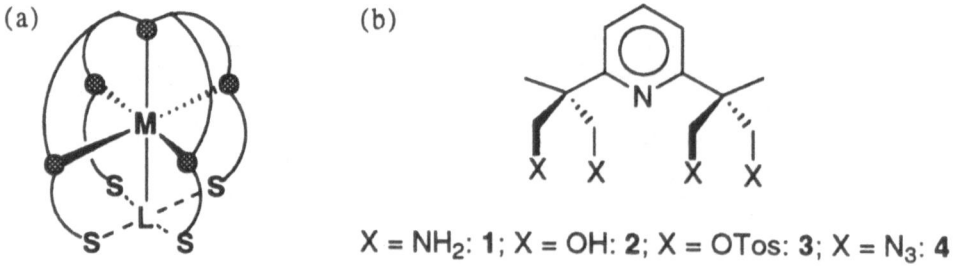

(a)   (b)

$X = NH_2$: **1**; $X = OH$: **2**; $X = OTos$: **3**; $X = N_3$: **4**

Figure 1.1 (a) General representation of a metal receptor derived from a square pyramidal coordination cap. (b) The pyridine-derived pentaamine ligand **1** and the intermediates of its preparation

ligand as initially prepared is contaminated with incompletely substituted side products. A straightforward route to pure ligand **1** utilises the facile complexation of unpurified product by Ni^{2+} ions and subsequent chromatographic separation of the nickel complex from non-complexing impurities.

Synthesis and Structures of Transition Metal Complexes of the Pentaamine Ligand

We have, so far, obtained complexes of the transition metals cobalt, rhodium and nickel in which the pentaamine ligand **1** does indeed form the desired square pyramidal coordination cap. About ten complexes have been structurally characterised (Figure 1.2 (a); M = Ni, L = [Ag(CN)$_2$]). The structural data obtained attest to a virtually strain-free coordination of the ligand, the pyridine nitrogen atom taking the apical position of a square pyramidal NN_4 arrangement of donor atoms, and the primary amino functionalities occupying the equatorial positions. This leads to the formation of 6 six-membered chelate rings, all of which adopt the energetically favourable boat conformation. In the case of nickel, ligand exchange experiments have given a range of compounds (L = Cl, Br, H_2O, ClO_4, N_3, [Ag(CN)$_2$]). Suitable bridging ligands allow the synthesis of polynuclear complexes, such as the dinuclear complex [(**1**)Ni(μ-Cl)Ni(**1**)](PF$_6$)$_3$ or the pentanuclear complex {[Fe(CN)$_6$]-[Ni(**1**)]$_4$}Cl$_4$. The latter bears a resemblance to systems which have been studied in connection with the search for "molecular magnetic materials". The magnetic and structural

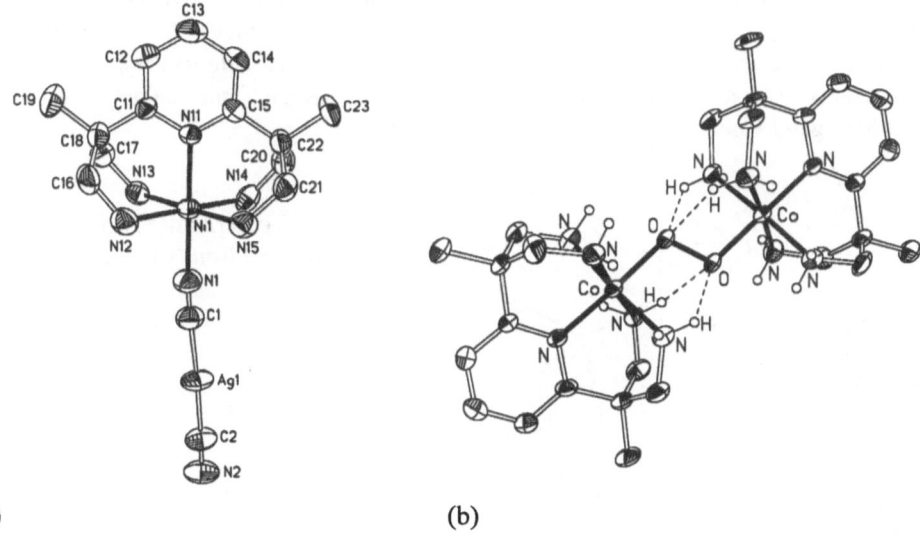

(a)   (b)

Figure 1.2 X-ray structures of the complexes (a) [Ni(**1**){Ag(CN)$_2$}]$^+$ and (b) [(**1**)Co-O-O-Co(**1**)]$^{4+}$, showing the virtually strain-free coordination of **1** in both cases. The local symmetry of both nickel and cobalt is octahedral, with all angles close to 90° and 180°, respectively

properties of the chloro-bridged dinuclear nickel complex have been studied in detail (antiferromagnetic coupling between the two paramagnetic Ni(II) ions in the ground state, S = 0) [8]. A mixture of **1** and Co(II) gives, upon aerial oxidation, the brown dinuclear peroxo complex shown in Fig 1.2 (b), in which intramolecular hydrogen bonds lock the Co(III)-O-O-Co(III) unit into a transoid arrangement with a dihedral angle of 180 ° [d(O-O) = 150.4(4) pm, $\angle$ (Co-O-O) = 109.8(2) ° and 110.6(2) °]. We are exploring the potential of this complex as a template for fourfold bridge formation.

Structural and spectroscopic data for M = Zn(II) and Ru(II) show that the pentaamine ligand **1** adopts two further coordination modes, depending on the nature of the starting material and the reaction conditions. In the zinc complex, only 3 of the 4 primary amino functionalities are together with the pyridine nitrogen atom coordinated to Zn^{2+}, while the fourth amino-substituted sidearm acts as a bridge to a second metal centre, which preliminary structural data indicate to be a trichlorozincate unit. By contrast, the only Ru(II) complex we have obtained so far is a symmtrical binuclear species; in it, each of the 1,3-diaminoproprane units of the ligand is coordinated to a $[RuCl_2(PPh_3)_2]$ fragment while the pyridine nitrogen atom remains uncoordinated.

Towards Functionalised Square Pyramidal Coordination Caps

The further derivatization of the pentaamine ligand **1** by Schiff base formation proceeds in excellent yields using aromatic aldehydes containing an *ortho*-hydroxyl group. The products may be isolated as crystalline solids, and reduction gives the respective secondary amine derivatives which are likewise solids and may be isolated in pure form. We are currently studying complexation reactions of the salicylaldehyde derivative shown in Figure 1.3. With cobalt(III), preliminary spectroscopic and mass spectrometric results point to the formation of a mononuclear complex, but we have not yet been able to achieve structural characterization. In a similar vein, we are currently studying the integration of the pentaamine ligand **1** into suitably substituted calix[4]arenes.

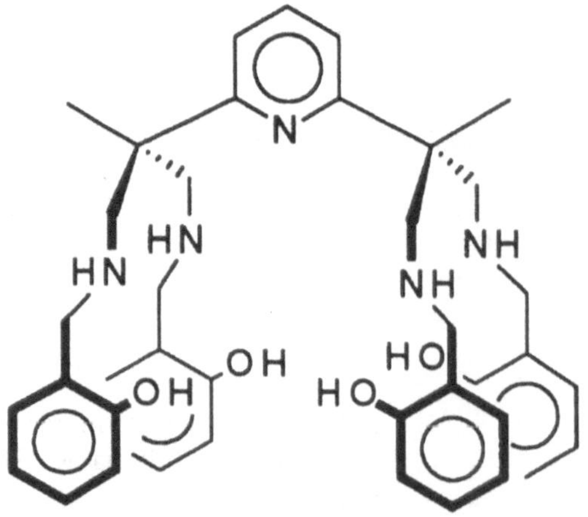

Figure 1.3 The derivatised coordination cap obtained from **1** and salicyl-aldehyde by Schiff base condensation and subsequent reduction

Acknowledgements

We wish to thank Professor D. Sellmann for support of this work, and Raschig AG and Chemetall GmbH for the donation of basic chemicals. We further thank the Fonds der Chemischen Industrie for financial support.

References

1. F. M. Raymo, J. F. Stoddart, *Chem. Ber.* **1996**, *129*, 981.
2. G. B. Jameson, J. A. Ibers in *Bioinorganic Chemistry*; I. Bertini, H. B. Gray, S. J. Lippard, J. S. Valentine (Hrsg.); University Science Books: Mill Valley, California, **1994**, p. 167.
3. J. E. Kickham, S. J. Loeb, *Inorg. Chem.* **1995**, *34*, 5656.
4. B. R. Cameron, S. J. Loeb, *J. Chem. Soc., Chem. Commun.* **1996**, 2003.
5. P. Scrimin, P. Tecilla, U. Tonellato, G. Valle, A. Veronese, *J. Chem. Soc., Chem. Commun.* **1995**, 1163.
6. A. Grohmann, F. Knoch, *Inorg. Chem.* **1996**, *35*, 7932.
7. S. Schmidt, C. Dietz, L. Omnès, A. Grohmann, W. Donaubauer, F. A. Knoch, F. W. Heinemann, J. Kuhnigk, C. Krüger, *J. Chem. Soc. Dalton Trans.* **1997**, submitted.
8. C. Dietz, A. Grohmann, F. W. Heinemann, M. Gerdan, A. X. Trautwein, manuscript in preparation.

Activation of Small Molecules at Transiton Metal Sites

Siegfried Schindler

Institute for Inorganic Chemistry, University of Erlangen-Nürnberg, Egerlandstraße 1, 91058 Erlangen, Germany

Activation of Dioxygen

Small molecules can be activated at transition metal sites and then further react with a substrate. An example of a basic reaction scheme for the oxidation of a substrate with the small molecule dioxygen is shown by equation 1. Dioxygen is reversibly bound to the metal complex forming a superoxo or peroxo adduct which then selectively oxidizes the substrate by transfering one or two oxygen atoms.

$$\text{Complex} \; \underset{}{\overset{O_2}{\rightleftharpoons}} \; \text{Oxo-Complex} \; \xrightarrow{\text{Substrate}} \; \text{Product} \qquad (1)$$

Nature uses enzymes for very selective reactions that are usually difficult to perform with catalysts in chemical industry; for example methane monooxygenase is responsible for the selective oxidation of methane to methanol. Copper and iron ions are found in the active sites of a large number of metalloproteins involved in important biological electron transfer reactions. Interest in preparing functioning model complexes for such proteins is twofold; first the models provide a better understanding of the biological molecules, and second they assist in the development of new homogeneous catalysts used for selective oxidations under mild conditions. Considerable effort has been directed towards the development of functional model compounds for copper enzymes. A fundamental step in metalloenzyme redox reactions is activation upon binding of dioxygen at the active site prior to the reaction with a substrate (first step in equation 1). Therefore, it is important to gain a better understanding of the reaction of dioxygen with copper(I) complexes.

The copper(I) complex of the tripodal ligand tmpa (tmpa = tris[(2-pyridyl)me-thyl]amine) proved to be a useful compound for performing such studies. [Cu(tmpa) (CH$_3$CN)]$^+$ reacts with dioxygen at low temperatures to form a copper peroxo complex which was structurally characterized by Karlin and coworkers. [1] An elegant kinetic investigation was performed on that system in propionitrile which allowed the spectroscopic observation of a superoxo complex prior to formation of the peroxo complex. [2]

We have started a detailed investigation on a series of tripodal copper(I) complexes to gain better understanding on their reactions with dioxygen. [3, 4] Therefore, we have studied how a) chelate ring sizes, b) nitrogen donor atoms and c) solvents influence these reactions. The ligands obtained modifying the parent ligand tmpa by increasing the pendant arm lengths are shown in Figure 1.

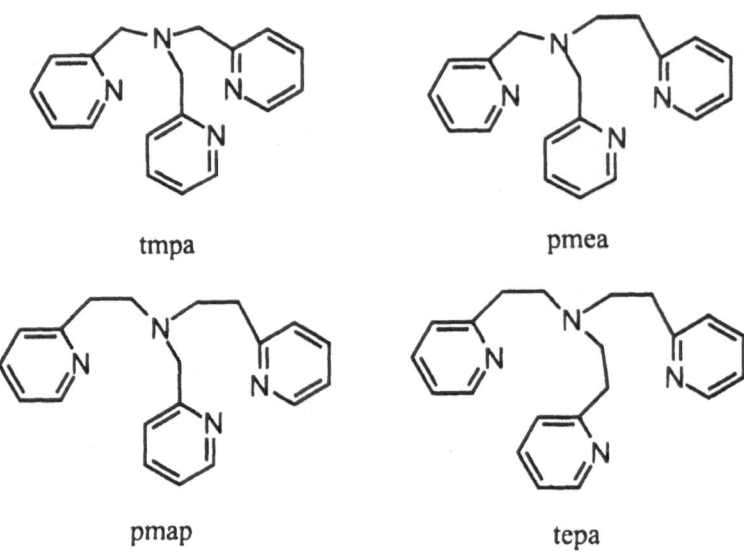

tmpa   pmea

pmap   tepa

Figure 1 Modification of the ligand tmpa by increasing the pendant arm lengths

In contrast to [Cu(tmpa)(CH$_3$CN)]$^+$ the copper(I) complexes of the ligands pmea, pmap and tepa did not form stable peroxo complexes at low temperatures. The reason for the different behavior towards dioxygen can be explained by the different geometries of the copper(I)/copper(II) complexes of the tripodal ligands. For the tmpa system it was found that the coordination geometry of the Cu(I) complex already closely resembled that of the peroxo product. [5] This is not the case for the other complexes. [4] Here the copper(I) complexes have a distorted tetrahedral geometry and are oxidized to copper(II) complexes with a distorted square pyramidal coordination geometry. Thus these complexes have to undergo a geometrical rearrangement during the reaction with dioxygen. Furthermore, the dioxygen ligand is much more kinetically labile to substitution in a square pyramidal complex compared to trigonal bipyramidal[6] which is most likely an important reason for the difference in behavior.

To investigate the influence of different nitrogen donors we replaced the pyridine arms in tmpa by aliphatic amines and used the ligands tris(2-dimethylaminoethyl)amine (= Me$_6$tren) and tris(2-benzylaminoethyl)amine (= Bz$_3$tren). The copper(I) complexes of both ligands react reversibly with dioxygen at low temperatures similar to [Cu(tmpa)(CH$_3$CN)]$^+$. So far X-ray structures of the copper(I) complexes could not be obtained but the geometry of the copper(II) complexes of Bz$_3$tren[3] and Me$_6$tren[7] was found to be trigonal bipyramidal in solution and in the solid state.

224

Propionitrile has proved to be a versatile solvent for reactions of dioxygen with copper(I) complexes of tripodal ligands. A positive feature about this solvent is that it stabilizes Cu(I) complexes but during the reaction with dioxygen it strongly competes as a ligand and therefore shifts the equilibrium to the educt as shown in eqations 2 and 3.

$$[Cu(Me_6tren)RCN]^+ + O_2 \rightleftharpoons [Cu(Me_6tren)O_2]^+ + RCN \quad (2)$$
$$[Cu(Me_6tren)O_2]^+ + [Cu(Me_6tren)RCN]^+ \rightleftharpoons [Cu_2(Me_6tren)_2O_2]_2^+ + RCN \quad (3)$$

As an alternative acetone proved to be the solvent of choice and the effect on the reaction of $[Cu(Me_6tren)(CH_3CN)]^+$ with dioxygen was very dramatic. At $-90\ °C$ the formation of the superoxo complex was much faster than in propionitrile and only a minor part of that reaction could be observed. In contrast the peroxo intermediate was dramatically stabilized compared to propionitrile. Therefore, it was possible for the first time to observe the peroxo intermediate at room temperature.[3] The spectral changes during the reaction of $[Cu(Me_6tren)(CH_3CN)]^+$ with dioxygen in acetone at $20\ °C$ are shown in Figure 2 and the absorbance time trace for the formation and decomposition of the peroxo species is shown as an insert.

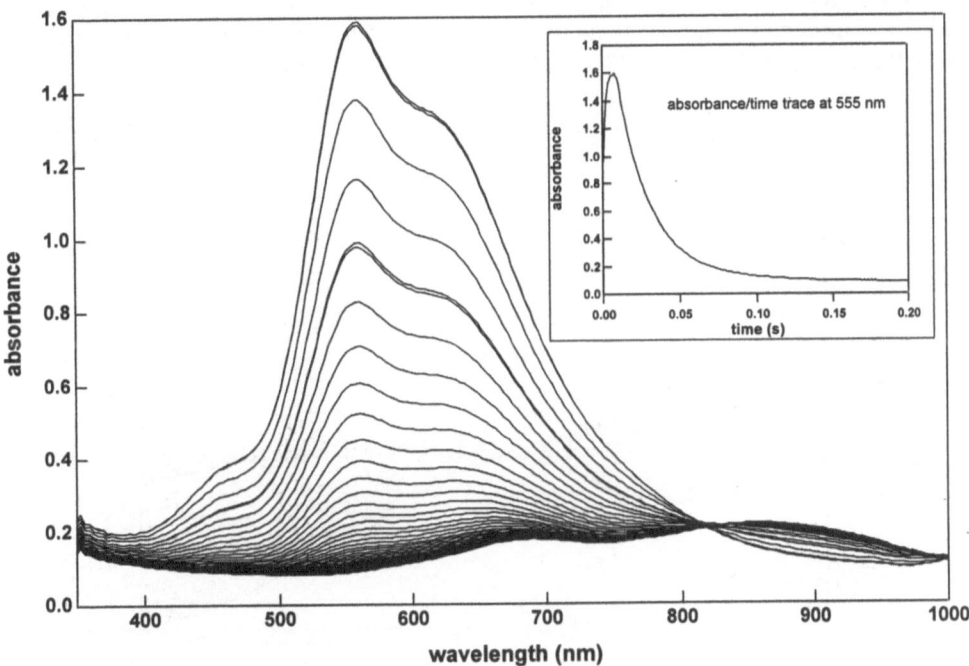

Figure 2 Spectral changes during the reaction of $[Cu(Me_6tren)(CH_3CN)]^+$ with dioxygen.

Activation of Substrate

A different reaction pathway (compared to equation 1) can occur, if the substrate itself is activated instead of the small molecule. This was found to be the case for the reaction of carbon dioxide with propionaldehyde at a Ni(0)-complex as shown in equation 4.[8]

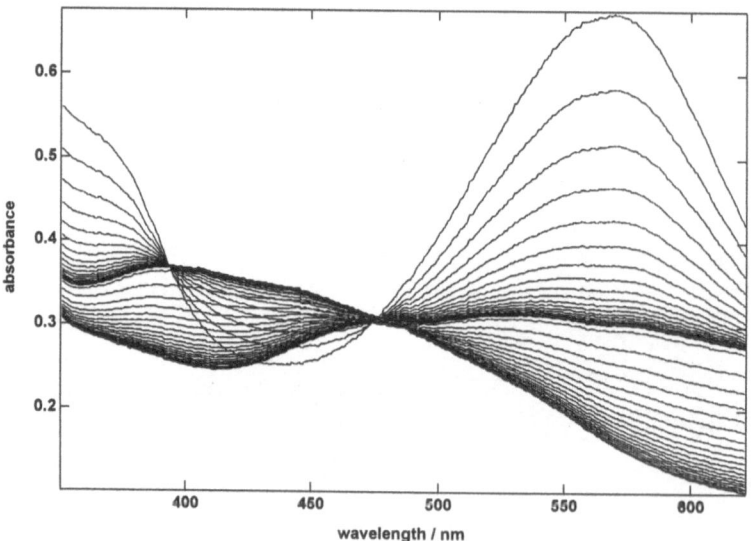

A detailed kinetic analysis revealed that Ni(bipy)(COD) reacts reversible with propionaldehyde and activation parameters for the forward and back reactions were calculated from a temperature dependence study; forward reaction (propionaldehyde) $\Delta H^{\#}=55 \pm 3$ kJ/mol, $\Delta S^{\#} = -58 \pm 9.0$ J/(mol K); back reaction (COD) $\Delta H^{\#} = 39 \pm 0.5$ kJ/mol, $\Delta S^{\#} = -58 \pm 2$ J/(mol K).[9] It could be clearly shown that carbon dioxide did not react with Ni(bipy)(COD) directly but with the nickel propionaldehyde complex to form a five membered cyclic nickel complex. The spectral changes during this reaction are shown in Figure 3. The reaction proceeded according to an associative mechanism, during which the carbon dioxide was inserted into a Ni-O bond by an oxidative addition step. The activation parameters for this process were: $\Delta H^{\#} = 43 \pm 6$ kJ/mol and $\Delta S^{\#} = -129 \pm 19$ J/(mol K).[9]

Figure 3 Spectral changes during the reaction of CO_2 with Ni(bipy)(COD) and propionaldehyde.

References

1. Jacobson, R. R.; Tyeklár, Z.; Farooq, A.; Karlin, K. D.; Liu, S.; Zubieta, J. *J. Am. Chem. Soc.* **1988**, *110*, 3690.
2. Karlin, K. D.; Wei, N.; Jung, B.; Kaderli, S.; Niklaus, P.; Zuberbühler, A., D. *J. Am. Chem. Soc.* **1993**, *115*, 9506.
3. Becker, M.; Liehr, G.; Schindler,S., manuscript submitted to *Inorganic Chemistry*.
4. Becker, M.; Hampel, F.; Schindler,S., manuscript submitted to *Inorganic Chemistry*.
5. Tyeklar, Z.; Jakobson, R., R.; Wei, N.; Murthy, N., N.; Zubieta, J.; Karlin, K. D. *J. Am. Chem. Soc.* **1993**, *115*, 2677.
6. Powell, D. H.; Merbach, A. E.; Fabian, I.; Schindler, S.; van Eldik, R. *Inorg. Chem.* **1994**, *33*, 4468.
7. Scott, M. J.; Lee, S. C.; Holm, R. H. *Inorg. Chem.* **1994**, *33*, 4651.
8. Dinjus, E.; Kaiser, J.; Sieler, J.; Walther, D. *Z. Anorg. Allg. Chem.* **1981**, *483*, 63.
9. Geyer, C.; Dinjus, E.; Schindler, S. manuscript submitted to *Organometallics*.

Organic Transformations with Early-Late Hetero-bimetallics: Synergism and Selectivity

Lutz H. Gade,* Stefan Friedrich, Harald Memmler, Uta Kauper, Martin Schubart, Bernd Findeis, Andreas Schneider, Sylvie Fabre and Dominique J. M. Trösch

Institut für Anorganische Chemie, Universität Würzburg, Am Hubland, D-97074 Würzburg, Germany

Summary

The coordination of novel types of chelating amido ligands has provided the key to the synthesis of stable early-late heterobimetallic complexes containing unsupported metal-metal bonds. This has been the prerequisite for the systematic development of their reactivity and their utilization in organic chemical transformations. Reactions with polar, unsaturated organic substrates have been studied in particular detail and are reported in this paper.

Introduction

Heterodinuclear complexes containing polar metal-metal bonds between early and late transition elements may display cooperative reactivity of the electrophilic and nucleophilic metal centres [1], although direct experimental evidence is still rare [2]. This is mainly due to the thermal lability and extreme chemical reactivity of most of the systems studied to date.

With the development of halide complexes containing tripodal amido ligands, *monofunctional* building blocks for the generation of directly metal-metal bonded M'-M heterobimetallic compounds (M' = Ti, Zr, Hf; M = late transition metal) of unprecedented stability have been made available [3]. Their structures have been investigated in detail by X-ray crystallographic studies of a whole range of such dinuclear complexes and the nature of their metal-metal bonds has been elucidated by modern *ab initio* quantum chemical and DFT methods [4].

More recently, this strategy has been extended to *difunctional* complexes of zirconium which provided the key to the synthesis of stable trinuclear complexes containing unsupported Zr-M bonds [5].

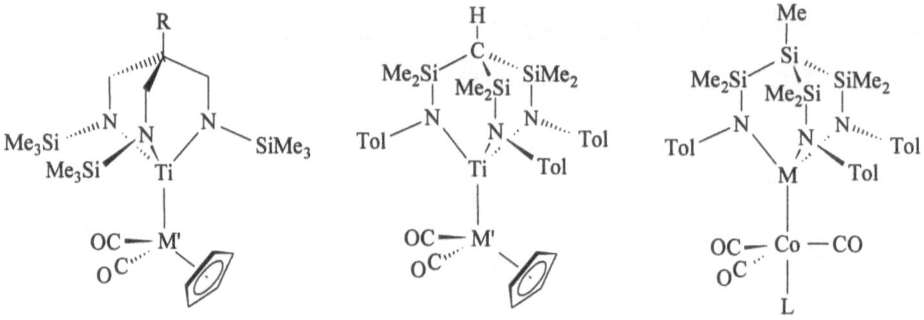

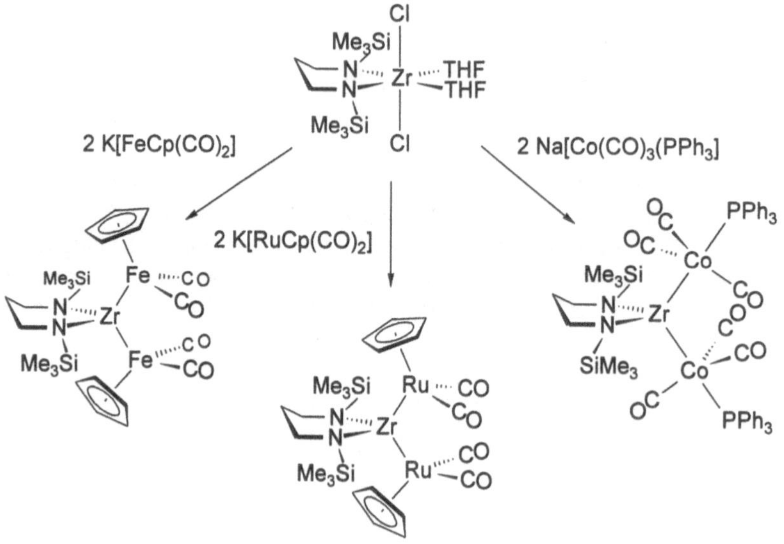

The thermal stability of early-late heterobimetallics containing a polydentate amido ligand at the early transition metal centre as depicted above enabled us to study systematically their reactivity towards organic substrates, in particular those containing a polar unsaturated functional group [6]. Their interaction with the substrates may, in principle, occur in a highly specific manner and induce patterns of reactivity which are not observed for the individual mononuclear or homodinuclear compound.

So far, three different reactivity patterns could be observed: *1.* cleavage of the metal-metal bond by the polar substrate X-Y, *2.* insertion of the unsaturated substrate X=Y into the highly polar metal-metal bond and *3.* insertion and ligand-substrate or substrate-substrate coupling. These three types of reactions may be followed by subsequent conversions.

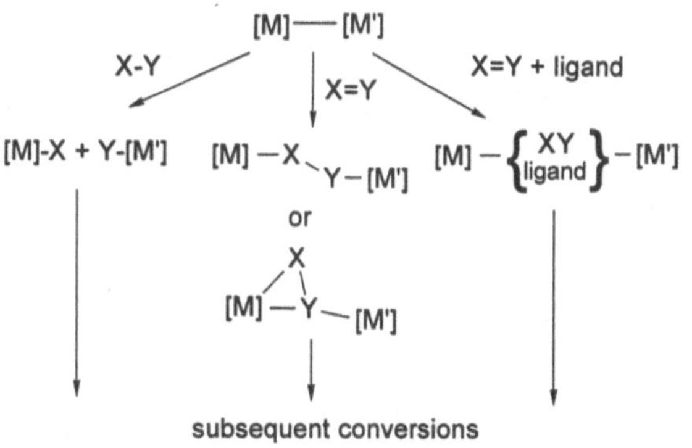

Simple Reactions Involving the Cleavage of a Substrate

This type of reaction is observed when reacting the highly polar metal-metal bond with polar, saturated functional groups such as R-OH or R-NH$_2$ or with element halides R$_3$EX (E = C, Si, Ge, Sn; X = Cl, Br, I). In each case, the more electronegative fragment will end up coordinated to the early transition metal while the remaining fragment combines with the late transition metal complex unit.

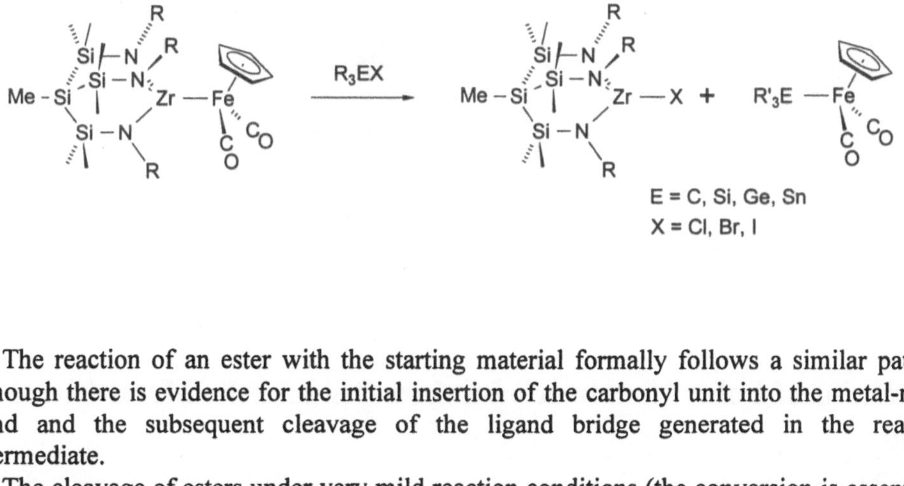

The reaction of an ester with the starting material formally follows a similar pattern, although there is evidence for the initial insertion of the carbonyl unit into the metal-metal bond and the subsequent cleavage of the ligand bridge generated in the reaction intermediate.

The cleavage of esters under very mild reaction conditions (the conversion is essentially spontaneous at -40°C) has been extended to include lactones, in particular the biaryl lactones studied by G. Bringmann and coworkers as precursor materials in the synthesis of biaryl alkaloids [7].

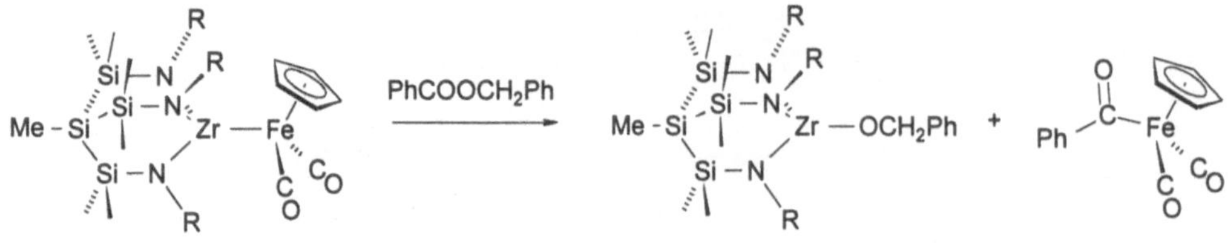

R = 2-FC$_6$H$_4$ or 2,3,4-F$_3$C$_6$H$_2$

The initial product of the reactions with lactones is the ring opened species, in which the metalloacyl oxygen atom coordinates to the Lewis acidic zirconium centre, thus generating a carbene type complex characterized by the ^{13}C-NMR resonance at ca. 300 ppm for the Fe and at ca. 280 ppm for the Ru complexes. These complexes slowly undergo decarbonylation generating reaction products in which the Cp(CO)$_2$M-unit is directly bonded to the naphthalin ring of the biaryl system [8].

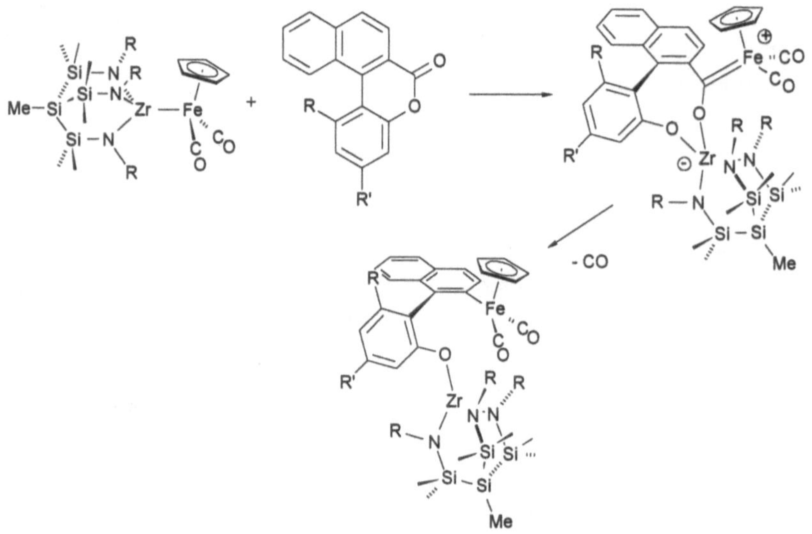

Insertions of Polar Unsaturated Substrates into the Metal-Metal Bond

Insertion into the polar metal-metal bonds is the dominating type of reaction with carbonyl derivatives, heteroallenes or related compounds. This provides a simple example of the *cooperative* reactivity of a metal electrophile and a metal nucleophile generated subsequently to the cleavage of the metal-metal bond [6].

232

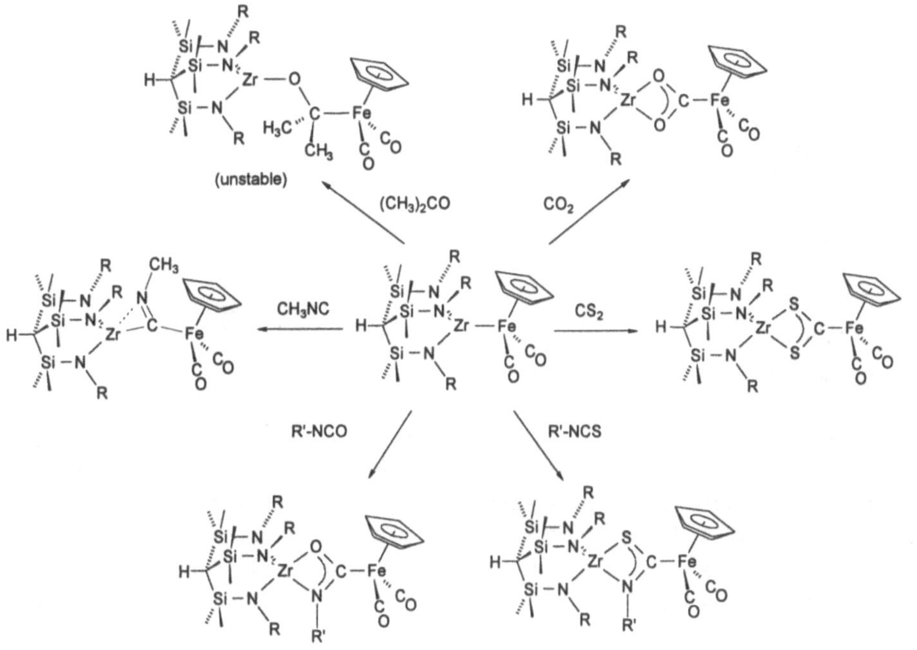

Atom Transfer Reactions

Insertion steps as discussed above appear to be involved in a number of atom transfer reactions taking place within the coordination sphere of the two metal centres. An example of such an atom transfer reaction which occurs extremely rapidly and at temperatures below -50°C is the Cannizzaro-type disproportionation of aryl aldehydes to yield zirconium alkoxides and acyliron or -ruthenium complexes [8].

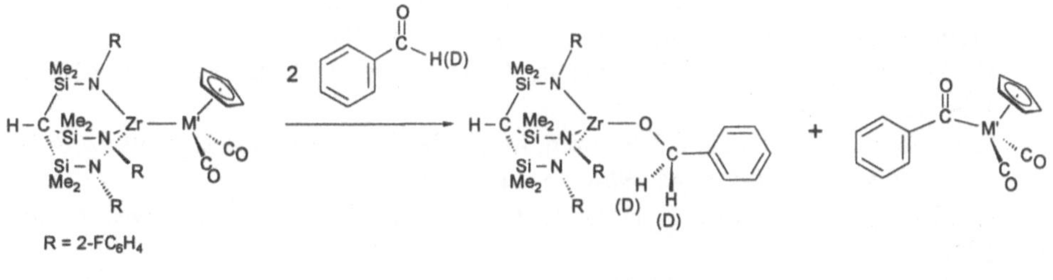

Another example of atom transfer is the deoxygenation of sulfoxides by oxygen transfer from the sulfur to a carbonyl ligand generating reaction products in which CO_2 acts as a bridge between the metal centres. The thioether which is produced in this conversion remains coordinated to the late transition metal centre. This type of reaction has been observed to occur with all Zr-M-complexes studied to date [9].

233

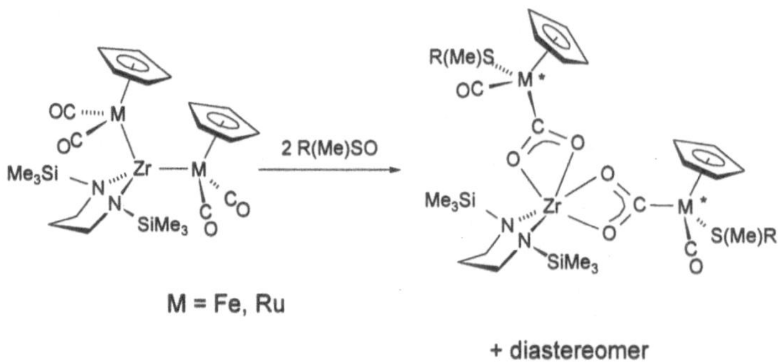

M = Fe, Ru

+ diastereomer

Conclusions

The early-late heterobimetallic complexes discussed in this paper display pronounced cooperativity in their reactions with polar organic substrates acting as pairs of metal electrophiles and nucleophiles. This cooperativity is not only manifested in the selectivity of their interaction with polar organics but in particular by the considerable enhancement of the reaction rates. In view of these first results they promise great potential as "molecular tools" in novel types of organic transformations.

References

[1]   D. W. Stephan, *Coord. Chem. Rev.* **1989**, *95*, 41.

[2]   (a) G. S. Ferguson, P. T. Wolczanski, L. Parkanyi, M. C. Zonneville, *Organometallics* **1988**, *7*, 1967. (b) C. P. Casey, *J. Organomet. Chem.* **1990**, *400*, 205. (c) F. Ozawa, J. W. Park, P. B. Mackenzie, W. P. Schaefer, L. M. Henling, R. H. Grubbs, *J. Am. Chem. Soc.* **1989**, *111*, 1319. (d) A. M. Baranger, R. G. Bergman, *J. Am. Chem. Soc.* **1994**, *116*, 3822.

[3]   (a) S. Friedrich, H. Memmler, L. H. Gade, W.-S. Li, M. McPartlin, *Angew. Chem.* **1994**, *106*, 705; *Angew. Chem. Int. Ed. Engl.* **1994**, *33*, 676. (b) B. Findeis, M. Schubart, C. Platzek, L. H. Gade, I. J. Scowen, M. McPartlin, *Chem. Commun.*, **1996**, 219. (c) S. Friedrich, H. Memmler, L. H. Gade, W.-S. Li, I. J. Scowen, M. McPartlin, C. E. Housecroft, *Inorg. Chem.* **1996**, *35*, 2433.

[4]   G. Jansen, M. Schubart, B. Findeis, L. H. Gade, I. J. Scowen, M. McPartlin, submitted for publication.

[5]   S. Friedrich, L. H. Gade, I. J. Scowen, M. McPartlin *Angew. Chem.* **1996**, *108*, 1440; *Angew. Chem. Int. Ed. Engl.* **1996**, *35*, 1338.

[6]   H. Memmler, U. Kauper, L. H. Gade, I. J. Scowen, M. McPartlin, *Chem. Commun.* **1996**, 1751.

[7]   A. Schneider, L. H. Gade, M. Breunig, G. Bringmann, I. J. Scowen, M. McPartlin, in preparation.

[8]   A. Schneider, *Diplomarbeit*, Universität Würzburg **1996**.

[9]   (a) H. Memmler, *Dissertation*, Universität Würzburg **1996**. (b) D. J. M. Trösch, *Diplomarbeit*, Universität Würzburg **1996**. (c) S. Fabre, L. H. Gade, unpublished results.

Design of Monoanionic Self Adapting N-Heteroaromatic Substituted Claw Ligands

Matthias Pfeiffer, Thomas Kottke, and Dietmar Stalke*

Institut für Anorganische Chemie der Universität Würzburg, Am Hubland,

D-97074 Würzburg, Germany

Introduction

Aromatic nitrogen heterocycles are prime examples for electronically active ligands widely used in metal organic chemistry [1]. Due to their potential to π-interact with or σ-coordinate to a metal center, they allow a high coordination flexibility. Multidentate ligand systems which are electronically *and* geometrically variable are created when single heteroaromatic rings are linked by suitable bridging groups via single bonds. The poly(pyrazolyl)borates represent well-known examples for such heteroaromatic substituted chelate ligands [2]. Monoanionic ligand systems implicitly emulating this structural motif are designed by bridging heteroaromatic rings with formal negatively charged p-block elements [3]. Bidentate ligands result when group 15 (a) or monosubstituted group 14 (b) elements (E) are used to bridge two heteroaromatic rings. Bridging of three heteroaryl groups by unsubstituted group 14 elements creates monoanionic tridentate ligands (c). Rotation of the heteroaromatic substituents about the central bonds allows to vary the grip of the ligand periphery. In addition, the heterocycles operate as charge spacers between the negatively charged bridging position and the metal cation. This provides the metal complexes with the potential to react as both a nucleophilic and an electrophilic reagent.

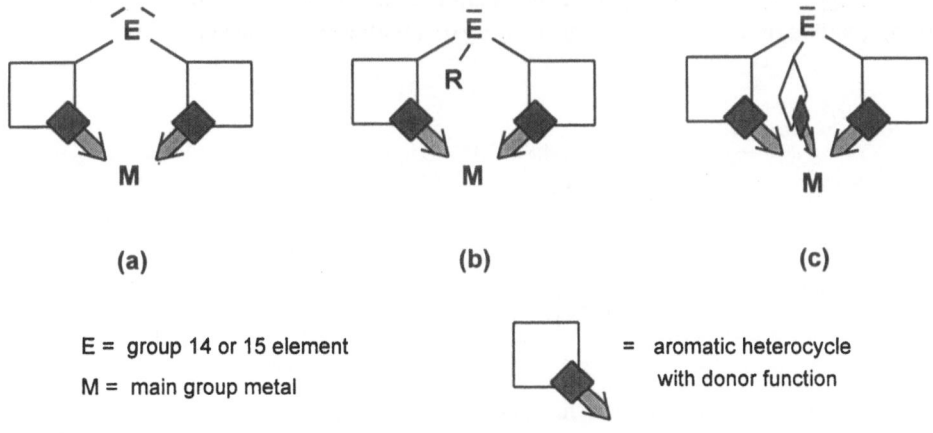

(a) (b) (c)

E = group 14 or 15 element
M = main group metal

= aromatic heterocycle with donor function

235

Monoanionic Chelate Ligands

Di(pyridyl)methan Py_2CH_2 has been already introduced as a neutral bidentate ligand in a variety of transition metal complexes [4]. The corresponding monoanionic ligand is accessible via deprotonation at the central carbon atom [5]. Metal complexes are formed only with small cations like Li^+. The lack of geometric flexibility of the di(pyridyl)methyl ligand is due to the complete conjugation of the anion and a partial double bond character of the bridging bonds. This allows stabilization of appropriately sized cations in an unusual environment like in the rarely observed lithium lithate (1) and the so far unprecedented sodium lithate (2) [5]. In all of the $Py_2C(H)M$ complexes (M= Li, $AlMe_2$, $GaMe_2$) the monoanionic ligand has to be classified as an amide rather than as a carbanion.

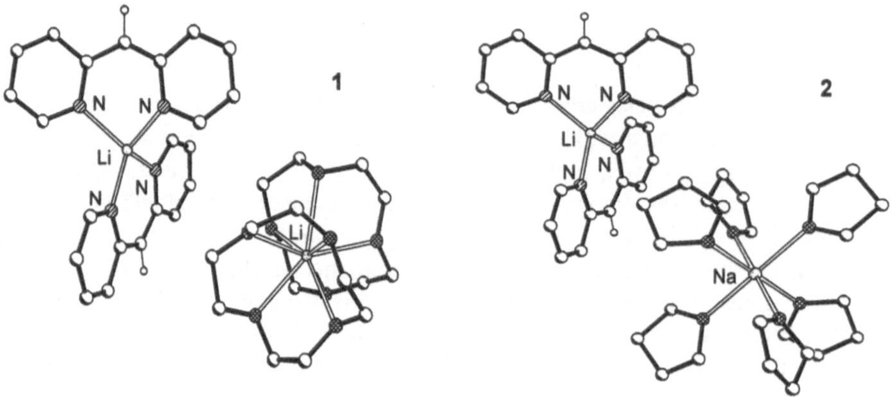

As depicted in the reaction scheme below, a synthetic route different to the preparation of the di(pyridyl)methyl derivatives is applied to introduce heavier group 15 elements in the bridging position [6]. Conversion of trisubstituted Py_3E (E= P, As) with lithium metal in THF yields the lithium precursor $[Py_2ELi \cdot 2THF]$ and dipyridine, which forms upon a ligand coupling reaction of the initial product PyLi with Py_3E. Complexes with group 13 metals are advantageously prepared by transmetallation using the corresponding metal dimethylchlorides, although the direct metallation of Py_2PH with Me_3M is also possible.

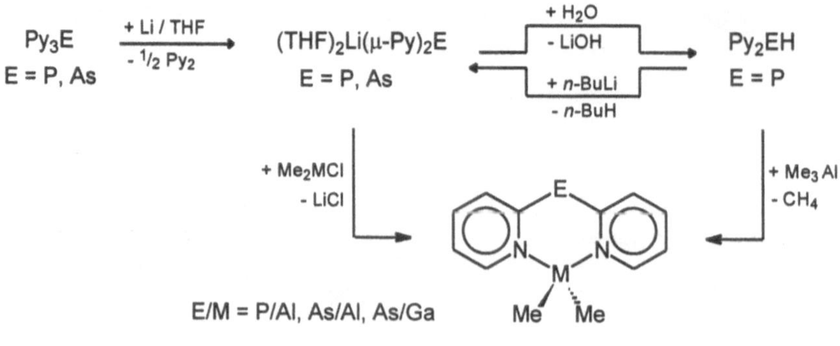

236

The Py_2P^--and the Py_2As^--ligand coordinate the lithium or the di(methyl)aluminum cation similar to the di(pyridyl)methyl ligand. The cation is always σ-coordinated via the two nitrogen atoms of the pyridyl substituents, while the formal negatively charged bridging atom (P or As) does not participate in metal coordination. Noticeable differences occur with respect to ligand planarity. In contrast to the planar lithium complex Py_2PLi, the conformation of Py_2PAlMe_2 and $Py_2AsAlMe_2$ increasingly deviates from planarity as a consequence of the decreasing conjugation of the anion.

An entirely different situation is crucial in complexes employing the monoanionic di(pyridyl)ammino-ligand Py_2N^-, which is accessible via deprotonation of the corresponding amine with suitable bases (e.g. $LiAlH_4$, $Sn[N(SiMe_3)_2]_2$, Me_2Zn, Et_3Al) [7]. Here, the heteroaromatic rings can rotate about the central bonds allowing the bridging nitrogen atom to be involved in coordinaton of the metal center in addition to or in competition with the pyridyl nitrogen atoms. Three distinct conformations of the Py_2N^--ligand arise characterized by the arrangement of the nitrogen atoms with respect to each other. While the all-cis conformation of the ligand coincides with a low degree of aggregation, the all-trans arrangement supports the formation of higher aggregated species. In the cis-trans conformation both, low and high aggregated complexes are observed.

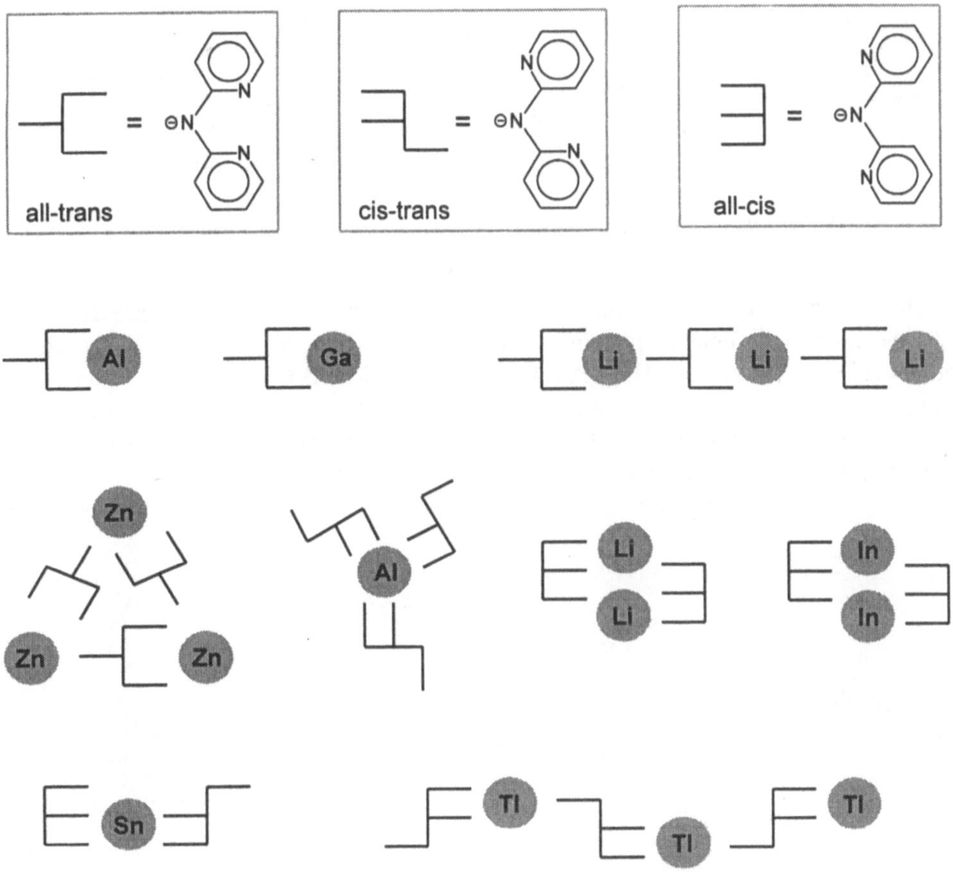

Monoanionic Tridentate Ligands

The formal replacement of the B(H)-unit in poly(pyrazolyl)borates with negatively charged group 14 elements affords pyramidal monoanionic tridentate ligand systems which – like the di(pyridyl)amino ligand – are potentially able to vary their coordination behaviour via substituent rotation about the bridging single bonds. A second active domain of the ligand is created when germanium(II) or tin(II) are introduced in the bridging position. Due to the known amphoteric character of these elements in their compounds [8] this side of the ligand remains active, i.e. attractive for nucleophiles *and* electrophiles, even after complexation of a metal cation by the ligand periphery.

A variety of coordination modes results when group 1, 2, and 14 metals are complexed with this ligand system [9]. A tridentate (a) or a monodentate (b) coordination fashion is the consequence of an exclusice N(σ) donation of the pyrazolyl ring nitrogen atoms to the metal center. A combination of the N(σ) donating and the π-interacting potential of the heteroaromatic rings results in a mixed coordination mode (c).

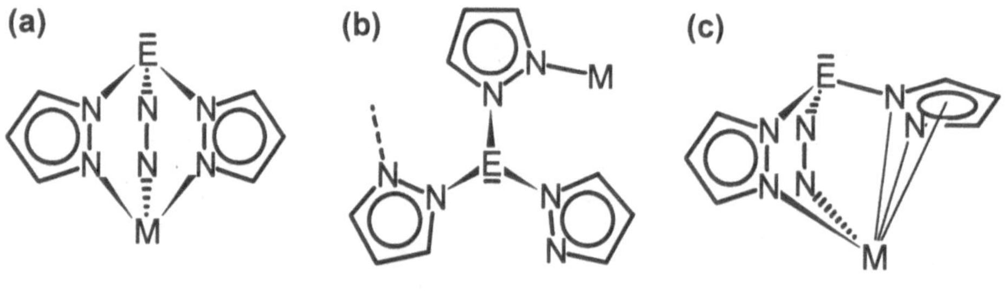

(a)   (b)   (c)

M/E= Na/Ge, Ge/Ge, Sn/Sn   M/E= Na/Sn   M/E= Ba/Ge

The trigonal symmetric coordination of cations in the type (a) complexes corresponds with the metal coordination generally observed in tri(pyrazolyl)borates. In the type (b) sodium stannate [SnPz$_3$Na(THF)$_2$(PzH)]$_2$ (Pz= pyrazol-1-yl) this symmetry is lost and the pyrazolyl substituents are arranged perpendicular to each other such that the nitrogen atoms are facing *outside* the ligand cone. A twelve-membered ring system results composed of alternating sodium and tin atoms which are connected via pyrazolyl rings.

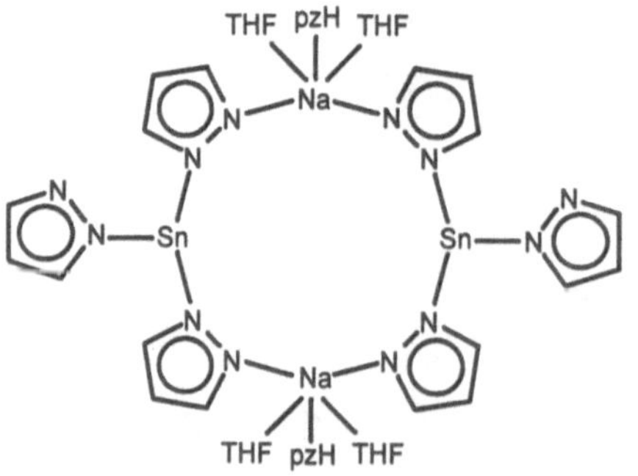

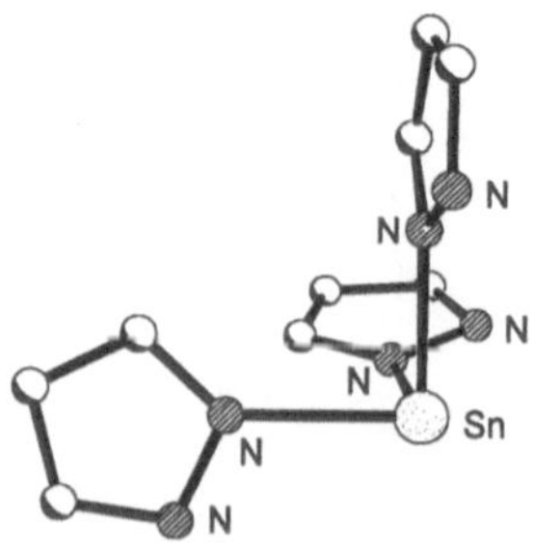

238

The central N-Sn-N' angle in the [Sn(Pz)$_3$]$^-$-unit is quite rigid such that the cavity in the type (**a**) form is too small to accomodate a sodium cation on condition of appropriate Na-N distances. A higher flexibility with respect to the bridging angle is realized in the [Ge(Pz)$_3$]$^-$-unit. This allows the complexation of cations which are by far larger than the sodium cation. Ba[Ge(Pz*)$_3$]$_2$ (Pz*= 3,5-dimethylpyrazol-1-yl) is formed by conversion of Ba(Pz*)$_2$ with germanium dichloride in the molar ratio 3:2. The reaction of Ba(Pz*)$_2$ with GeCl$_2$ or SnCl$_2$ in the molar ratio 1:2 yields the homo-bimetallic type (**a**) complexes [E(Pz*)$_3$E]$^+$ (E= Ge, Sn) with ECl$_3^-$ as counterion.

Results of the oxidation studies

The oxidation reactions were carried out with equimolar amounts of hydroperoxide and substrate (0.2 mmol) in 0.5 ml solvent under air. The data for conversion, product distribution, and mass balance were obtained by ^{1}H-NMR measurements (error ±5%). Naphthalene was used as internal standard. The ee data were obtained by HPLC with a Daicel Chiracel OD column (UV-detector; λ = 254 nm; mobile phase hexane/2-propanol 9:1; 0.8 ml/min). The results are compiled in Tables 1 - 3.

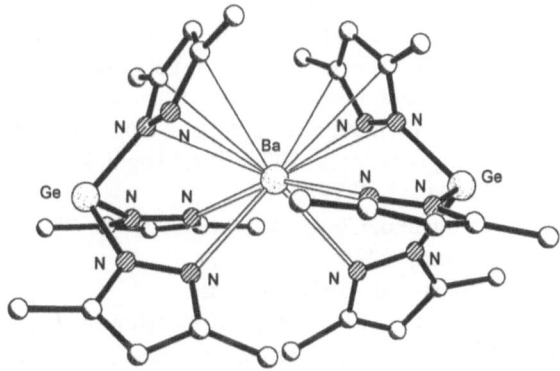

The coordination sphere of barium consists of four σ-donating nitrogen atoms and two π-interacting Pz* substituents, in total resembling the type (**c**) coordination mode. To our knowledge, Ba[Ge(Pz*)$_3$]$_2$ is the first example of a *side-on* coordination by a pyrazolyl ligand toward an alkaline earth metal which demonstrates the versatility of the E(Pz*)$_3^-$ ligand system. In the electronically isovalent poly(pyrazolyl)borate Ba[BH(Pz*)$_3$]$_2$, all six coordinating nitrogen atoms interact N(σ) with the metal center, exclusively, and a S$_6$-symmetry of the complex results [10].

Conclusion

The connection of heteroaryl rings by formal negatively charged bridging groups creates monoanionic ligand systems which are potentially able to adapt to the steric and electronic requirements of the complexed metal center. Within most complexes discussed the heteroaromatic substituents operate as charge spacers between the formal anionic center and the metal cation without encapsulating either site. This allows the complexes to react as nucleophiles via the lone pair at the bridging element or as Lewis acids via the cation. As an application, soft/hard bimetallic reagents with specific characteristics may be designed by introducing d-block metals as complex centers. Multinuclear linear arrays are created when soft main group elements in the bridging position connect two complex units via $p\pi$-$d\pi$ interactions. A different aspect refers to the application in CVD processes. As the ligand systems are composed of volatile but stable substituents the monomeric binuclear complexes may prove to constitute valuable precursors particularly en route to III/V-semiconducting thin films.

Acknowledgement

Funding was kindly provided by the Deutsche Forschungsgemeinschaft and the Stiftung Volkswagenwerk.

References

[1]   for example: a) J. Reedijk, in G. Wilkinson, R. D. Gillard, J. A. McCleverty (Eds.), *Comprehensive Coordination Chemistry, Vol. 2,* Pergamon Press, Oxford, 1987, 73; b) A. R. Katritzky, C. W. Rees, *Comprehensive Heterocyclic Chemistry*, Pergamon Press, Oxford, 1984; c) T. Eichner, S. Hauptmann, *Chemie der Heterocyclen*, Georg Thieme Verlag, Stuttgart, New York, 1994; d) A. Togni, L. M. Venanzi, *Angew. Chem. Int. Ed. Engl.* **1994**, *33*, 497-526.

[2]   for example: a) S. Trofimenko, *Chem. Rev.* **1993**, *93*, 943-980; b) N. Kitajima, W. B. Tolman, *Prog. Inorg. Chem.* **1995**, *43*, 419-531.

[3]   T. Kottke, D. Stalke, *Chem. Ber.* **1997**, in press.

[4]   a) A. J. Canty, G. Hayhurst, N. Chaichit, B. M. Gatehouse, *J. Chem. Soc., Chem. Commun.* **1980**, 316-318; b) A. J. Canty, N. Chaichit, B. M. Gatehouse, E. E. George, G. Hayhurst, *Inorg. Chem.* **1981**, *20*, 2414-2422; c) E. Spodine, J. Manzur, M. T. Garland, M. Kiwi, O. Pena, D. Grandjean, L. Toupet, *J. Chem. Soc., Dalton Trans.* **1991**, 365-369.

[5]   a) H. Gornitzka, D. Stalke, *Angew. Chem. Int. Ed. Engl.* **1994**, *33*, 693-695; b) H. Gornitzka, D. Stalke, *Organometallics* **1994**, *13*, 4398-4405.

[6]   a) A. Steiner, D. Stalke, *J. Chem. Soc., Chem. Commun.* **1993**, 444-446; b) A. Steiner, D. Stalke, *Organometallics* **1995**, *14*, 2422-2429.

[7]   H. Gornitzka, D. Stalke, *J. Am. Chem. Soc.* submitted.

[8]   a) M. Veith, *Angew. Chem. Int. Ed. Engl.* **1987**, *26*, 1-14; b) M. Veith, *Chem. Rev.* **1990**, *90*, 3-16; c) W. P. Neumann, *Chem. Rev.* **1991**, *91*, 311-334; d) J. T. B. H. Jastrzebski, G. van Koten, *Adv. Organomet. Chem.* **1993**, *35*, 241-294.

[9]   a) A. Steiner, D. Stalke, *J. Chem. Soc., Chem. Commun.* **1993**, 1702-1704; b) A. Steiner, D. Stalke, *Inorg. Chem.* **1995**, *34*, 4846-4853.

[10] S. G. Dutremez, D. B. Leslie, W. E. Streib, M. H. Chisholm, K. G. Caulton, *J. Organomet. Chem.* **1993**, *462*, C1-C2.

Enantioselective Organic Syntheses Using Chiral Transition Metal Complexes - The Search for Highly Dissymetric Templates

W. A. Schenk*, N. Burzlaff, M. Hagel

Institut für Anorganische Chemie, Universität Würzburg, Am Hubland, D-97074 Würzburg, Germany

1   New Chiral, Enantiomerically Pure Ruthenium Complexes

Chiral phosphines should preferably be accessible in simple reactions as pure enantiomers. The (S,S)-bis(dibenzophospholyl)butane **A** was seen as a ligand electronically similar to the well-known (S,S)-CHIRAPHOS, but sterically more rigid.

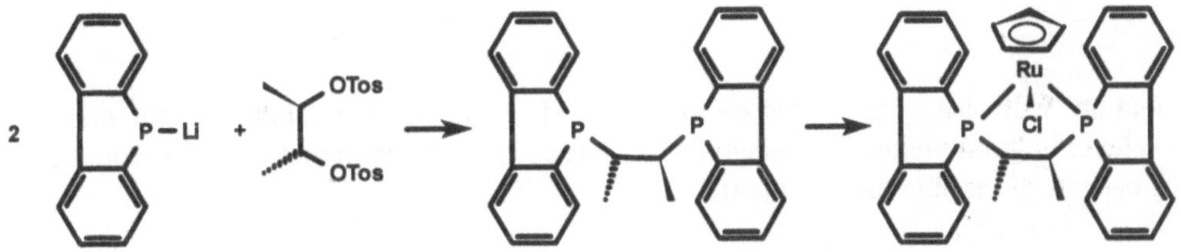

A

A was synthesized from (R,R)-2,3-butanediol and dibenzophospholyllithium and converted to the desired ruthenium complex [1]. A structure determination of the latter revealed that, due to steric congestion, the two dibenzophosphole "wings" are forced into a common plane. Therefore, little asymmetric induction is to be expected from this complex. Nevertheless, **A** should have some potential in asymmetric hydrogenations using square-planar complexes.

The binaphthol-derived phosphinite **B** [2] is sterically as demanding as (S,S)-CHIRAPHOS, but less electron-donating. The ruthenium complex obtained thereof was converted to thiolate, thioether, thioaldehyde, or thionolactone complexes. Addition and oxidation reactions of those complexes proceeded with similar or even better diastereoselectivity than those of the corresponding (S,S)-CHIRAPHOS complexes.

241

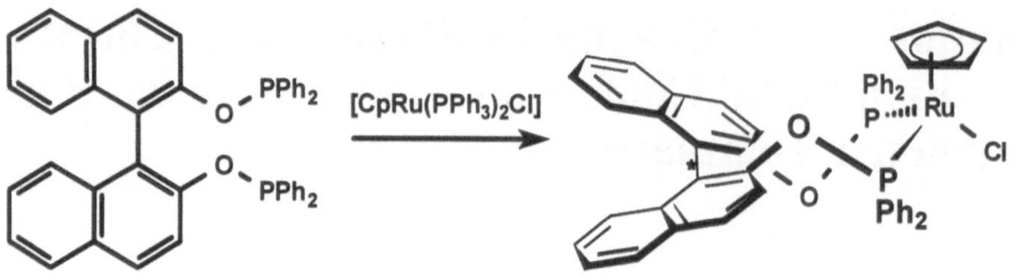

2 Chiral Rhenium Thioaldehyde Complexes

Chiral rhenium thiolate complexes have been synthesized by either acid demethylation or ligand exchange.

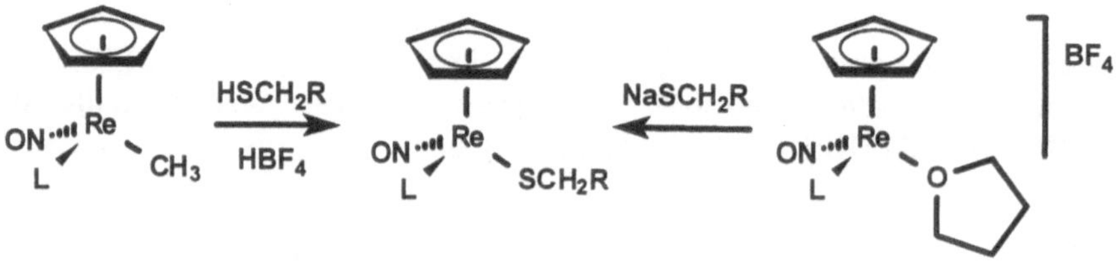

Oxidation with trityl or ferrocenium salts [3] gave the corresponding thioaldehyde complexes in good yields. Thioaldehyde complexes bearing functionalized side groups have thus become accessible for the first time.

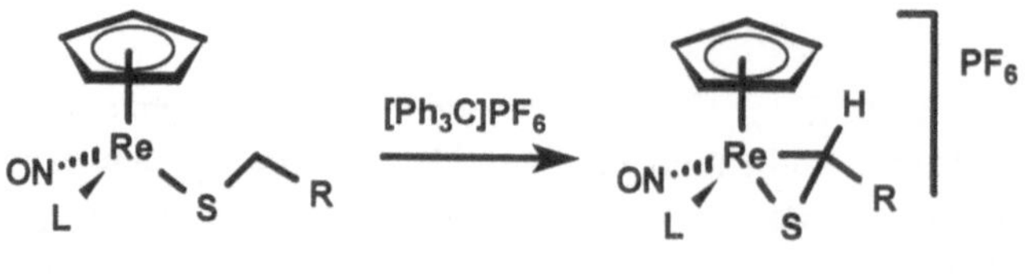

Nucleophilic additions and cycloadditions with these complexes proceed with moderate diastereoselectivity.

242

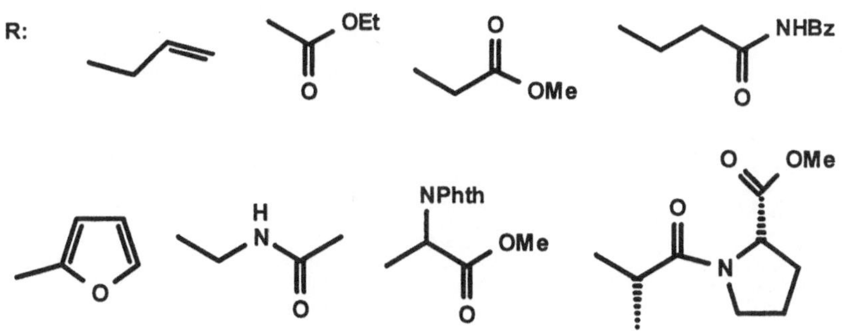
R:
OEt
O
O
OMe
NHBz
O
O
OMe
NPhth
H
N
OMe
O
O
N
O
O

Acknowledgements

This work was supported by the *Deutsche Forschungsgemeinschaft* (SFB 347 "Selektive Reaktionen Metall-aktivierter Moleküle", Project B-3).

References

[1] M. Stubbe, *Dissertation,* Universität Würzburg, **1996**.
[2] R. H. Grubbs, R. A. DeVries, *Tetrahedron Lett.* **1977**, *22*, 1879.
[3] W. A. Schenk, N. Burzlaff, H. Burzlaff, *Z. Naturforsch. B* **1994**, *49*, 1633.

Enantioselective Oxidation of Thioethers Using Ruthenium Complexes as Chiral Auxiliaries

W. A. Schenk*[a)], M. Dürr[a)], B. Steinmetz[a)], W. Adam*[b)], C. R. Saha-Möller[b)]

[a)] Institut für Anorganische Chemie, Universität Würzburg, Am Hubland,
D-97074 Würzburg, Germany
[b)] Institut für Organische Chemie, Universität Würzburg, Am Hubland,
D-97074 Würzburg, Germany

1 Introduction

The halfsandwich complex [CpRu(chir)]$^+$ is a highly efficient chiral auxiliary for the enantioselective oxidation of thioethers [1]. This method has *inter alia* been used to synthesize (*R*)-sulforaphane, a natural product occurring in broccoli [2].

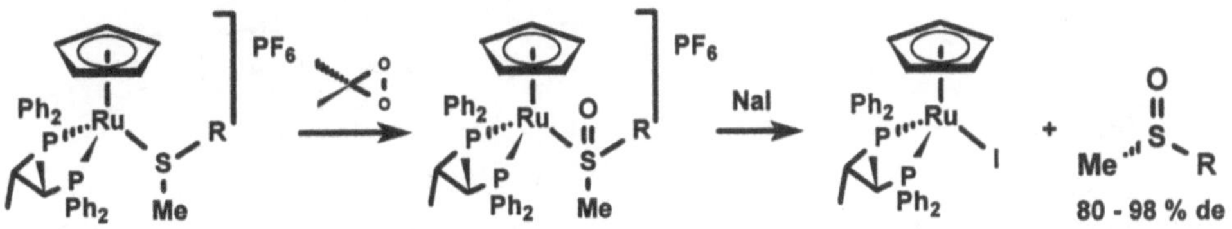

2 Variation of the Oxidant

Besides dimethyldioxirane (DMD), N-tosylphenyloxaziridine and caroate (KHSO$_5$) also could be used. The oxaziridine is less reactive but occasionally gave cleaner products. KHSO$_5$ in some cases gave even higher diastereoselectivities than DMD. This also is proof that in the rate-determining step the oxidant directly attacks the sulfur atom.

R	DMD	KHSO$_5$	oxaziridine
Ph	46	98	8
CH$_2$Ph	98	64	98
i-Pr	86	98	n.r.
Cy	84	98	n.r.

3 Variation of the Complex

Two other kinds of halfsandwich complexes were included in this study. The oxidation of complexes **A** is slower but occasionally more selective (R = Ph, 66 % de with DMD). As expected, the selectivity of oxidation of **B** is much lower (8 - 38 % de) due to the large spatial separation of the chiral neomenthyl and the prochiral thioether groups.

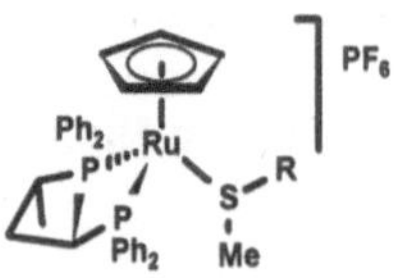

A 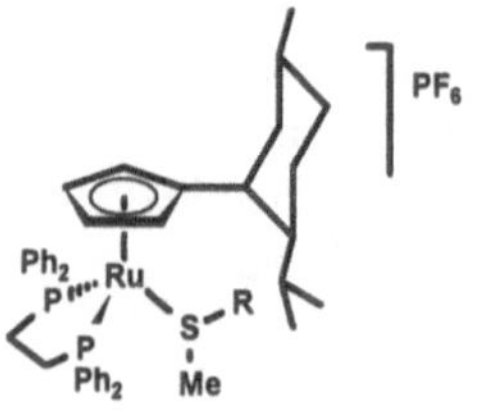   B

4 Oxidation of Remote Functionalities

A variety of allylthioether complexes was synthesized and oxidized with DMD to the corresponding (sulfinylmethyl)epoxides.

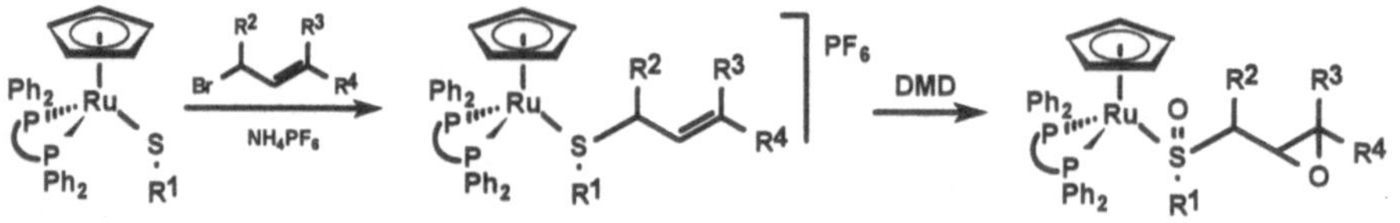

Monitoring the reaction by NMR revealed oxidation of the sulfur atom prior to the C-C double bond. The diastereoselectivity of the second step is only moderate (10 - 58 %) [3]. Similar experiments with the corresponding (*S,S*)-CHIRAPHOS complexes gave mixtures of four diastereoisomers with as yet unsatisfactory selectivity.

Acknowledgements

This work was supported by the *Deutsche Forschungsgemeinschaft* (SFB 347 "Selektive Reaktionen Metall-aktivierter Moleküle", Project B-3).

References

[1]   W. A. Schenk, J. Frisch, W. Adam, F. Prechtl, *Angew. Chem.* **1994**, *106*, 1699; *Angew. Chem. Int. Ed. Engl.* **1994**, *33*, 1609; W. A. Schenk, J. Frisch, M. Dürr, N. Burzlaff, D. Stalke, R. Fleischer, W. Adam, F. Prechtl, A. Smerz, *Inorg. Chem.* **1997**, *36*, 2372.
[2]   W. A. Schenk, M. Dürr, *Chem. Eur. J.* **1997**, *3*, 713.
[3]   W. A. Schenk, B. Steinmetz, M. Hagel, W. Adam, C. R. Saha-Möller, *Z. Naturforsch. B*, submitted.

Stereo- and Enantioselective Reactions of Thioaldehydes, Thioketones, Thioketenes, and Thionolactones Mediated by Ruthenium Complexes

W. A. Schenk*[a], T. Beucke[a], J. Kümmel[a], F. Servatius[a], N. Sonnhalter[a], G. Bringmann*[b], A. Wuzik[b]

[a] Institut für Anorganische Chemie, Universität Würzburg, Am Hubland, D-97074 Würzburg, Germany
[b] Institut für Organische Chemie, Universität Würzburg, Am Hubland, D-97074 Würzburg, Germany

1   Thioaldehyde and Thioketone Complexes

Halfsandwich-type ruthenium complexes of thioaldehydes and thioketones are obtained by acid-catalyzed condensation of Ru-SH complexes with the respective carbonyl compounds. Thioaldehyde complexes may also be generated by a formal hydride abstraction reaction [1].

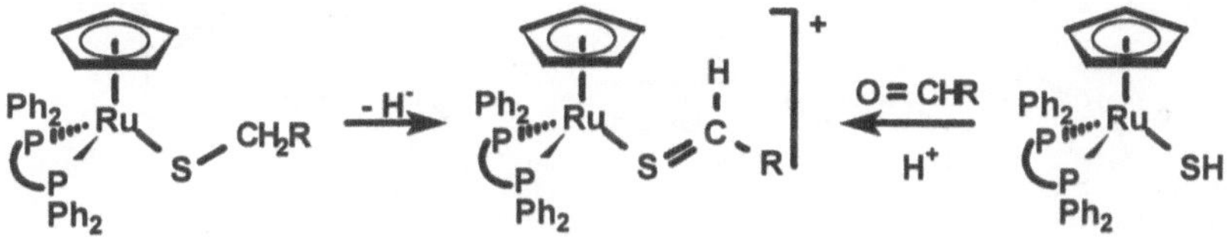

Thioaldehyde complexes undergo [2+4] cycloadditions with a variety of dienes. With chirally modified complexes, moderate to good diastereoselectivities can be achieved.

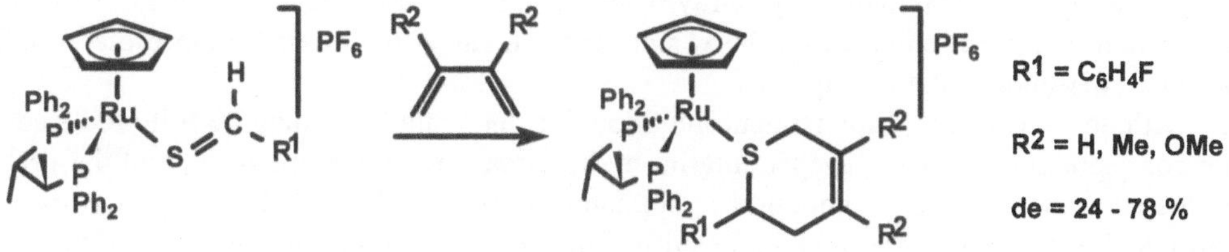

$R^1 = C_6H_4F$

$R^2 = H, Me, OMe$

de = 24 - 78 %

[4+2] cycloadditions of α,β-unsaturated thioaldehyde complexes with dienophiles proceed with high regioselectivity. Kinetic analysis revealed domination of the transition state of this reaction by a HOMO(dienophile)-LUMO(diene) interaction. The organic part is readily released from the metal by ligand substitution with iodide ion.

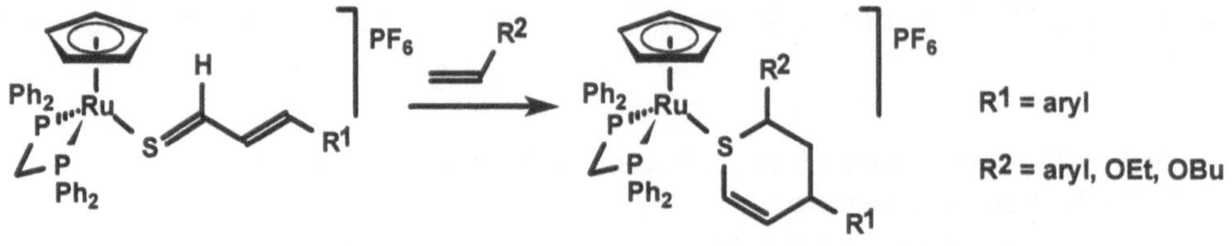
PF6
R2
Ph2
Ru
P
S
H
R1
R1 = aryl
R2 = aryl, OEt, OBu

2 Thioketene Complexes

Complexes of thioketenes are readily accessible from thiocarboxylate complexes by an acylation-elimination route [2]. This method is particularly suited for thioketenes which cannot be isolated in the free state.

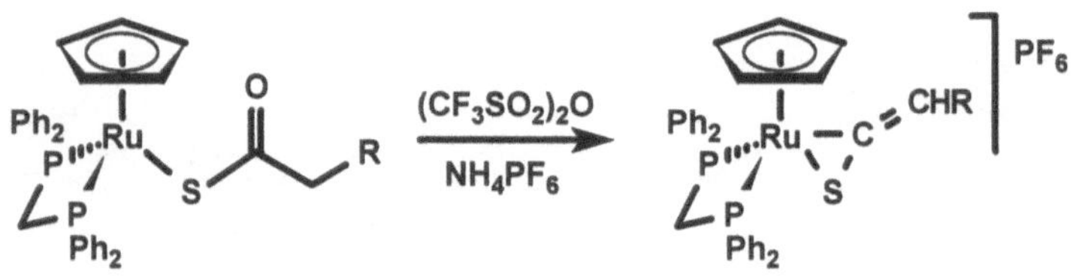
(CF3SO2)2O
NH4PF6
PF6
CHR
Ph2
Ru
P
S
O
R

3 Thionolactone Complexes

Chiral, enantiomerically pure complexes of thionolactones were synthesized from thiophene complexes by ligand exchange [3]. X-ray structure determination (R = Me) revealed that the thiocarbonyl group is shielded on the *si* side. Indeed, single hydride addition preferentially produces (*S*)-thiolactolate complexes.

With an excess of hydride reagent the thionolactone complexes may be fully reduced to the corresponding thiolates. Depending on hydride reagent, solvent polarity, and the group R, the ring-opened biaryl compound is obtained in up to 76 % de [4]. Again, the organic part may be released by methylation at sulfur followed by ligand exchange with iodide.

R	de (%)
H	86
OMe	86
Me	14
Et	16
i-Pr	20
t-Bu	4

Acknowledgements

This work was supported by the *Deutsche Forschungsgemeinschaft* (SFB 347 "Selektive Reaktionen Metall-aktivierter Moleküle", Project B-3).

References

[1]   W. A. Schenk, T. Stur, E. Dombrowski, *Inorg. Chem.* **1992**, *31*, 723; idem, *J. Organomet. Chem.* **1994**, *472*, 257.

[2]   W. A. Schenk, N. Sonnhalter, N. Burzlaff, *Z. Naturforsch. B* **1997**, *52*, 117.

[3]   G. Bringmann, B. Schöner, O. Schupp, W. A. Schenk, I. Reuther, K. Peters, E. M. Peters, H. G. von Schnering, *J. Organomet. Chem.* **1994**, *472*, 257.

[4]   G. Bringmann, M. Breuning, S. Busemann, J. Hinrichs, T. Pabst, R. Stowasser, S. Tasler, A. Wuzik, W. A. Schenk, J. Kümmel, D. Seebach, G. Jaeschke, *this Volume*.

Salen-type Oxo Vanadium Complexes as Catalysts for Sulfoxidation and Epoxidation Reactions with Hydroperoxides

Horst Elias[a, *], Frank Stock[a], Waldemar Adam[b], Catherine Mitchell[b], Margareta Neuburger[c] and Markus Neuburger[c]

[a] Institut für Anorganische Chemie, Technische Universität Darmstadt, Petersenstr. 18, D-64287 Darmstadt, Germany; [b] Institut für Organische Chemie, Universität Würzburg; Germany[c] Institut für Anorganische Chemie, Universität Basel, Switzerland

1 Introduction

V_2O_5 was probably one of the first transition metal compounds applied for catalytic epoxidation reactions with organic hydroperoxides [1]. Oxo vanadium complexes, such as $OV^{IV}(acac)_2$, turned out to be rather poor catalysts for the epoxidation of olefins such as cyclohexene [2]. It was found, however, that vanadium and oxo vanadium complexes catalyze the epoxidation of allylic alcohols remarkably well [3,4].

Oxo vanadium salen complexes, reported to be poor catalysts for the oxidation and epoxidation of olefins [5,6], have been prepared enantiomerically pure and applied as catalysts for the asymmetric sulfoxidation of thioethers with organic hydroperoxides [7,8].

In contrast to the extensive literature on transition metal *salen* complexes, rather little is known about the corresponding *tetrahydrosalen* complexes. The latter are easily accessible and, in contrast to Schiff base complexes, stable towards hydrolysis of the C=N bond. Pecoraro et al. [9] prepared OV^{3+} and O_2V^+ tetrahydrosalen-type complexes but, to our knowledge, the catalytic potential of such complexes has not been explored so far.

We report on the synthesis and characterization of an oxo vanadium(V) and an oxo vanadium(IV) tetrahydrosalen-type complex, prepared from enantiomerically pure 1,2-diaminocyclohexane and substituted salicylaldehydes. In these complexes the tetradentate N_2O_2-type ligand was N-methylated to avoid oxidative dehydrogenation (i.e., C=N bond formation [10]) in the presence of peroxides. The two oxo vanadium complexes were used as catalysts for various sulfoxidation and epoxidation reactions and the results are reported.

2 Preparation and Properties of the Vanadium Catalysts

2.1 Ligands

The synthesis of the hydrogenated ligands is described elsewhere [11]. The preparation of the corresponding N,N′-dimethylated derivatives H_2L^1 and H_2L^2 (see Fig. 1), as accomplished by a modified procedure published by Borch and Durst, was reported recently [12].

2.2 Complexes

A solution of 1 mmol $OV^{IV}(acac)_2$ in dry EtOH was added to a stirred and heated solution of 1 mmol H_2L^1 or H_2L^2 in dry EtOH. The mixture was refluxed for 12 h. After cooling, the product $OV^{IV}L^2$ was separated by filtration. Yield: 73 %; mp.: dec; μ_{eff} (295 K): 1.70 BM; MS(FD): $m/z(\%)$ = 620 [M^+ - H] (100); Vis(CH_2Cl_2): λ_{max} 572 nm (ε = 320 l M^{-1} cm^{-1}). Correct elemental analysis.

In the case of the system $OV^{IV}(acac)_2/H_2L^1$ the black-violet ethanol solution did not yield a crystalline product upon cooling. The solution was therefore taken to dryness and the residue dissolved in dry MeOH. Upon boiling for about 2 h, the complex $OV^VL^1(OMe)$ precipitated. Yield: 54 %; mp 190 °C; MS(FD): $m/z(\%)$ = 590 [M^+] (100); Vis(CH_2Cl_2): λ_{max} 554 nm (ε = 6600 l M^{-1} cm^{-1}). Characterized by ^{1}H NMR (200 MHz, $CDCl_3$). Correct elemental analysis.

2.3 X-ray structure of $OV^VL^1(OMe)$

Suitable crystals for an X-ray structure determination were grown from a solution of $OV^VL^1(OMe)$ in $CH_2Cl_2/MeOH$. The details of the structure analysis are to be published [13]. Figure 2 gives a view of the coordination geometry.

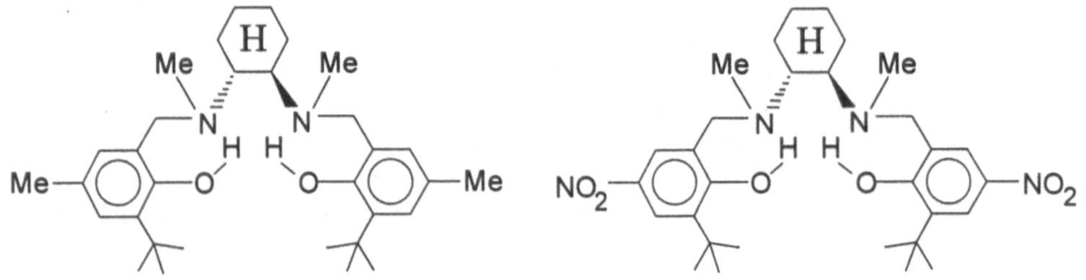

Figure 1 Structure formulae of the ligands H_2L^1 and H_2L^2.

3 Results of the oxidation studies

The oxidation reactions were carried out with equimolar amounts of hydroperoxide and substrate (0.2 mmol) in 0.5 ml solvent under air. The data for conversion, product distribution, and mass balance were obtained by ^{1}H-NMR measurements (error ±5%). Naphthalene was used as internal standard. The ee data were obtained by HPLC with a Daicel Chiracel OD column (UV-detector; λ = 254 nm; mobile phase hexane/2-propanol 9:1; 0.8 ml/min). The results are compiled in Tables 1-3.

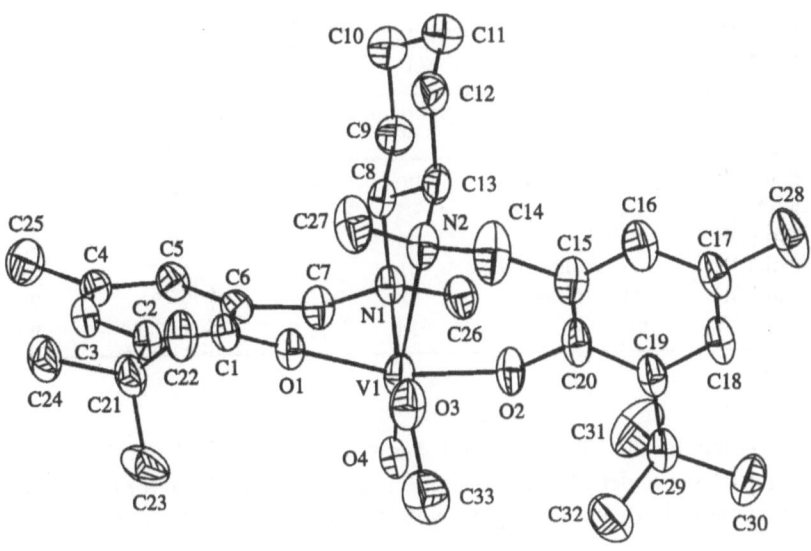

Figure 2 View of the coordination geometry of the complex $OV^VL^1(OMe)$.

Table 1. Sulfoxidation of thioethers by hydroperoxides in d-chloroform at ambient temperature in the presence of 5 mol percent of $OV^VL^1(OMe)$ (t = 10-18 h).

sulfide	oxidant	conversion of sulfide (%)	product distribution (%)		ee of R-sulfoxide (%)
			sulfoxide	sulfone	
thioanisole	tBuOOH	87	95	5	1
thioanisole	H_2O_2	$\geq$95	$\geq$95	$\leq$5	15
4-NO$_2$-thioanisole	H_2O_2	$\geq$95	90	10	6*
thioanisole	H_2O_2	73	96	4	1**

* Analyzed by HPLC using a Daicel Chiracel OB-H column; λ = 220 nm; mobile phase hexane/2-propanol 1:1; 0.5 mL/min; ** Catalyst: $OV^{IV}L^2$.

Table 2. Sulfoxidation of thioanisole by H_2O_2 at ambient temperature in the presence of $OV^VL^1(OMe)$ (t = 10-27 h).

mol% catalyst	solvent	conversion of sulfide (%)	product distribution (%)		ee of R-sulfoxide (%)
			sulfoxide	sulfone	
0.5	CDCl$_3$	24	$\geq$95	$\leq$5	3
0.5	CD$_3$CN	88	$\geq$95	$\leq$5	0
5	CDCl$_3$	$\geq$95	$\geq$95	$\leq$5	15
20	CDCl$_3$	95	$\geq$95	$\leq$5	19

Table 3. Epoxidation of cyclic olefins by hydroperoxides in d-chloroform at ambient temperature in the presence of 5 mol percent of vanadium catalyst (t = 18-24 h).

olefin	oxidant	catalyst	conversion of olefin (%)	yield of epoxide (%)	mass balance (%)
c-octene	tBuOOH	$OVL^1(OMe)$	20	12	93
c-octene	H_2O_2	$OVL^1(OMe)$	≤ 5	-	93
c-octene	tBuOOH	OVL^2	25	17	92
c-octene	tBuOOH	OVL^2 (40°C)	61	59	≥ 95
c-hexene	tBuOOH	$OVL^1(OMe)$	38	9	73
c-hexene	tBuOOH	OVL^2	28	7	83

The epoxidation of the allylic alcohol 4-methylpent-3-en-2-ol by $OV^VL^1(OMe)$ and $OV^{IV}L^2$ was performed under the same conditions as described above for olefin epoxidation and gave the epoxide in quantitative yield with the threo isomer being the favored diastereomer.

4 Discussion

Under aerobic conditions, the reaction of $OV^{IV}(acac)_2$ with H_2L^2 yields the oxo vanadium(IV) complex $OV^{IV}L^2$, whereas in the case of H_2L^1 the oxo vanadium(V) complex $OV^VL^1(OMe)$ is obtained. This difference in the stability of the V(IV) state reflects the electronic effect of the nitro group (H_2L^2) and methyl group (H_2L^1), respectively. It follows from the X-ray structure analysis that the metal center in $OV^VL^1(OMe)$ is coordinated in a distorted octahedral fashion with the tetradentate N_2O_2-ligand being folded. The oxygens of the oxo group and of the methoxide ligand are cis-orientated and disordered.

In the presence of TBHP, the V(V) complex $OV^VL^1(OMe)$, dissolved in chloroform, decomposes slowly at ambient temperature. When TBHP is added to the suspension of $OV^{IV}L^2$ in chloroform, the complex dissolves rapidly and a dark brown solution is obtained. It is important to note that the solution is paramagnetic for hours. This might mean that $OV^{IV}L^2$ forms a relatively stable adduct with TBHP, which is slowly oxidized to V(V).

Both $OV^VL^1(OMe)$ and $OV^{IV}L^2$ catalyze the sulfoxidation of thioethers (such as thioanisole) with TBHP and H_2O_2. The asymmetric induction obtained is better for $OV^VL^1(OMe)$ than for $OV^{IV}L^2$ and better for H_2O_2 than for TBHP. In both cases, however, the ee is lower than 20 % and depends on the ratio catalyst/substrate. Details of the results are summarized in Tab. 1 and Tab. 2.

The epoxidation of cyclohexene and cyclooctene with TBHP or H_2O_2 is again catalyzed by both complexes (see Tab. 3). The main results are that (i) at ambient temperature the conversion of olefin and yield of epoxide is low and higher for TBHP than for H_2O_2, (ii) at 40°C, $OV^{IV}L^2$ is a reasonably good catalyst for epoxidation, whereas $OV^VL^1(OMe)$ decomposes at 40°C, and (iii) the catalytic activity of both complexes for the epoxidation of allylic alcohols by TBHP is very good and comparable to that of $OV^{IV}(acac)_2$ [14,15].

Acknowledgment

The authors thank the *Deutsche Forschungsgemeinschaft* (Schwerpunktprogramm „Peroxidchemie: Mechanistische und präparative Aspekte des Sauerstofftransfers"), the *Verband der Chemischen Industrie e. V.* and the *Otto-Röhm-Stiftung*.

References

[1] E. G. E. Hawkins, *J. Chem. Soc.* **1950**, 2169.

[2] G. L. Linden, M. F. Farona, *J. Catal.* **1977**, *48*, 284.

[3] M. N. Sheng, J. G. Zajacek, *J. Org. Chem.* **1970**, *35*, 1839.

[4] K. B. Sharpless, T. R. Verhoeven, *Aldrichim. Acta* **1979**, *12*, 63.

[5] H. Mimoun, M. Mignard, P. Brechot, L. Saussine, *J. Am. Chem. Soc.* **1986**, *108*, 3711.

[6] D. D. Agarwal, R. Rastogi, P. K. Sangha, *Indian. J. Chem.* **1995**, *34B*, 254.

[7] K. Nakajima, M. Kojima, J. Fujita, *Chem. Lett.* **1986**, 1483.

[8] K. Nakajima, K. Kojima, M. Kojima, J. Fujita, *Bull. Chem. Soc. Jpn.* **1990**, *63*, 2620.

[9] G. J. Colpas, B. J. Hamstra, J. W. Kampf, V. L Pecoraro, *Inorg. Chem.* **1994**, *33*, 4669.

[10] A. Böttcher, H. Elias, E.-G. Jäger, H. Langfelderova, M. Mazur, L. Müller, H. Paulus, P. Pelikan, M. Rudolph, M. Valko, *Inorg. Chem.* **1993**, *32*, 4131.

[11] A. Böttcher, H. Elias, J. Glerup, M. Neuburger, C. E. Olsen, H. Paulus, J. Springborg, M. Zehnder, *Acta Chem. Scand.*, **1994**, *A 48*, 967.

[12] H. Elias, F. Stock, C. Röhr, *Acta Cryst.* **1997**, *C 53*, 862.

[13] F. Stock, *Dissertation*, Technische Universität Darmstadt, in preparation.

[14] B. E. Rossiter, T. R. Verhoeven, K. B. Sharpless, *Tetrahedron Lett.* **1979**, 4733.

[15] W. Adam, B. Nestler, *Tetrahedron Lett.* **1993**, *34*, 611.

Metal Mediated Synthesis of Chiral Secondary Phosphane Ligands via the Organophosphenium Complexes Cp(OC)(L)M=P(H)R (M = Mo, W; L = OC, Me₃P; R = t-Bu, Mes)[1]. Regiospecific Functionalization of Iron and Ruthenium Substituted Disilanes[2]

Wolfgang Malisch*, Klaus Grün, Heinrich Jehle, Joachim Reising, Stephan Möller, Oliver Fey and Christa Abdelbaky

Institut für Anorganische Chemie der Universität Würzburg, Am Hubland, D-97074 Würzburg, Germany

1 Introduction

The transition metal mediated synthesis of organic compounds transferring heteroatoms from an organometallic reagent to organic substrates represents a field of increasing interest[3,4]. In this context metal phosphorus compounds found only occasional application[5] due to the limited access to complexes containing reactive organophosphorus units. A promissing approach is offered by diorganophosphenium complexes $Cp(OC)_2M=PR_2$ (M = Mo, W; R = alkyl, aryl) for which in a series of experiments the coupling activity of the M=P-unit towards various unsaturated organic compounds has been demonstrated[6]. Moreover, the cycloadducts are produced in general with extraordinary high regio- and diastereoselectivity[7].

Analogous reactions with the PH-functionalized phosphenium complexes $Cp(OC)_2M=P(H)R$ [M = Mo (1a), W (1b)][6] are considered to be even more attractive, since additional coupling or hydrogen transfer reactions involving the PH-function can be expected. However, the short lifetime of the organophosphenium complexes creates severe experimental problems. These are circumvented by the intermediate generation from the precursor complexes $Cp(OC)_2[R(H)_2P]M$-Cl (M = Mo, W; R = alkyl, aryl) in the presence of the organic substrate. The same approach is necessary to get the novel phosphenium complex $Cp(OC)(Me_3P)Mo=P(H)Mes$ (1c) additionally characterized by a chiral metal centre, involved in controlled coupling processes.

This communication reports about the interaction of 1a,b with diverse dienes as well as the regio- and stereoselectivity of the [2+4] cycloadduct formation. In the second part, first coupling reactions of the chiral phosphenium complex 1c with heteroallenes are described.

2 [2+4] Cycloaddition Reactions of the PH-functionalized Phosphenium Complexes $Cp(OC)_2M$-P(H)R (M = Mo, W; R = t-Bu, Mes) with 1,3-Dienes: A New Route to Alkenylphosphane Ligands

The Diels-Alder reaction is a well-established procedure in organic chemistry for the formation of cyclohexene derivatives from alkenes and 1,3-dienes[8]. Several groups have performed Diels-Alder reactions of silicon- or phosphorus-containing double bonded systems, for example phospha-alkenes[9], thiophosphoranes[10], disilylenes[11] and silenes[12]. In the last case a concerted Diels-Alder mechanism is ruled out due to primary interaction of the sp^2-hybridized silicon with the π-bond of one C=C unit.

The transfer of Diels-Alder reactions to multiple bonded transition metal main group element systems M=E is limited to primary carbene complexes of the type $(OC)_5M$=C(H)Ph (M = Cr, W). However, the M=C unit reacts with 1,3-dienes exclusively via [2+1] cycloaddition yielding metal-coordinated vinylcyclopropanes[13].

In contrast, the analogous reaction of the diorganophosphenium complexes $Cp(OC)_2M$=PR_2 (M = Cr, Mo, W; R = alkyl, aryl) affords access to the phosphametallacyclohexenes $Cp(OC)_2\overline{W\text{-}PR_2\text{-}CH_2\text{-}CH=CH\text{-}CH_2}$ in excellent yields. Figure 1 shows the structure of the o-tolyl derivative (R= o-Tol)[14].

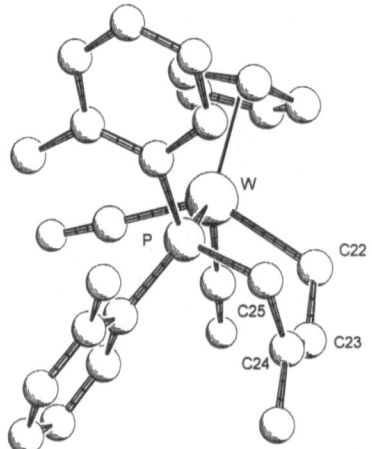

Figure 1 Crystal structure of

$Cp(OC)_2\overline{W\text{-}P(o\text{-}Tol)_2CH_2C(Me)=C(H)CH_2}$
Selected bond lengths [Å] and angles [°]: W-P 2.469, W-C22 2.28, P1-C25 1.87, C23-C24 1.31; P-W-C22 77.2, W-P-C25 113.9, C22-C23-C24 123.2, C23-C24-C25 119.2, C25-C24-C26 117.2, P-C25-C24 106.0.

In the case of the mesityl- and tert-butylphosphenium complexes $Cp(OC)_2W$=P(H)R [R = t-Bu (**1a**), Mes (**1b**)] the regioselectivity of the [2+4] cycloaddition is studied using 2-methyl-1,3-butadiene and 2-trimethylsiloxy-1,3-butadiene. The phosphametallacycles **2** and **3** are obtained nearly quantitatively as a mixture of both regioisomers (**2a** : **2b** = **3a** : **3b** = 80 : 20) (eq. 1).

NOE-NMR experiments concerning the dominating isomer **a** prove the Cp-ligand in a *syn* position to the phosphorus bound mesityl group. While the mesityl species **2a,b** are formed diastereomerically pure, the tert.-butyl substituted derivatives **3a,b** were isolated as a mixture of diastereoisomers (**3a** : **3a′** = 80 : 20; **3b** : **3b′** = 89 : 11).

258

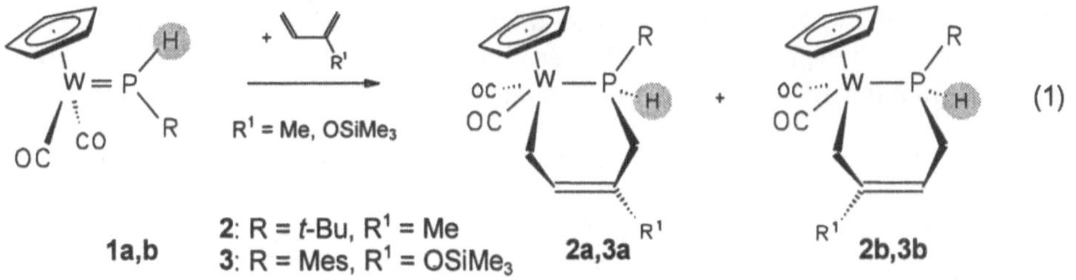
R'
R^1 = Me, OSiMe$_3$
R
H
R
H
R
H
(1)
1a,b
2: R = *t*-Bu, R^1 = Me
3: R = Mes, R^1 = OSiMe$_3$
2a,3a
R^1
R^1
2b,3b

The use of α-subsituted 1,3-dienes leads to [2+4] cycloadducts with a stereogenic carbon centre. Treatment of **1b** with *trans*-1,3-pentadiene, 2-methyl-1,3-pentadiene or the ethyl ester of sorbic acid at room temperature yields regiospecifically the phosphametallacycles **4a-c** with the methyl group next to the phosphorus. Moreover, the cycloadducts **4a-c** were generated diastereospecifically.

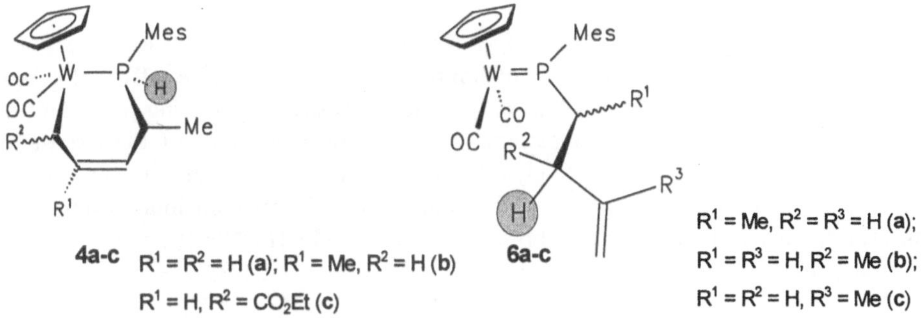
Mes
H
Me
R^2
R^1
4a-c R^1 = R^2 = H (a); R^1 = Me, R^2 = H (b)
R^1 = H, R^2 = CO$_2$Et (c)
Mes
R^1
R^2
R^3
H
6a-c
R^1 = Me, R^2 = R^3 = H (a);
R^1 = R^3 = H, R^2 = Me (b);
R^1 = R^2 = H, R^3 = Me (c)

With the exception of **4c** the six-membered phosphametallacycles are characterized by the high mobility of the P-bound hydrogen. As a result **2,3** and **4a,b** rearrange in toluene at room temperature via formal P→C-hydrogen transfer to the alkenylphosphenium complexes **6**. Tautomerization to **6** is complete within 30 min (**4a**) to 21 d (**3b**). This method offers a novel stereocontrolled access to special alkenylphosphorus fragments.

6a-d shows the typical behaviour of phosphenium complexes, e. g. [2+1] cycloaddition with sulfur to give the three-membered phosphametallacycles **7a-d** (eq. 2).

The same reaction can be conducted with **4c** showing no direct isomerization at room temperature to yield Cp(OC)$_2$W-S-P(Mes)[C(H)Me-CH$_2$-CH=C(H)CO$_2$Et] (**7e**). This result indicates that intermediate formation of the corresponding phosphenium complex Cp(OC)$_2$W=P(Mes)[C(H)Me-CH$_2$-CH=C(H)CO$_2$Et] can be forced by subsequent cycloaddition.

Addition of HCl to the W=P bond of **6a-d** leads to the secondary phosphane complexes **8a-d** (eq. 2), which represent promising candidates for the release of the novel alkenyl phosphanes from the metal fragment via ligand exchange.

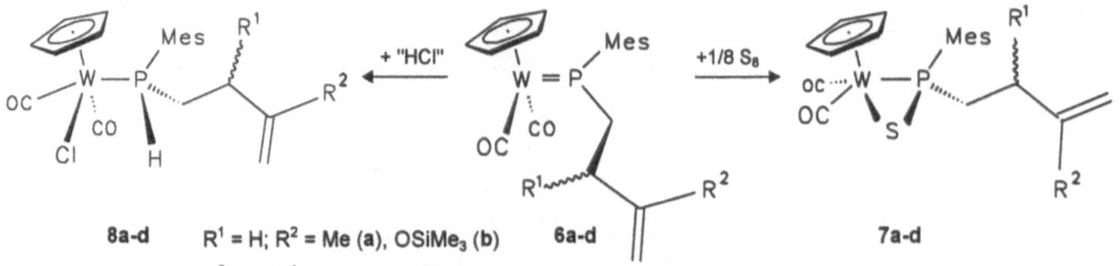

8a-d   R^1 = H; R^2 = Me (a), OSiMe$_3$ (b)   **6a-d**   **7a-d**
            R^2 = H; R^1 = Me (c), OSiMe$_3$ (d)

3   Functionalized Chiral Phosphane Ligands via Stereoselective Coupling of the Phosphenium Complex Cp(OC)(Me$_3$P)Mo=P(H)Mes with Heteroallenes

The phosphenium complex Cp(OC)(Me$_3$P)Mo=P(H)Mes (**1c**) derived from the corresponding mesitylphosphane complex **9** is of limited existence. Therefore, efficient coupling of **1c** with the organoisocyanates **10a,b** is only accomplished by treating a mixture of both reactants with Et$_3$N. The P-H-insertion products **11a,b** are formed diastereoselectively (de 76 %) (eq. 3). Most likely **11a,b** originate from the [2+2] cycloadduct Cp(OC)(Me$_3$P)Mo-P(H)(Mes)-C(=O)NR, formed from the phosphenium complex **1c** and **10**, via ring cleavage with hydrogenbromide, transferred by [Et$_3$NH]Br.

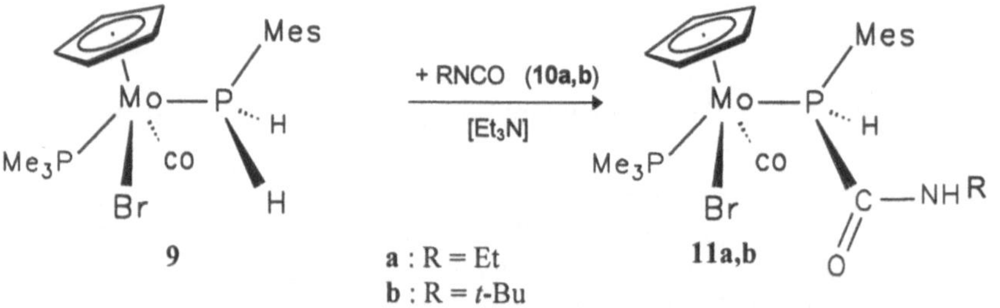

9   a : R = Et   **11a,b**
                         b : R = t-Bu

With DBU **11b** can be converted to the functionalized phosphenium complex **12** bearing a stereogenic metal- and a prostereogenic phosphorus center. Reaction of **12** with phenylisothiocyanate leads regio-, chemo- and stereospecifically to the phosphametallacycle

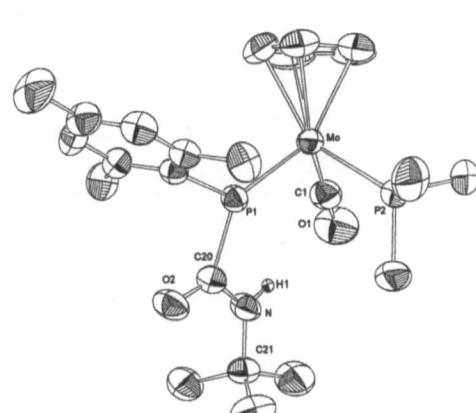

$$12 \xrightarrow{\text{+ PhNCS}} 13 \qquad (4)$$

12   **13**

12 is determined by X-ray crystal structure analysis (Figure 2). The most important findings are the planarity of the phosphorus, the *syn* position of the mesityl group to the Cp ligand and the eclipsed conformation of the phosphenium group plane to the carbonyl ligand.

Figure 2 Crystal structure ofCp(OC)(Me₃P)Mo=P(Mes)[C(O)NH*t*Bu] (**12**)

Selected bond lengths, bond and torsion angles: Mo-P1 2.2528(1), Mo-P2 2.4177(1), O2-C20 1.217(3), N-C20 1.337(3); C20-P1-Mo 131.16(9); Mo-P1-C20-N -2.3(3), Cp(Z)-Mo-P1-C10 15.44.

4. Regiospecific Oxygenation and Chlorination of the Iron and Ruthenium Substituted Disilanes C₅R₅(OC)₂M-Si₂X₅ (M = Fe, Ru; X = H, Cl; R = H, Me)

Investigations concerning metallo-disilanes L$_n$M-Si₂X₅ (X = H, Cl) bearing exclusively hydrogen or chlorine substituted silicon are focused to the "transition metal effect" on the chemical properties and reactivity of the silicon atoms in the α- and β-position. After preliminary studies concerning derivatives with C₅R₅(OC)₂(Me₃P)M-fragments (M = Mo, W; R = H, Me)[15], this series was extended to iron and ruthenium substituted disilanes C₅R₅(OC)₂M-Si₂X₅ (M = Fe, Ru; R = H, Me).

The metallo-pentachlorodisilanes C₅R₅(OC)₂M-Si₂Cl₅ [R = H, M = Fe (**14a**), Ru (**14b**); R = Me, M = Fe (**14c**), Ru (**14d**)], attractive precursors for transformations, are obtained via heterogenous metalation of Si₂Cl₆ with the corresponding alkali-metalates Na[M(CO)₂C₅R₅]

261

(**15a-d**). Subsequent Cl/H-exchange with LiAlH$_4$ yields the metallo-pentahydridodisilanes (**16a-d**)[16]. The IR-, ^{1}H- and ^{29}Si-nmr spectra reveal two extremely different kinds of Si-H units, characterizing the α-silicon as very electron-rich, while the β-silicon is very similar to that of free silanes.

The expected difference in the reactivity is documented by the regioselective oxygenation of **16a-d** with dimethyldioxirane to yield the metallo-dihydroxy-disilanes **17a,d**, or the chlorination with CCl$_4$ to generate the metallo-dichloro-disilanes **18a-c** respectively (eq. 5).

These products clearly indicate considerable activation of the α-position by the metal towards the attack of electrophilic reagents. On the other hand, the transition metal fragment deactivates the α-silicon with respect to nucleophilic exchange. An impressive example is given by the hydrolysis of the pentachloro-disilanyl complexes **14a,d**, resulting in the introduction of three hydroxy groups exclusively at the β-silicon. The structure of the formed metallodisilanetriols **18a,b** is determined spectroscopically and by the controlled base-assisted condensation of **19b** with Me$_2$Si(H)Cl to give the metallo-trisiloxy-disilanes **20** (eq. 6).

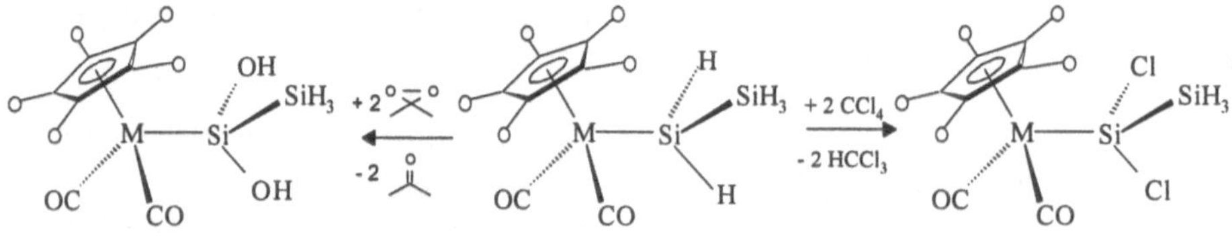

(5)

| 17a,d | | 16a-d | | 18a-c |

	a	b	c	d
M	Fe	Ru	Fe	Ru
-o	H	H	Me	Me

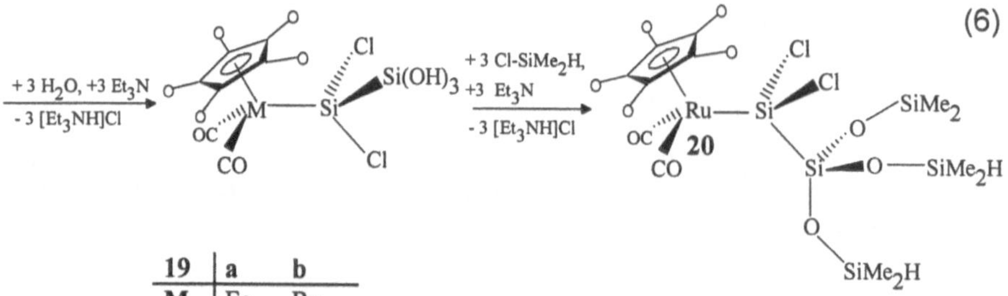

(6)

19	a	b
M	Fe	Ru
-o	H	Me

Acknowledgement

This work was gratefully supported by the Deutsche Forschungsgemeinschaft (SFB 347 "Selektive Reaktionen Metall-aktivierter Moleküle") and by the Fonds der Chemischen Industrie.

References

[1] Phosphenium Complexes, Part 35. - Part 34: W. Malisch, C. Hahner, K. Grün, J. Reising, R. Goddard, C. Krüger, *Inorg. Chim. Acta* **1996**, 2, 147.

[2] Synthesis and Reactivity of Silicon Transition Metal Complexes, Part 43. - Part 42: S. Möller, H. Jehle, W. Malisch, W. Seelbach in *Organosilicon Chemistry III, From Molecules to Materials* (Eds.: N. Auner, J. Weis), VCH, Weinheim, **1997**, in press.

[3] J. P. Collman, L. S. Hegedus, J. R. Norton, R. Finke, *Principles and Applications of Organotransition Metal Chemistry*, Science Books, Mill Valley, CA, **1987**.

[4] P. L. McGrane, M. Jensen, T. Livinghouse, *J. Am. Chem. Soc.* **1992**, *114*, 5459. - J. M. Burn, M. G. Fickes, F. J. Hollander, R. G. Bergman, *Organometallics* **1995**, *14*, 137. - J. Du Bois, C. S. Tomooka, J. Hong, E. M. Carreira, *Acc. Chem. Res.* **1997**, *30*, 364.

[5] T. L. Breen, D. W. Stephan, *Organometallics* **1997**, *16, 365*. - P. A. T. Hoye, P. G. Pringle, M. B. Smith, K. Worboys, *J. Chem. Soc., Dalton Trans.* **1993**, 269. - S. Nielsen-Marsh, R. J. Crowte, P. G. Edwards, *J. Chem. Soc., Chem. Commun.* **1992**, 699.

[6] W. Malisch, U.-A. Hirth, T. A. Bright, H. Käb, T. S. Ertel, S. Hückmann, H. Bertagnolli, *Angew. Chem.* **1992**, *104*, 1537; *Angew. Chem. Int. Ed. Engl.* **1992**, *31*, 1486.- W. Malisch, K. Grün, U.-A. Hirth, M. Noltemeyer, *J. Organomet. Chem.* **1996**, *513*, 31. - W. Malisch, U.-A. Hirth, K. Grün, M. Schmeußer, O. Fey, U. Weis, *Angew. Chem.* **1995**, *107*, 2717; *Angew. Chem. Int. Ed. Engl.* **1995**, *34*, 2500.

[7] W. Malisch, U.-A. Hirth, A. Fried, H. Pfister, *Phosphorus, Sulfur, Silicon and Related Elements* **1993**, *77*, 17. - H. Lang, M. Leise, L. Zsolnai, *Organometallics* **1993**, *12*, 2393. - H.-U. Reisacher, W. F. McNamara, E. N. Duesler, R. T. Paine, *Organometallics* **1997**, *16*, 449.

[8] H. Wollweber in *Methoden der organischen Chemie* (Ed.: E. Müller), Georg Thieme Verlag, Stuttgart, **1970**, Vol. *V/1c*, p. 977ff.

[9] R. Hussong, H. Heydt, G. Maas, M. Regitz, *Chem. Ber.* **1987**, *120*, 1263.

[10] J. Born, G. Huttner, O. Orama, L. Zsolnai, *J. Organomet. Chem.* **1985**, *282*, 53.

[11] H. Sakarai, K. Sakamoto, M. Kira, *Chem. Lett.* **1984**, 1379. - A. Marchand, P. Gerval, F. Dubondin, P. Mazerolles, *J. Organomet. Chem.* **1984**, *267*, 93.

[12] N. Wiberg, G. Fischer, S. Wagner, *Chem. Ber.* **1991**, *124*, 769 and literature therein.

[13] H. Fischer, J. Hofmann, *Chem. Ber.* **1991**, *124*, 981.

[14] W. Malisch, K. Grün, A. Fried, W. Reich, M. Schmeußer, U. Weis, C. Krüger, *unpublished results*.

[15] W. Malisch, S. Möller, R. Lankat, J. Reising, S. Schmitzer, O. Fey, in *Organosilicon Chemistry* (Eds.: N. Auner, J. Weis), VCH, Weinheim, **1996**, 575. - W. Malisch, R. Lankat, W. Seelbach, J. Reising, M. Noltemeyer, R. Pikl, U. Posset, W. Kiefer, *Chem. Ber.* **1995**, *128*, 1109.

[16] B. Stadelmann, P. Lassacher, H. Stueger, E. Hengge, *J. Organomet. Chem.* **1994**, *482*, 201.

Chapter II

Spectroscopical and Theoretical Studies on the Structure and Dynamics of Metal Complexes

Light-Induced Radical Formation from Organometallic Complexes

D. J. Stufkens

Anorganisch Chemisch Laboratorium, J. H. van 't Hoff Research Institute, University of Amsterdam, Nieuwe Achtergracht 166, 1018 WV Amsterdam, The Netherlands

1 Introduction

Organometallic complexes may possess different low-energy electronic transitions which determine their photophysical and photochemical behaviour in solution. Best known are the ligand-field (LF) and metal-to-ligand charge transfer (MLCT) transitions. MLCT states are normally not reactive and long lived and complexes having such a lowest MLCT state are used as photosensitizers for energy- and electron transfer processes. Best known is the complexion $Ru(bpy)_3^{2+}$. Organometallic analogues are the complexes $Re(L)(CO)_3(\alpha\text{-diimine})^{n+}$ (n=0,1) and $Ru(L_1)(L_2)(CO)_2(\alpha\text{-diimine})$ (α-diimine=bpy, etc.). These latter complexes have the special property that variation of L (or L_1 and L_2) may not only influence the energy of the lowest-excited state but also change its character and reactivity. Thus, if L represents an organic donor molecule, excitation into the MLCT band may be followed by an intramolecular electron transfer from the donor to the Re centre [1,2]. The complex then arrives in an LLCT (organic donor to α-diimine charge transfer) state from which it normally decays to the ground state without decomposition. A similar excited state process occurs when L represents a metal fragment or alkyl group. MLCT excitation is then followed by electron transfer from a high-lying σ orbital giving rise to a $\sigma(\text{M-L})\pi^*(\alpha\text{-diimine})$ (LLCT) state [2]. However, contrary to most complexes with an organic donor ligand L, this $\sigma\pi^*$ (LLCT) state is normally reactive; the complex decomposes into radicals by homolysis of the metal-metal or metal-alkyl σ bond. The efficiency of this homolysis reaction will first of all depend on the relative energies of the ^{3}MLCT and $^3\sigma\pi^*$ states, which may be influenced by varying the energy of the σ bond. It will also depend on the reactivity of the $^3\sigma\pi^*$ state, which may be affected by varying the strength and delocalisation of the σ bond. Finally, occupation of the $^3\sigma\pi^*$ state, and as a result radical formation, may be prevented if the MLCT states themselves are reactive.

In recent years we have studied in detail the photophysical and photochemical properties of the bi- and trinuclear metal-metal bonded complexes $Re(ML_n)(CO)_3(\alpha\text{-diimine})$ [2-10], $Ru(ML_n)(M'L'_n)(CO)_2(\alpha\text{-diimine})$ (ML_n, $M'L'_n$=$Mn(CO)_5$, $Re(CO)_5$, $Co(CO)_4$, $SnPh_3$ or $FeCp(CO)_2$) [11-15] and $Os_3(CO)_{10}(\alpha\text{-diimine})$ [16,17], and the metal-alkyl (R) complexes $M(R)(CO)_3(\alpha\text{-diimine})$ (M=Mn, Re) [18-20], and $Ru(X)(R)(CO)_2(\alpha\text{-diimine})$ (X=halide) [21-23]. With the exception of the $Ru(ML_n)(M'L'_n)(CO)_2(\alpha\text{-diimine})$ complexes we found no evidence for an allowed σ → π* transition and irradiation took place into the metal-to-α-diimine charge transfer (MLCT) transitions in the visible region. In this article we present the properties of the $^3\sigma\pi^*$ states, their structure, lifetime and reactivity, and the formation and fate of the radicals. The schematic structures of the complexes and of the α-diimine ligands used are presented in Figure 1.1.

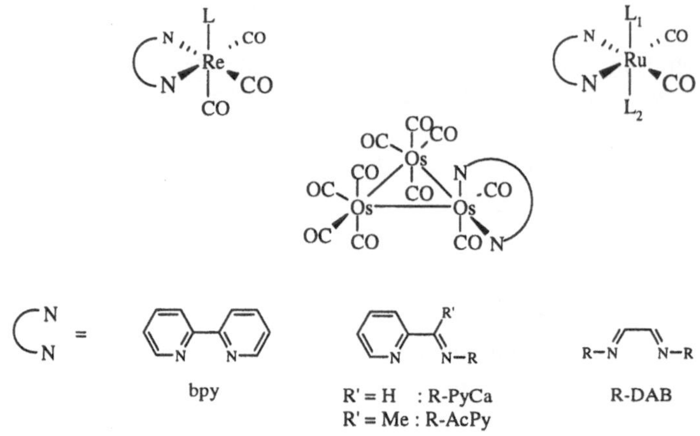

Figure 1.1 Structures of the complexes and α-diimine ligands used.

2 Metal-Alkyl Complexes

The metal-alkyl complexes M(R)(CO)$_3$(α-diimine) (M=Mn,Re) and Ru(X)(R)(CO)$_2$(α-diimine) (X=halide) showed a strong dependence of their photoreactivity on the metal, R and the α-diimine ligand. Thus, the complex Re(Me)(CO)$_3$(iPr-DAB) is only weakly photoreactive ($\phi=10^{-2}$) [19,20], whereas Re(Me)(CO)$_3$(4,4'-Me$_2$-bpy) photodecomposes with $\phi=0.4$ [24], Re(Et)(CO)$_3$(iPr-DAB) and Re(Bz)(CO)$_3$(iPr-DAB) with a quantum yield close to unity [19,20]. This difference in behaviour was ascribed to a change in the relative energies of the ^{3}MLCT and $^3\sigma\pi^*$ states. At least in the case of Re(Me)(CO)$_3$(iPr-DAB) this interpretation could be confirmed by the observation in the UV-Photoelectron spectrum of a higher ionization potential for the σ(Re-Me) electrons compared to those occupying the d$_\pi$(Re) orbitals. This places the σπ* state at higher energy than the MLCT states. Replacing the iPr-DAB ligand by 4,4'-Me$_2$-bpy (bpy') changes this picture completely since radical formation becomes much more efficient ($\phi=0.4$). The nanosecond time-resolved absorption spectra of Re(Me)(CO)$_3$(bpy') in toluene (Fig. 2.1) clearly show that MLCT excitation gives rise to the formation of two transients, absorbing close to each other [24]. The transient species absorbing at longer wavelength (ca. 525 nm) has a lifetime of 40 nanoseconds, which is typical for a Re(L)(CO)$_3$(bpy') complex having a lowest ^{3}MLCT state. The transient species absorbing at ca. 500 nm is still present when the ^{3}MLCT state has decayed to the ground state.

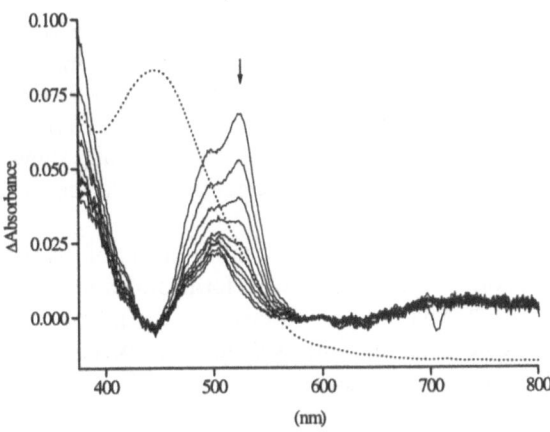

Figure 2.1 Nanosecond time-resolved absorption spectra of [Re(Me)(CO)$_3$(bpy')] measured in toluene at 293 K. (- - - = Ground state absorption spectrum and ——— = transient absorption spectra; τ_d = 10 ns, 20 ns, 30 ns, 40 ns, 50 ns, 60 ns, 70 ns, 80 ns, 90 ns and 100 ns, respectively)

It has a lifetime of a few microseconds and represents the [Re(CO)$_3$(bpy')]• radical. An analysis of the spectra shows that the radicals are formed directly after irradiation into the ^{1}MLCT state and not out of the ^{3}MLCT state. Excitation takes place into the Franck Condon levels of the ^{1}MLCT state. About 40% of the complexes cross to the reactive $^3\sigma\pi^*$ state, from which they decompose into radicals; the remaining 60% relax to the nonreactive ^{3}MLCT state from which they decay to the ground state. From the temperature dependence of the quantum yield of the reaction a value of ca. 2000 cm^{-1} was derived for the barrier between the ^{1}MLCT and $^3\sigma\pi^*$ states.

The transient absorption spectra of the corresponding benzyl and isopropyl complexes only showed the presence of the [Re(CO)$_3$(bpy')]• radicals directly after the 7 ns laser pulse, in agreement with the very high quantum yield of their formation [24]. The repulsive character of the $^3\sigma\pi^*$ state was evident from the fact that for none of these bpy' complexes this state could be observed with ns transient absorption spectroscopy. However, the situation changed when bpy' was replaced by iPr-DAB. The $^3\sigma\pi^*$ states of the Re(R)(CO)$_3$(iPr-DAB) (R= Et, iPr, Bz) complexes could be observed with transient absorption spectroscopy and appeared to have a lifetime in apolar solvents at room temperature of ca. 250 ns [20]. From this state the complexes decomposed completely into radicals. Apparently, the $^3\sigma\pi^*$ state of these complexes is not a repulsive but a bound state, from which they slowly, but still completely decompose into radicals. We could even observe such a complex in its $^3\sigma\pi^*$ state and its decomposition into radicals with ns time-resolved IR spectroscopy in the CO-stretching region [20]. This bound character of the $^3\sigma\pi^*$ state disappeared completely when the complex was instead irradiated in THF. The lifetime of the $^3\sigma\pi^*$ state then dropped to less than 20 picoseconds, presumably because of the efficient cleavage by the solvent molecules of the weakened Re-alkyl bond.

269

Similar light-induced homolysis reactions have been observed for the complexes Ru(X)(R)(CO)$_2$(α-diimine) (X=halide) [23]. These complexes do, however, not photodecompose into radicals for R=Me and Et, since the $^3\sigma\pi^*$ state is then too high in energy to become occupied by MLCT (Ru-to-α-diimine) excitation. They are photolabile for R=iPr and Bz and their reaction is even photocatalytic. A detailed study showed that, in this case, homolysis of the Ru-R bond is followed by an electron transfer chain reaction initiated by the [Ru(I)(S)(CO)$_2$(iPr-DAB)]$^\bullet$ radicals [23].

An important aspect of these reactions is the involvement of the $^3\sigma\pi^*$ state. In order to prove the triplet character of this state we have performed a ns time-resolved EPR study of the ethyl-, isopropyl-, and benzyl-radicals formed on irradiation of these Re- and Ru-complexes [25]. These studies indeed confirmed that the alkyl radicals originate from an excited biradical state having triplet character.

3 Binuclear Metal-Metal Bonded Complexes

From the great variety of metal-metal bonded compounds we have studied the excited state properties of the complexes Re(ML$_n$)(CO)$_3$(α-diimine), in which ML$_n$ represents a Mn(CO)$_5$, Re(CO)$_5$, Co(CO)$_4$, SnPh$_3$, or FeCp(CO)$_2$ group [2-10]. The UV-Photoelectron spectra of these complexes showed that the σ(Re-M) orbital is the HOMO, which implies that their $^3\sigma\pi^*$ state is lower in energy than the MLCT states. In fact, the $^3\sigma\pi^*$ state of several of these complexes could be detected both in emission in a glass at 77K and with ns time-resolved absorption spectroscopy [10]. The latter spectra nicely showed the decomposition of the $^3\sigma\pi^*$ excited complexes into radicals just as for the above mentioned Re(Bz)(CO)$_3$(iPr-DAB) complex. An exceptional behaviour was observed for the complex Re(SnPh$_3$)(CO)$_3$(bpy'), since it remained completely stable in the $^3\sigma\pi^*$ state due to its strong σ(Re-Sn) bond. A comparison of the $^3\sigma\pi^*$ state lifetime of this complex with that of a related complex having a lowest ^{3}MLCT state showed that $^3\sigma\pi^*$ states are much longer lived than ^{3}MLCT states, provided that the former ones are stable. Thus, the $^3\sigma\pi^*$ state of Re(SnPh$_3$)(CO)$_3$(bpy') has a lifetime at room temperature in toluene of 1100 ns, whereas the ^{3}MLCT state of Re(Cl)(CO)$_3$(bpy') lives only 50 ns under these circumstances [10].

These observations are very important since they demonstrate that these σ-bonded α-diimine complexes vary from very reactive with production of radicals to extremely long lived. Both aspects of these complexes deserved more attention and we have focused in recent years on the following two issues:

1. How should a complex be modeled in order to make it completely photostable and extremely long lived so that it can be used as an efficient photosensitizer or infrared emitter?

2. Which complexes produce a biradical species instead of two separate radicals and what are the chemical and catalytic properties of such species?

In order to tackle the first question we extended our investigation to the complexes Ru(X)(R)(CO)$_2$(α-diimine) in which both axial ligands X and R represent a metal fragment (ML$_n$, M'L'$_n$). Such a trinuclear metal-metal bonded complex Ru(ML$_n$)(M'L'$_n$)(CO)$_2$(α-diimine) should have a delocalised σ(M-Ru-M') bond, which might be very strong for a proper choice of ML$_n$ and M'L'$_n$. For the second project we started an investigation of the triangular clusters $Os_3(CO)_{10}(\alpha$-diimine), which might give rise to the formation of biradicals when the $^3\sigma\pi^*$ state of the Os_3-skeleton is lower in energy than the MLCT states. The results of these investigations will shortly be discussed in the next two chapters.

4 Linear Trinuclear Metal-Metal Bonded Complexes

First of all a series of complexes was prepared in which the halide ligand in $Ru(X)(Me)(CO)_2(\alpha$-diimine) was replaced by a metal fragment such as $Mn(CO)_5$ or $SnPh_3$. According to the density functional (DFT) MO calculations on the model complex $Ru(SnH_3)(Me)(CO)_2$(H-DAB) [13] the HOMO and LUMO are strongly delocalised and consist of contributions from the lowest π^* orbital of H-DAB as well as from Me, SnH_3 and Ru orbitals. As a result the $\sigma \rightarrow \pi^*$ transition is allowed and does not give rise to large changes in population. The latter effect is certainly responsible for the rather long lifetime of the $^3\sigma\pi^*$ state of the complex $Ru(SnPh_3)(Me)(CO)_2$(iPr-DAB) in a glass at 77K [13]. Unfortunately, this complex is still not photostable since the weaker, σ(Ru-Me), bond is broken photochemically [15]. In order to improve the photostability, the methyl ligand was replaced by a $SnPh_3$ group, thus creating two strong Ru-Sn bonds in the complex. In fact, the complex $Ru(SnPh_3)_2(CO)_2$(iPr-DAB) appeared to be nearly photostable at room temperature [13,15]. Moreover, the $^3\sigma\pi^*$ state of this complex has a lifetiime of 1μs at room temperature and of 264 μs in a glass at 77K [13]. The latter lifetime is extremely long for a charge transfer excited state. For comparison, the lowest ^{3}MLCT state of the related complex $Ru(Cl)(Me)(CO)_2$(iPr-DAB) has a lifetime of only 300 ns, although its energy is nearly the same as that of the $^3\sigma\pi^*$ state of the $Ru(SnPh_3)_2(CO)_2$(iPr-DAB) complex. Thus, the emission properties of these complexes do not follow energy gap law behaviour, which proves that the character of the lowest-excited state changes when Cl$^-$ and Me are both replaced by a $SnPh_3$ group. Going from a ^{3}MLCT ($^3d_\pi\pi^*$) to a $^3\sigma\pi^*$ state the complex becomes less distorted with respect to the ground state. This is manifested by a much smaller Stokes shift and by a drastic decrease of the rate constant of nonradiative decay to the ground state. The lack of distortion in the $^3\sigma\pi^*$ state is also evidenced by the ns time-resolved IR (TRIR) spectra [13]. Figure 4.1 shows the TRIR spectra of the complexes $Ru(Cl)(SnPh_3)(CO)_2$(iPr-DAB) and $Ru(SnPh_3)_2(CO)_2$(iPr-DAB) in the CO-stretching region. The spectra were measured in a PrCN glass at 77K. The lower spectra are FTIR ground state spectra, the upper spectra are difference TRIR spectra obtained immediately after the 355 nm excitation. The spectrum of $Ru(Cl)(SnPh_3)(CO)_2$(iPr-DAB) shows a shift of the v(CO) vibrations to higher frequencies by 18 and 29 cm^{-1}, respectively, which is characteristic of complexes having a lowest-excited state with (partial) ^{3}MLCT character. The corresponding spectrum of $Ru(SnPh_3)_2(CO)_2$(iPr-DAB) shows only very small differences between the wavenumbers of the CO-stretching vibrations of the complex in its ground and $^3\sigma\pi^*$ state.

In order to further increase the photostability of these complexes at room temperature, we strengthened their metal-metal bonds by replacing the central ruthenium atom by osmium. Quite recently, we succeeded in the synthesis of the complex $Os(SnPh_3)_2 (CO)_2$(bpy'), which appeared to be completely photostable at room temperature and has a $^3\sigma\pi^*$ emission at ca. 750 nm with a lifetime of 3 μs [26].

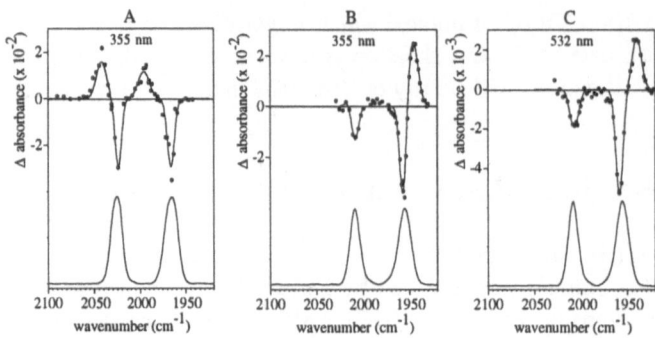

Figure 4.1 TRIR (top) and FTIR (bottom) spectra of Ru(Cl)(SnPh$_3$) (CO)$_2$ (iPr-DAB) (left) and Ru(SnPh$_3$)$_2$ (CO)$_2$(iPr-DAB) (right) at 77K in nPrCN. The difference TRIR spectra were obtained immediately after the 355 nm excitation. Adapted from ref. [13].

The energy of the π^* orbital and as a result that of the $^3\sigma\pi^*$ state can easily be lowered by replacing bpy' by another α-diimine ligand. This will shift the emission into the near infrared while it will still have an appreciable lifetime. Work is in progress to synthesize such complexes and to investigate their suitability as infrared emitters.

5 Trinuclear Clusters

The results obtained for the binuclear and trinuclear metal-metal bonded complexes prompted us to study also the photochemistry of the substituted triangular clusters Os$_3$(CO)$_{10}$(α-diimine) (Figure 1.1). In accordance with the behaviour of the complexes discussed above, MLCT excitation was expected to be followed by occupation of a $^3\sigma\pi^*$ state. In the case of the binuclear complexes, radicals are formed from this state, which were found to undergo the following reactions [7,8]. In noncoordinating solvents they dimerize or regenerate the parent compound; in coordinating solvents (S) they undergo an electron transfer reaction (e.g. [Mn(CO)$_5$]$^\bullet$ + [Mn(S)(CO)$_3$(α-diimine)]$^\bullet$ $\rightarrow$ Mn(CO)$_5^-$ + Mn(S) (CO)$_3$(α-diimine)$^+$), in viscous media the two radicals produce a ligand-bridged complex, provided that the α-diimine ligand possesses a reactive imine bond (R-DAB or R-PyCa). A detailed investigation has shown that very similar reactions occur for these Os-clusters, although the products are quite different and novel [17].

Irradiation of a cluster in a coordinating solvent such as acetonitrile or pyridine caused the formation of a transient zwitterion $^-$Os(CO)$_4$-Os(CO)$_4$-Os$^+$(S)(CO)$_2$(α-diimine), which regenerated the parent cluster [17]. The lifetimes of these zwitterions varied from a few seconds in acetonitrile to several minutes in pyridine and they could be stabilised in these solvents at low temperatures. In noncoordinating and weakly coordinating solvents (toluene, THF) no such zwitterions were formed. However, according to the time-resolved absorption spectra, short-lived species were still produced, their lifetimes varying from a few nanoseconds to one microsecond depending on the α-diimine and the solvent. These transients could be quenched by nitrosodurene and the resulting adducts were detected with EPR spectroscopy. They were therefore assumed to be the biradical species •Os(CO)$_4$-Os(CO)$_4$-Os(CO)$_2$(α-diimine)•, and this assumption was confirmed by the nanosecond transient absorption (TA) spectra [17]. Figure 5.1 presents these TA spectra for the photoreaction of Os$_3$(CO)$_{10}$(iPr-AcPy) in THF at room temperature. The spectra show a strong bleach at ca. 540 nm, which nearly coincides with the ground state absorption, represented by the dotted curve. There are transient absorptions in the whole spectral region and the lifetimes of transient and bleach are the same, viz. 111± 5 ns. This is in line with our observation that also this transient species mainly regenerates the parent cluster. The long-wavelength absorption of the transient is characteristic of an α-diimine radical anion, which confirms the biradical formation.

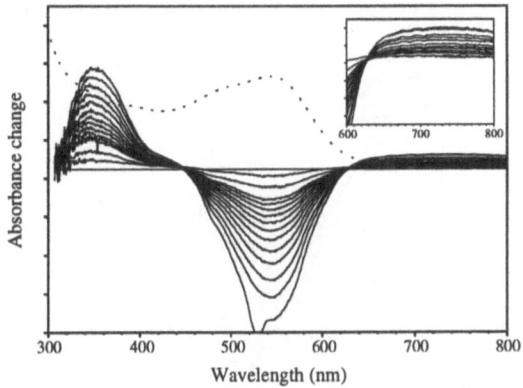

Figure 5.1 Nanosecond transient absorption spectra of Os$_3$(CO)$_{10}$(iPr-AcPy) in THF at 298K. Time delay between the first twelve spectra is 10 ns, after that it is 30 ns (--- = ground state absorption). Adapted from ref. [17].

Based on these observations it was concluded that these Os$_3$(CO)$_{10}$(α-diimine) clusters undergo a solvent dependent photochemistry, as indicated in Scheme 1.

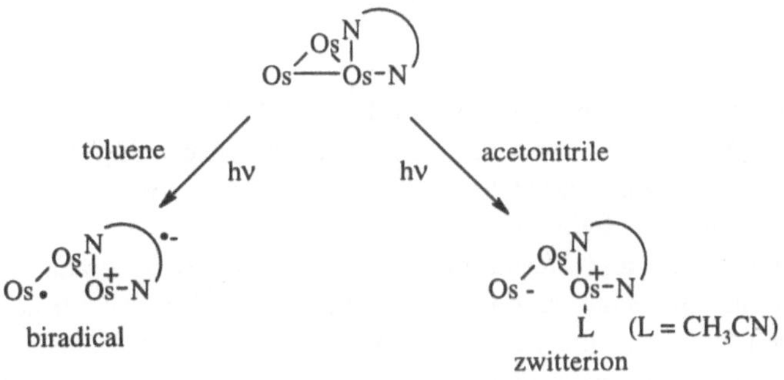

Scheme 1

The biradicals do not only regenerate the parent cluster. When the cluster is e.g. irradiated in THF in the presence of 0.5 M acetonitrile, the primarily formed biradical is converted into the long-lived acetonitrile-stabilised zwitterion. Secondly, a minor part (ca. 5%) of the biradicals does not regenerate the parent cluster, but transforms into a ligand-bridged isomer if the α-diimine possesses a reactive imine bond [17]. The formation of this isomer in which the α-diimine becomes σ,σ bonded to one Os atom and π bonded to another after cleavage of an Os-Os bond, is quite similar to the formation of ligand-bridged complexes in the case of e.g. the $Mn(Mn(CO)_5)(CO)_3$(α-diimine) (α-diimine= R-DAB, R-PyCa) complexes in a viscous medium such as paraffin [7].

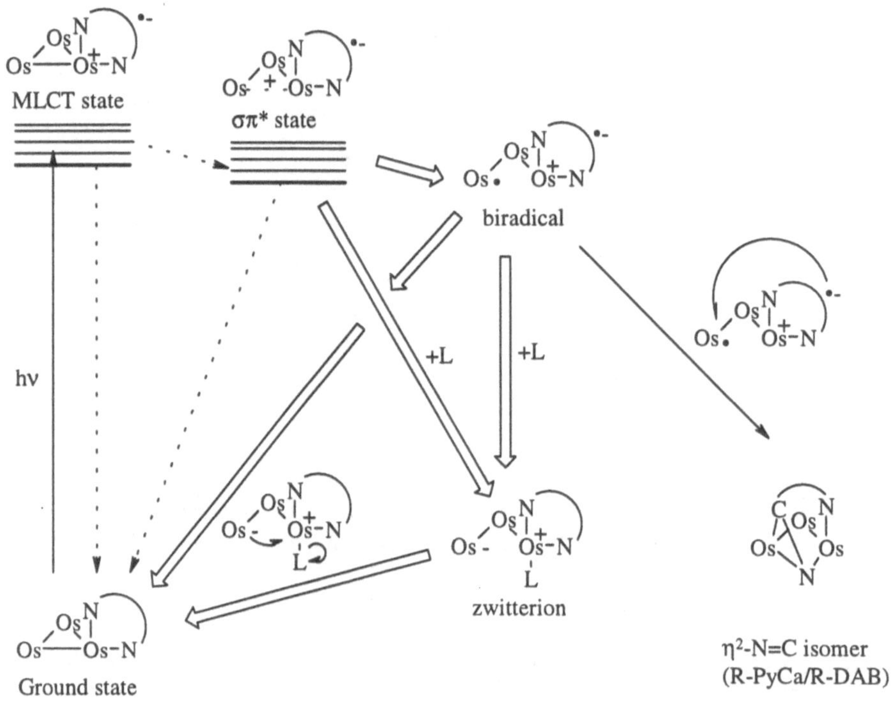

Scheme 2

The above results are summarized in Scheme 2 [17]. MLCT excitation is followed by surface crossing to the $^3\sigma\pi^*$ state just as for the binuclear complexes. From this state the cluster may decay to the ground state or transform into a biradical. A zwitterion may then be formed from the biradical by the uptake of a coordinating solvent molecule such as acetonitrile. Alternatively, such a zwitterion may be formed directly by attack of a solvent molecule to the cluster in its $^3\sigma\pi^*$ state. A minor part of the biradicals reacts further to give a stable α-diimine bridged isomer in the case of the R-DAB and R-PyCa clusters. Finally, both the biradicals and the zwitterions regenerate the parent cluster, and in both cases this backreaction is retarded in coordinating solvents.

The processes leading to the formation of biradicals, zwitterions and ligand-bridged complexes are analoguous to those of the binuclear metal-metal bonded complexes. Because of their triangular structure the products of these clusters are, however, different. An interesting aspect of these biradical and zwitterionic species is their chemical reactivity and catalytic potential, especially in view of the presence of two reactive sites. Work is in progress to further develop this field; the preliminary investigations show that olefins are activated by the zwitterions and undergo secondary reactions.

6 Conclusions

Metal-metal and metal-alkyl bonded α-diimine complexes have a $^3\sigma\pi^*$ excited state which is often lower in energy than the MLCT states. Such a $^3\sigma\pi^*$ state may vary from very reactive giving rise to the formation of radicals to stable and extremely long lived. The reactive metal-alkyl complexes are good photoinitiators of polymerisation reactions; the photostable metal-metal bonded complexes having a very long lived $^3\sigma\pi^*$ state may be good candidates for the development of novel near infrared emitters. Clusters of the type $Os_3(CO)_{10}(\alpha$-diimine) produce zwitterions in coordinating solvents, which activate olefins and may be of use in catalytic applications.

7 Acknowledgment

The results described in this article have been obtained by the PhD students M.P. Aarnts, C.J. Kleverlaan, J. Nijhoff and B.D. Rossenaar, whose devotion and skill I very much appreciate. Many thanks are due to my colleagues Prof. A. Oskam, Dr F. Hartl, Prof. A. Vlček, Jr for their valuable contribution and to the Netherlands Foundation for Chemical Research (SON) for financial support.

References

[1]   K. S. Schanze, D. B. MacQueen, T. A. Perkins, L. A. Cabana, *Coord. Chem. Rev.* **1993** *122* 63.
[2]   D. J. Stufkens, *Comments Inorg. Chem. 13* **1992** 359.
[3]   M. W. Kokkes, D. J. Stufkens, A. Oskam **1995**, *Inorg. Chem. 24* 4411.
[4]   R. R. Andréa, W. G. J. de Lange, D. J. Stufkens, A. Oskam, *Inorg. Chem.* **1989** *28* 318.
[5]   H. K. van Dijk, J. van der Haar, D. J. Stufkens, A. Oskam, *Inorg. Chem.* **1989** *28* 75.
[6]   P. C. Servaas, G. J. Stor, D. J. Stufkens, A. Oskam, *Inorg. Chim. Acta* **1990** *178* 185.
[7]   T. van der Graaf, D. J. Stufkens, A. Oskam, K. Goubitz, *Inorg. Chem.* **1991** *30* 599.
[8]   T. van der Graaf, A. van Rooy, D. J. Stufkens, A. Oskam, *Inorg. Chim. Acta* **1991** *187* 133.
[9]   J. W. M. van Outersterp, D. J. Stufkens, A. Vlček, Jr., *Inorg. Chem.* **1995** *34* 5183.

[10]   B. D. Rossenaar, E. Lindsay, D. J. Stufkens, A. Vlček Jr, *Inorg. Chim. Acta* **1996** *250* (1996) 5.

[11]   H. A. Nieuwenhuis, A. van Loon, M. A. Moraal, D. J. Stufkens, A. Oskam, K. Goubitz, *J. Organomet. Chem.* **1995** *492* 165.

[12]   M. P. Aarnts, M. P. Wilms, K. Peelen, J. Fraanje, K. Goubitz, F. Hartl, D. J. Stufkens, E. J. Baerends, A. Vlček Jr, *Inorg. Chem.* **1996** *35* 5468.

[13]   M. P. Aarnts, D. J. Stufkens, M. P. Wilms, E. J. Baerends, A. Vlček Jr, I. P. Clark, M. W. George, J. J. Turner, *Chem. Eur. J.* **1996** *2* 1556.

[14]   M. P. Aarnts, M. P. Wilms, D. J. Stufkens, E. J. Baerends, A. Vlček Jr, *Organometallics* **1997** *16* 2055.

[15]   M. P. Aarnts, D. J. Stufkens, A. Vlček Jr, *Inorg. Chim. Acta,* in press.

[16]   J. W. M. van Outersterp, M. T. Garriga Oostenbrink, H. A. Nieuwenhuis, D. J. Stufkens, F. Hartl, *Inorg. Chem.* **1995** *34* 6312.

[17]   J. Nijhof, M. J. Bakker, F. Hartl, D. J. Stufkens, W.-F. Fu, R. van Eldik, submitted for publication.

[18]   B. D. Rossenaar, D. J. Stufkens, A. Oskam, J. Fraanje, K. Goubitz, *Inorg. Chim. Acta* **1996** *247* 215.

[19]   B. D. Rossenaar, M. W. George, F. P. A. Johnson, D. J. Stufkens, J. J. Turner, A. Vlček, Jr., *J. Am. Chem. Soc.* **1995** *117* 11582.

[20]   B. D. Rossenaar, C. J. Kleverlaan, M. C. E. van de Ven, D. J. Stufkens, A. Vlček Jr, *Chem. Eur. J.* **1996** *2* 228.

[21]   H. A. Nieuwenhuis, D. J. Stufkens, A. Oskam, *Inorg. Chem.* **1994** *33* 3212.

[22]   H. A. Nieuwenhuis, D. J. Stufkens, A. Vlček, Jr., *Inorg. Chem.* **1995** *34* 3879.

[23]   H. A. Nieuwenhuis, M. C. E. van de Ven, D. J. Stufkens, A. Oskam, K. Goubitz, *Organometallics* **1995** *14* 780.

[24]   C. J. Kleverlaan, to be published.

[25]   C. J. Kleverlaan, D. M. Martino, H. v. Willigen, D. J. Stufkens, A. Oskam, *J. Phys. Chem.* **1996** *100* 18607.

[26]   J. van Slageren, to be published.

The Luminous and the Dark Side of Singlet Oxygen : Comparison Between Photochemical and Chemical Sources of Singlet Oxygen ($^1O_2$, $^1\Delta_g$) in Organic Synthesis

J.M. Aubry,[‡,*] V. Nardello,[‡] S. Bouttemy,[‡] T. Wirth,[¶] T. Linker,[¶] W. Adam[¶]

[‡] Equipe de Recherches sur les Radicaux Libres et l'Oxygène Singulet, CNRS 351, Faculté de Pharmacie, Laboratoire de Physique, B.P. 83, 59006 Lille Cedex, France.
[¶] Institut für Organische Chemie, Universität Würzburg, Am Hubland, 97074 Würzburg, Germany.

Singlet oxygen, $^1O_2$ ($^1\Delta_g$), exhibits an opposite chemical reactivity compared to that of ground state oxygen. This powerful oxidant has found considerable synthetic utility since it can undergo selective reactions with a wide variety of electron-rich molecules (olefins [1], conjugated dienes, polycyclic aromatic hydrocarbons, phenols, sulfides and heterocycles) [2,3]. It is usually generated by photosensitization, this method is very versatile because it can be conducted in a wide range of solvents and at low temperature when unstable oxidized products have to be prepared. However it requires photochemical reactors which are not always available in research laboratories or industrial plants, therefore, it has found very little industrial applications [4]. In 1985, we discovered a mild chemical generator of $^1O_2$ involving the disproportionation of hydrogen peroxide catalyzed by molybdate ions [5,6]:

$$2 H_2O_2 \xrightarrow[\text{water}]{MoO_4{}^{2-}} 2 H_2O + {}^1O_2 \ (\,100\,\%)$$

This reaction proceeds in aqueous environment at room temperature and generates a huge flux of $^1O_2$ from a readily available oxygen source, H_2O_2. It could constitute an attractive alternative to the usual photochemical method provided that a number of conditions are fulfilled. Various examples are presented below to illustrate the ability of the system $H_2O_2/MoO_4{}^{2-}$ to oxidize hydrophilic or hydrophobic substrates and to compare it to the usual photochemical method.

Hydrophilic Substrates

Poorly Reactive Substrates

This system releases efficiently $^1O_2$ only in aqueous alkaline media and hence, it is particularly relevant to perform the oxidation of hydrophilic organic substrates. At first

277

sight, water does not appear suitable to sustain $^1O_2$ reactions since the lifetime of $^1O_2$ in this solvent is very short (4.4 μs). However, it has been recognized recently that the reactivity of organic substrates towards $^1O_2$ is greatly increased in water [7].

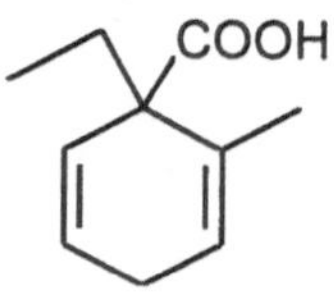

Thus, the following 1,4-cyclohexadiene derivative is 240 times more reactive with $^1O_2$ in water than in CCl_4. This accelerating effect has found an application in the oxidation, on the preparative scale, of the endoperoxide $NDPO_2^{2-}$. This compound behaves as an « $^1O_2$ carrier » since it binds $^1O_2$ at low temperature (25 °C) and releases $^1O_2$ on incubation (37°C / 2h 30 min). It has thus been used in biological media to mimic the photodynamic effect without light [8]. Unfortunately, the starting naphthalene derivative NDP is poorly reactive towards $^1O_2$ and 3 h of irradiation at 5°C with a 500W high pressure Hg lamp were required to convert 1 g of this compound into $NDPO_2$ with a 95 % yield but with simultaneous formation of side products (1%). On the other hand, 10 g or more can be readily oxidized within 1 h at room temperature and with a 98% yield by the system H_2O_2/MoO_4^{2-} without formation of side products [9].

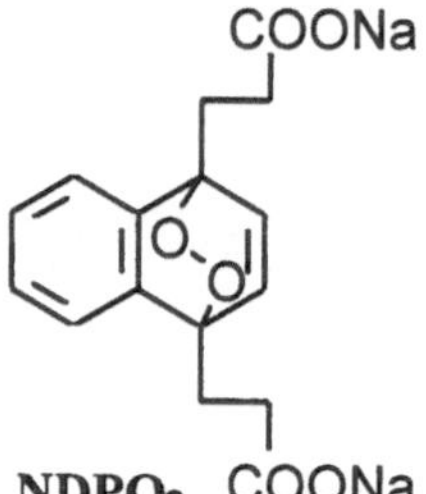

NDPO$_2$

Chemoselectivity of the System H_2O_2/MoO_4^{2-}

To take advantage of this chemical source of $^1O_2$ in organic synthesis, substrates and products must be resistant to experimental conditions (pH 9-12, 20°C) and to other oxidizing species present in the solution (H_2O_2, peroxomolybdates). In basic solutions, only the mono- , di-, tri- and tetra- peroxomolybdates $MoO_{4-n}(O_2)_n^{2-}$ are formed, among which the triperoxomolybdate is the precursor of $^1O_2$ [6], whereas in more acidic media a strong epoxidizing agent $Mo_2O_3(O_2)_4^{2-}$ is produced. Therefore, the chemoselectivity of the system

H_2O_2/MoO_4^{2-} can be tuned easily by varying the pH-value. Thus, an olefin such as tiglic acid is selectively epoxidized in slightly acid conditions or peroxidized through the ene-reaction of $^1O_2$ in basic media [10] :

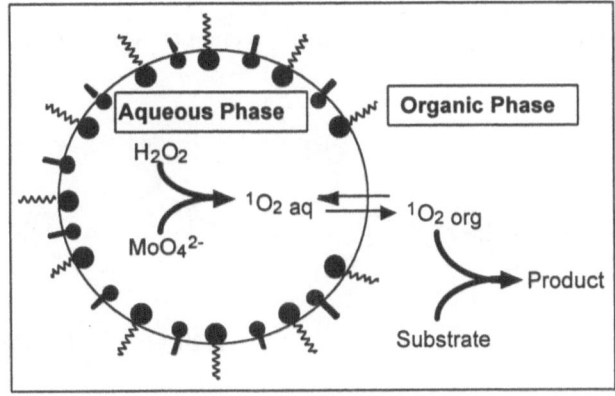

Hydrophobic Substrates

Peroxidation of Hydrophobic Substrates in Microemulsions

Hydrophobic substrates cannot be oxidized by the system H_2O_2/MoO_4^{2-} in a simple biphasic medium H_2O/organic solvent since most of $^1O_2$ would be deactivated by water molecules before diffusing into the organic phase. One effective means to overcome this problem is to resort to microemulsions. A microemulsion, which consists of water, organic solvent, surfactant and in most cases, cosurfactant, is defined as a transparent, thermodynamically stable, isotropic dispersion of two immiscible liquids. One important feature of these media is their ability to dissolve simultaneously huge amounts of hydrophilic compounds that are confined in the aqueous microdroplets and non-polar organic molecules that are localized in the continuous organic phase.

The microemulsion used in our study was constituted of H_2O / SDS / BuOH / CH_2Cl_2. Thus, H_2O_2 and MoO_4^{2-} are compartmentalized in an aqueous microreactor where they generate $^1O_2$. This small and uncharged molecule can then diffuse freely in the organic phase before deactivation since the typical size of microdroplets ($\approx$ 10 nm) is much smaller than the mean travel distance of $^1O_2$ in water ($\approx$ 200 nm). Once in the organic phase, $^1O_2$ reacts with the substrate. Typical hydrophobic substrates such as rubrene 1, 1,3-diphenyl-isobenzofuran 2, adamantylidene-adamantane 3, citronellol 4, 3-tert. butyl-6-methyl-thiophenol 5 have been readily oxidized in microemulsions, on the preparative scale and with excellent yield [11].

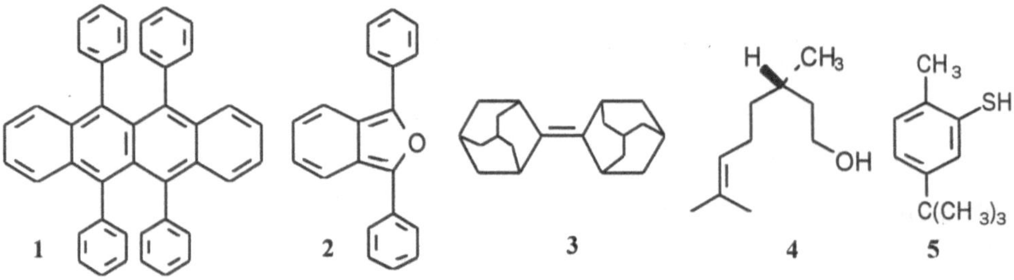

1   2   3   4   5

Diastereoselectivity in the Peroxidation of Chiral Allylic Alcohols

Photooxygenation of mesitylol leads to the expected hydroperoxide (entries 1 and 2) whereas chemical oxidation by H_2O_2/MoO_4^{2-} affords the epoxide besides the hydroperoxide with relative ratios depending on the nature of the solvent (Table 1).

In water, mesitylol was selectively epoxidized into E-threo whatever the pH was (entry 3) whereas in CH_2Cl_2-based microemulsion the expected allylic hydroperoxide was solely obtained (entry 5) because the direct interaction of mesitylol with the peroxomolybdates was avoided. In all cases the formation of the threo diastereoisomers were favored due to steric hindrance of the substituents and to hydrogen-bonding between the OH group of mesitylol and the two intermediates: perepoxide (ene reaction) and peroxomolybdates (epoxidation).

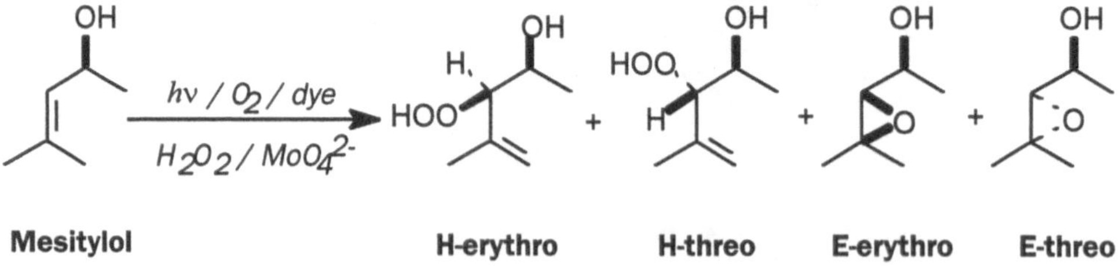

Mesitylol   H-erythro   H-threo   E-erythro   E-threo

Table 1 Chemo- and diastereoselectivity in the oxidation of mesitylol by photosensitization and by the system H_2O_2 / MoO_4^{2-}

Oxidizer	Solvent	H-erythro	H-threo	E-erythro	E-threo
hv / O_2 / dye	D_2O	20	80	0	0
	CCl_4	4	96	0	0
H_2O_2 / MoO_4^{2-}	D_2O	ε	ε	15	85
	μemulsion CCl_4	5	50	5	40
	μemulsion CH_2Cl_2	8	92	ε	ε

Conclusion

The catalytic system H_2O_2/MoO_4^{2-} constitutes an attractive alternative to the photochemical method since it can generate high flux of $^1O_2$ at room temperature (1 mol L^{-1} in 6 min) starting from a cheap and readily available source of oxygen, H_2O_2, and a reusable catalyst, Na_2MoO_4. It leads to the same products than those obtained by photooxygenation and does not modify the diastereoselectivity in the oxidation of chiral substrates. This chemical source of $^1O_2$ is particularly suitable for large scale oxidations since it does not require photochemical reactors. Aqueous solutions of hydrophilic substrates are preferably oxidized by this system rather than by photosensitization which induces side reactions of the substrate and degradation of the photosensitizer through electron transfer. However, the substrates must be inert to H_2O_2 and to the peroxomolybdates. Hydrophobic substrates can be oxidized on the preparative scale in microemulsions but the recovering of the products is more tedious than from a photosensitized solution.

References

[1]   Prein, M.; Adam, W. *Angew. Chem., Int. Ed. Engl.* **1996**, *35*, 477-494.
[2]   Wasserman, H.H.; Ives, J.L. *Tetrahedron* **1981**, *37*, 1825-1852.
[3]   Ohloff, G. *Pure Appl. Chem.* **1975**, *43*, 481-501.
[4]   Gollnick, K. *Chim. Ind.* **1982**, *64*, 156-166.
[5]   Aubry, J.M., *J. Am. Chem. Soc.* **1985**, *107*, 5844-5849.
[6]   Nardello, V.; Marko, J.; Vermeersch, G.; Aubry, J.M. *Inorg. Chem.* **1995**, *34*, 4950-4957.
[7]   Aubry, J.M.; Mandard-Cazin, B.; Rougee, M.; Bensasson, R.V. *J. Am. Chem. Soc.* **1995**, *117*, 9159-9164.
[8]   Dewilde, A.; Pellieux, C.; Hajjam, S.; Wattré, P.; Pierlot, C.; Hober, D.; Aubry, J.M. *J. Photochem. Photobiol. B: Biology*, **1996**, *36*, 23-29.
[9]   Aubry, J.M.; Cazin, B.; Duprat, F. *J. Org. Chem.* **1989**, *54*, 726-728.
[10] Nardello, V.; Bouttemy, S.; Aubry, J.M. *J. Mol. Catal.* **1997**, *117*, 439-447.
[11] Aubry, J.M.; Bouttemy, S. *J. Am. Chem. Soc.* **1997**, *119*, 5286-5294.

Raman Spectroscopy on Transition Metal Complexes

P. Scholz[a], C. Fickert[a], A. Gbureck[a], K. Nielsen[b], W. A. Schenk[b], G. Wahl[c], J. Sundermeyer[c], M. E. Schneider[b], H. Werner[b], A. Materny[a], and W. Kiefer[a]*

a) Institut für Physikalische Chemie, Universität Würzburg, Am Hubland, D-97074 Würzburg, Germany.
b) Institut für Anorganische Chemie, Universität Würzburg, Am Hubland, D-97074 Würzburg, Germany.
c) Fachbereich Chemie der Universität Marburg, Hans-Meerwein-Str., D-35042 Marburg, Germany.

1 Introduction

Solid-state effects in the vibrational spectra of crystals are sources for a vast amount of information concerning crystal dynamics and electrostatic interactions. Factor group splittings in the ν(CO) region of transition metal complexes can be used to get information about the crystal structure of the complexes.

Complexes of the type $MoO(\eta^2-O_2)_2LL'$ are catalytically active compounds in the epoxidation of olefines [1]. In a single phase system the catalytic activity depends only on the Lewis acidity of the d^0-metal fragment and therefore on the σ donor strength of the ligands L and L' [2]. Raman spectroscopy should be used to analyse the influence of the ligands L or L' on the bonding situation and on the other hand for structural characterisation of unsaturated "$MoO(O_2)_2L$" complexes.

Raman spectroscopy is also a very common tool for studying the chemistry of surfaces. Catalytically active rhodium(I) complexes $[Rh(F_x-acac)(C_2H_4)_2]$ (x = 0, 3, 6) immobilized on a SiO_2-surface yield heterogeneous catalysts, which were investigated by Fourier-Transform (FT)-Raman spectroscopy. This technique with an excitation in the near infrared (NIR) region has some advantages over conventional Raman spectroscopy: on one hand photochemical reactions in the laser focus are avoided, and on the other hand fluorescence (which often obscures the whole Raman spectrum) in most cases can be avoided.

2 Experimental

The compounds have been prepared by established methods [2-4]. Raman spectra were excited with the 647.1 nm line of a krypton ion laser (Spectra Physics model 2025). The scattered light was dispersed by means of a SPEX model 1404 double monochromator and detected with a Charge Coupled Device (CCD) camera system (photometrics, model RDS 2000) employing the scanning multichannel technique [5]. Temperature dependend measurements have been carried out using the surface-scanning technique developed by Zimmerer and Kiefer [6] and a cryocooler (CTI-Cryogenics, model 22C). Raman spectra of the peroxo complexes were measured with a microscope setup and sample materials were handled under argon atmosphere.

FT-Raman spectra were recorded with a Bruker IFS 120 instrument equipped with the FT-Raman-module FRA 106. A Nd:YAG laser operating at 1064 nm was used with a power of 500 mW. 180°-scattering geometry was employed. Typically 10000 scans were averaged. The spectra were recorded with a resolution of 2 cm^{-1}.

3 Results and Discussion

3.1 Factor group splitting in the ν(CO) region of polycrystalline (dppe)M(CO)₃(SO₂)

Considering the bidentate ligands as point masses, the *mer*-(dppe)M(CO)$_3$(SO$_2$) complexes (M = Mo, W) are build up by twelve atoms and can be regarded to have local symmetry C_s. As a consequence three ν(CO) modes of 2 A' + A'' species can be observed in the solution spectra. In order to discuss solid spectra, factor group analysis [7] has been carried out (both complexes have space group P2$_1$/n (C_{2h}^5), Z=4) [3]. Factor group splittings into two Raman-active modes (A$_g$ + B$_g$) and two IR-active modes (A$_u$ + B$_u$) are expected.

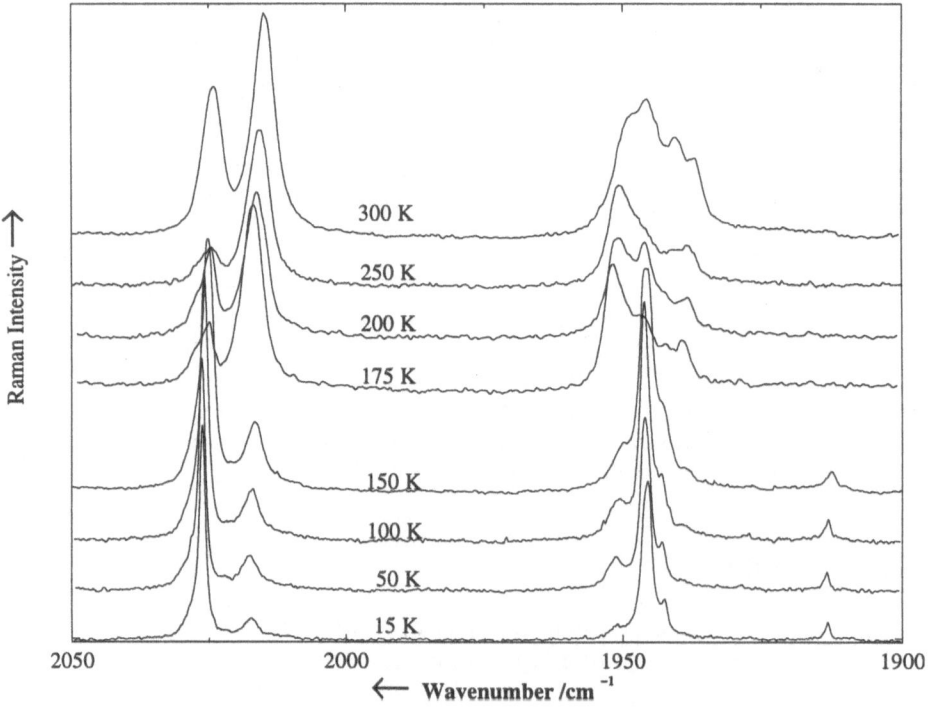

Figure 1 Raman spectra of the ν(CO) stretching region of polycrystalline *mer*(dppe)W (CO)$_3$(SO$_2$), at different temperatures as indicated.

The predicted splitting does occur for the molybdenum derivative and for the tungsten only at lower temperatures. At higher temperatures the tungsten complex shows a different factor group splitting (Figure 1).

The differences in factor group splittings may be caused by a temperature dependent change of the crystal structure. It is concluded that the tungsten complex in its microcrystalline form has a hitherto unknown structure which, at low temperature, undergoes a phase change to the structure known from single-crystal measurements.

3.2 Raman spectroscopic studies of molybdenum bis(peroxo)oxo complexes

Characteristic vibrations of $MoO(\eta^2-O_2)_2LL'$ are the $v(Mo=O)$ mode located between 900 - 1000 cm^{-1}, the symmetric and asymmetric $v(O-O)$ modes around 880 cm^{-1}, and the $v(Mo(O_2)_2)$ modes between 500 - 600 cm^{-1} [8]. The σ donor ligand in *trans* position to the oxo function extremely affects the Mo=O bonding leading to characteristic shifts for $v(Mo=O)$. The different donor ability of the ligands $OP(n\text{-dodec})_3$, HMPA, and collidin-N-oxide are reflected by the $v(Mo=O)$ values given in Table 1. An increasing Lewis acidity of the metal fragment produces an increasing bonding order of the oxo function. Therefore, the shifts in Table 1 refer to a decrease of the σ donor strength in the series collidin-N-oxide > HMPA > $OP(n\text{-dodec})_3$ [2].

Figure 2 shows the Raman spectra of three different HMPA complexes. Spectra A and B possess an equal band pattern with small shifts depending on the different ligands L' (H_2O or HMPA) in *trans* position to the [Mo=O] unit. Spectrum C of the unsaturated complex "$MoO(O_2)_2(HMPA)$" shows a complicated band pattern in the $v(Mo(O_2)_2)$ region, which could be explained by a dimerisation of two metal fragments. A dimer (see Figure 2) possesses two different peroxo ligands [2].

Table 1 Characteristic Raman data of $MoO(\eta^2-O_2)_2L_2$.

$MoO(\eta^2-O_2)_2L_2$ L=			assignment
$OP(n\text{-dodec})_3$ [cm^{-1}]	HMPA	collidin-N-oxide	
954 s	949 s	935 s	$v(Mo=O)$
877 m	876 m	880 m	$v(O-O)$
866 sh	866 sh	868 vw	

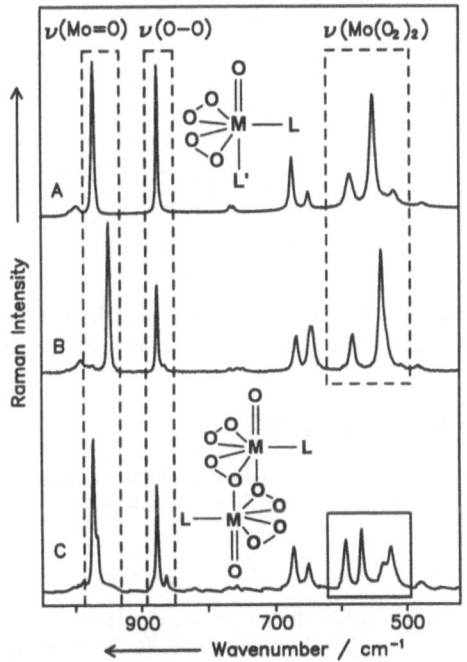

Figure 2 Raman spectra of $MoO(\eta^2O_2)_2$ (HMPA)L' [(A): L' = H_2O, (B): L' = HMPA] and (C): "$MoO(O_2)_2(HMPA)$".

3.3 FT-Raman spectroscopic investigations on silica-supported rhodium complexes

The FT-Raman-spectra of the catalytically active rhodium(I) complexes [$Rh(F_x\text{-acac})(C_2H_4)_2$] (x = 0, 3, 6) [4] were first recorded in benzene solutions and then after immobilization on a SiO$_2$-surface. Activation of the SiO$_2$-immobilized complexes with SiCl$_4$ or TiCl$_4$ improved the catalytic activity. Therefore, the FT-Raman spectroscopic investigations were also extended to this case.

An interesting feature of heterogeneous catalysts in general is the structure of the catalyst on the surface. By studying the characteristic Raman bands of the ethene as well as the acetylacetonato ligand we could show that immobilization on SiO$_2$ does not change the structure of the catalytically active complex. A substitution of the anionic ligand by surface hydroxyl groups does not occur like in the case of the immobilization of the tris(allyl)rhodium(III) complex [Rh(η_3-C$_3$H$_5$)$_3$] on SiO$_2$. However, a substitution of the anionic ligand as well as a conversion of the ethene ligand is very likely after treatment with the Lewis acids SiCl$_4$ or TiCl$_4$. The behaviour of the anionic ligand is demonstrated in Figure 3, which shows the FT-Raman spectra of [Rh(acac)(C$_2$H$_4$)$_2$].

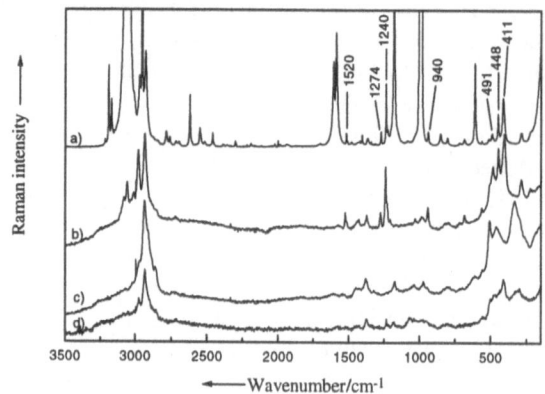

Figure 3 FT-Raman spectra of [Rh(acac)(C$_2$H$_4$)$_2$] in benzene solution (a), heterogenized on SiO$_2$ (b), after treatment with SiCl$_4$ (c) or TiCl$_4$ (d).

Acknowledgements

Financial support from the Deutsche Forschungsgemeinschaft (SFB 347, projects C-2, B-3, A-4, and D-1) is highly acknowledged.

References

1. K. A. Jørgensen, *Chem. Rev.* **1989**, *89*, 431.
2. G. Wahl, J. Sundermeyer, R. Stowasser, G. Bringmann, C. Fickert, W. Kiefer, in preparation.
3. K. Nielsen, *Diploma thesis*, Universität Würzburg **1995**.
4. M. E. Schneider, *Ph. D. thesis*, Universität Würzburg, **1996**.
5. V. Deckert, W. Kiefer, *Appl. Spectrosc.* **1992**, *46*, 322.
6. N. Zimmerer, W. Kiefer, *Applied Spectrosc.* **1974**, *28*, 279.
7. S. Bhagavantam, T. Venkatarayudu, *Therory of Groups and its Application to Physical Problems*, 3. Edition, Andhra University, Waltair **1962**.
8. N. J. Campbell, A. C. Dengel, C. J. Edwards, W. P. Griffith, *J. Chem. Soc. Dalton Trans.* **1989**, 1203.

Photochemistry Studies by Matrix Isolation Raman Spectroscopy

C. Fickert[a], V. Nagel[a], S. Möller[b], W. Malisch[b], R. Stowasser[c], G. Bringmann[c], A. Materny[a], and W. Kiefer[a]*

[a]Institut für Physikalische Chemie, [b]Institut für Anorganische Chemie, [c]Institut für Organische Chemie, Universität Würzburg, Am Hubland, D-97074 Würzburg, Germany.

1 Introduction

The matrix isolation technique has been used to stabilize reactive species for spectroscopic identification in order to analyse reaction pathways. This is demonstrated here for two different examples. The photochemistry of $[CpFe(CO)_2]_2$ (1) (Cp = η^5-C_5H_5) is dominated by two different photochemical channels, either a CO loss to form the triply-bridged intermediate $Cp_2Fe_2(CO)_3$ (1') or homolysis of 1 into the radicals •$FeCp(CO)_2$. Intermediate 1' possesses a strong electronic absorption centered at 510-550 nm. Therefore, Raman measurements with λ_0 = 514.5 nm excitation should be a sensitive detection tool for 1'. Due to resonance enhancement an analysis of 1' in low concentrations should be possible [1].

Metallosilanes show photochemical activity and, as a consequence of the induced CO loss, they are predicted to undergo intramolecular insertion, in which silylene complexes have been proposed to occur as intermediates. In irradiation experiments of $Cp(CO)_2$ FeSi-Me_3 a short living 16 electron species $[Cp(CO)FeSiMe_3]$ was detected by IR spectroscopy, but no intramolecular stabilization by the $SiMe_3$ unit could be observed [2]. Further investigations with $Cp(CO)_2FeSi_2Me_5$ allowed the detection of a CO loss product, which was stabilized by the Si_2Me_5 unit [2]. An intramolecular stabilization of a CO loss species is also expected for hydridosilyl complexes like $Cp(CO)_2FeSiH_3$ (2) (see Figure 1), due to the higher reactivity of the Si-H bond.

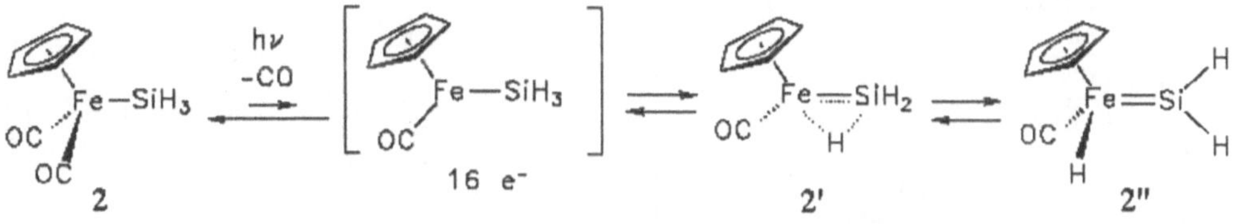

Figure 1 Possible reaction pathway for the photochemistry of 2.

2 Experimental

1 and 2 have been synthesized by standard procedures [3, 4]. A description of the matrix isolation apparatus is given in Ref. [5]. Sample material was heated in a Knudsen cell at 90°C (for 1) and at 35°C (for 2). The gas was directly mixed with krypton or nitrogen and

deposited on the target at 20 K by the "slow-spray-on" technique. For excitation of the Raman spectra the 647.1 nm line of a krypton or the 514.5 and 457.9 nm lines of an argon ion laser were used. Photochemistry of **2** was initiated by UV irradiation with an argon ion laser, operating multiline in the wavelength range between 333 and 364 nm. Density functional theory (DFT) calculations were performed on CRAY T90 computers by means of DGauss 3.0 using BP functional with DZVP as basis set [6].

3 Results and Discussion

3.1 Resonance Raman study of matrix isolated *trans*-[CpFe(CO)$_2$]$_2$

The UV/VIS spectrum of **1** shows several absorption bands between 300 and 650 nm. In the 647.1 nm Raman spectra of **1** no resonance enhancement can be observed. Characteristic Raman data of **1** are listed in Table 1. Due to the low concentration of **1** in the matrix layer only the strongest Raman bands below 1100 cm^{-1} could be observed in the spectra for off resonance excitation. The $v(CO)$ values in Table 1 are derived from Raman spectra of polycrystalline samples, which show factor group splittings [7].

The Raman spectra as displayed in Figure 2 are completely different from the λ_0 = 647.1 nm spectrum. Spectrum A was measured resonantly exciting **1'** at an

Table 1 Characteristic Raman data of **1** (λ_0 = 647.1 nm).

$\tilde{v}$ [cm^{-1}]	assignment
1972 w / 1960 w	A$_g$ $v(CO)_{term}$
1793 w / 1784 vw	A$_g$ $v(CO)_{brid}$
414 m	A$_g$ $v(Fe_2C_2)_{brid}$
231 s	A$_g$ $v(FeFe)$

Table 2 Characteristic Raman data of **1'**.

$\tilde{v}$ [cm^{-1}]		assignment
514.5	457.9 nm	
	1867 s	A$_1$' $v(CO)_{brid}$
520 vvs	522 vs	A$_1$' $v(Fe_2C_3)_{brid}$
221 vvs	220 vs	A$_1$' $v(FeFe)$

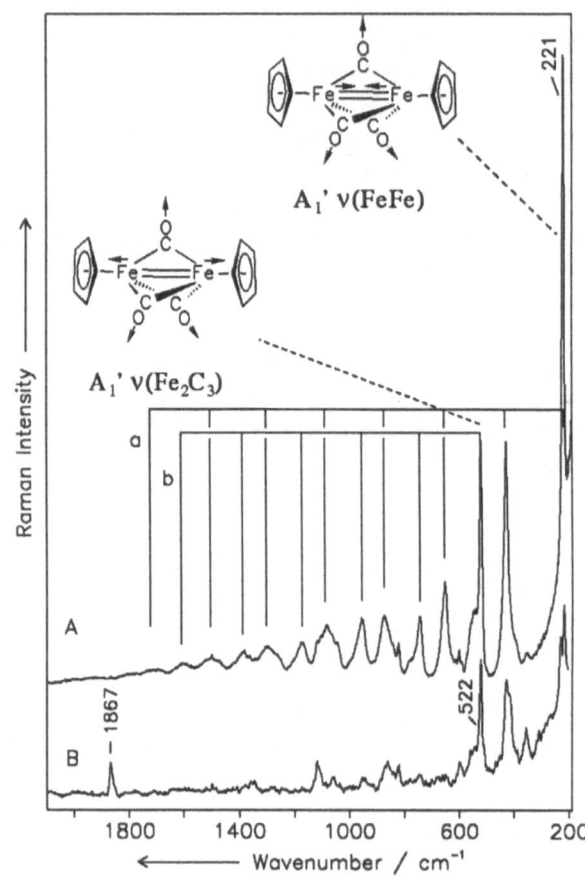

Figure 2 Raman spectra of matrix isolated **1** in a 9:1 mixture of Kr and N$_2$;
A: λ_0 = 514.5 nm and B: λ_0 = 457.9 nm.

288

excitation wavelength of $\lambda_0 = 514.5$ nm. It shows two characteristic Franck Condon progressions (a and b) starting with the two totally symmetric modes of the Fe_2C_3 unit of **1'**. Some fundamental vibrations of **1'** are listed in Table 2. The A_1' $v(CO)$ mode could be observed in spectrum B ($\lambda_0 = 457.9$ nm) at 1867 cm^{-1} [1, 7].

3.2 Photochemistry of Cp(CO)₂FeSiH₃ in low temperature matrices

The photochemical CO loss in **2** (Figure 1) is correlated with characteristic changes in the Fe-Si bonding behavior. Earlier investigations of silyl complexes have shown that Raman spectroscopy is a sensitive detection tool for the Fe-Si unit [8, 9]. The $v(FeSi)$ mode of **2** occurs as strong Raman band at 322 cm^{-1} (Figure 4, spectrum A).

The UV/VIS spectrum of **2** in benzene shows only one strong broad absorption around 325 nm. Therefore, the irradiation experiments were carried out within the UV multiline range (333-364 nm) of an argon ion laser. Figures 3 and 4 show the Raman spectra of matrix isolated **2** before (A) and after UV irradiation (B). The Raman excitation was varied between 647.1 and 514.5 nm, which results in characteristic differences in the spectra after UV irradiation (see Figure 4 B). The 647.1 nm spectrum is dominated by the Raman signals of educt **2**. These are also detectable in the 514.5 nm spectrum besides several additional bands, which can be attributed to a photochemical intermediate. During the irradiation experiment the color of the matrix layer changed to orange, so resonance enhancement of a new photogenerated intermediate is possible using $\lambda_0 = 514.5$ nm excitation. By this, new signals in B could be assigned to $v(FeSi)$, $v(FeCp)$, $v(FeC)$, or $\delta(SiH)$ modes. A possible

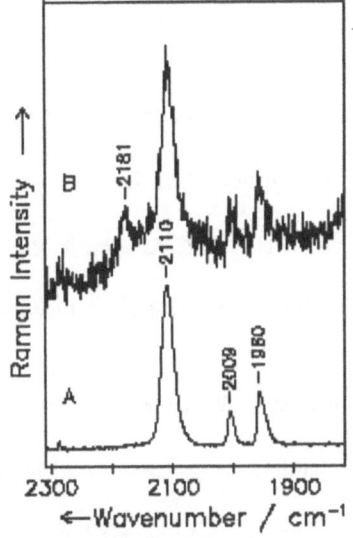

Figure 3 Raman spectra of **2** in N₂ matrix; (A) before UV irradiation and (B) after UV irradiation ($\lambda_0 = 514.5$ nm).

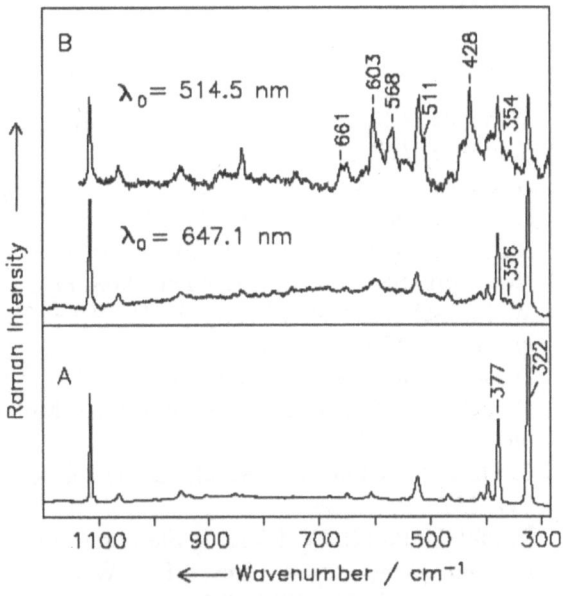

Figure 4 Raman spectra of **2** in N₂ matrix; (A) before UV irradiation and (B) after UV irradiation.

289

Table 3 Comparison of experimental Raman data and DFT calculations of **2** and **2'**.

2		2'		assign-ment
DFT BP/DZVP	Raman	DFT BP/DZVP	Raman 514.5 nm	
2003	2009 w	1963		$v_s(CO)$
1966	1960 w			$v_{as}(CO)$
531	523 m	591	603 s	$v(FeC)$
			568 s	
367	377 vs	340	354 vw	$v_s(FeCp)$
316	322 vs	459	428 vw	$v(FeSi)$

interpretation is given in Table 3 by comparison with DFT calculated vibrational data of **2** and **2'**, which are in good agreement with the experimental data. Due to the higher FeSi bonding order in **2'** the $v(FeSi)$ mode is shifted to higher wavenumbers. Therefore, an assignment to the strong enhanced signal at 428 cm^{-1} is possible [6]. A shift to higher wavenumbers is also expected for the $v(SiH_2)$ modes, which is confirmed by the new signal at 2181 cm^{-1} between the $v(SiH_3)$ band of **2** at 2110 cm^{-1} in spectrum B of Figure 3. Due to the low signal/noise ratio an additional $v(CO)$ band for **2'** could not be observed, which is expected in the region of the $v_{as}(CO)$ of **2**. Equal results are found in analogous matrix isolation experiments of Cp(CO)FeSiH$_2$Me [6, 10].

Acknowledgement

Financial support from the Deutsche Forschungsgemeinschaft (SFB 347, Teilprojekte C-2, B-1, and B-2) as well as from the Fonds der Chemischen Industrie is gratefully acknowledged.

References

1. M. Vitale, K.K. Lee, C.F. Hemann, R. Hille, T.L. Gustafson, B.E. Bursten, *J. Am. Chem. Soc.* **1995**, *117*, 2286.
2. A. Haynes, M.W. George, M.T. Haward, M. Poliakoff, J.J. Turner, N.M. Boag, M. Green, *J. Am. Chem. Soc.* **1991**, *113*, 2011.
3. R.B. King in *Organometallic Syntheses*, Eds. J.J. Eisch, R.B. King, Academic Press, New York, San Francisco, London, **1965**, 114.
4. W. Malisch, S. Möller, O. Fey, H.-U. Wekel, R. Pikl, U. Posset, W. Kiefer, *J. Organomet. Chem.* **1996**, *507*, 117.
5. D. Gernet, W. Kiefer, in preparation.
6. C. Fickert, R. Stowasser, G. Bringmann, S. Möller, W. Malisch, W. Kiefer, in preparation.
7. C. Fickert, P. Günther, P. Scholz, D. Gernet, R. Pikl, W. Kiefer, *Inorg. Chim. Acta* **1996**, *251*, 157.
8. R. Pikl, U. Posset, W. Kiefer in *Stereoselective Reactions of Metal-Activated Molecules (Second Symposium)*, Eds. H. Werner, J. Sundermeyer, Vieweg, Braunschweig/Wiesbaden, **1995**, 221.
9. R. Pikl, C. Fickert, W. Kiefer, *Trends in Organometallic Chemistry* **1997**, *2*, 71.
10. C. Fickert, R. Pikl, D. Gernet, S. Möller, W. Malisch, W. Kiefer, *Fresenius J. Anal. Chem.* **1996**, *355*, 340.

Sol-Gel-Processed Functionalized Metalla-Phospha-Adamantanes. Highly Mobile Reaction Centers for Chemistry in Interphases

Hermann A. Mayer* and Joachim Büchele

Institut für Anorganische Chemie der Universität Tübingen, Auf der Morgenstelle 18, 72076 Tübingen, Germany

To combine the advantages of the homogeneous with those of the heterogeneous catalysis efforts to transfer the principles of polymer-bound reagents to transition metal complexes have been made for many years. The strategy to immobilize transition metal complexes on a matrix like silica gel suffers from the loss of homogeneity due to minor changes of the structure of the reactive centers. This leads to a reduced reactivity and selectivity of the immobilized catalysts. Among others, a further drawback are the short lifetimes of these catalysts caused by the leaching. In many cases, the influence of the matrix on the final outcome of the reactivity and selectivity is not known. This is due to the lack of structural knowledge about both the reactive centers and the polymer matrix.

Optimal results in the performance of heterogenized homogeneous catalysts should be obtained, if the reactive center is in a state, which is able to simulate homogeneous reaction conditions. Homogeneous catalysts have uniform and well-defined reactive centers, which lead to high and reproducible selectivities. This leads to the concept of interphases.

1 Concept of Interphases

Interphases (Figure 1) are particular regions within a material, in which a stationary component and a mobile component penetrate each other on a molecular level. In these regions a reactive center becomes highly mobile simulating the properties of a solution. In interphases no homogeneous mixtures are formed. The stationary phase is composed of an inert carrier matrix, a flexible spacer and a reactive center, whereas the mobile phase is a solvent, a gaseous or liquid reactant.

The aim is to prepare highly cross-linked polymers, in which the functional groups are securely incorporated in the hybrid-polymers. At the same time, sufficient swelling abilities maintain the accessibility and mobility of the reactive centers. The resulting advantages of interphases are (i) very similar accessibility of the reactive centers in the interphase compared to that in the homogeneous case, (ii) tuneable densities and distances of the reactive centers and therefore adjustable reactivities in the interphase, (iii) reduced leaching of functional groups due to the high degrees of cross-linkage, (iv) easy and complete separation of the reaction products from the polymer by a simple filtration or centrifugation process.

291

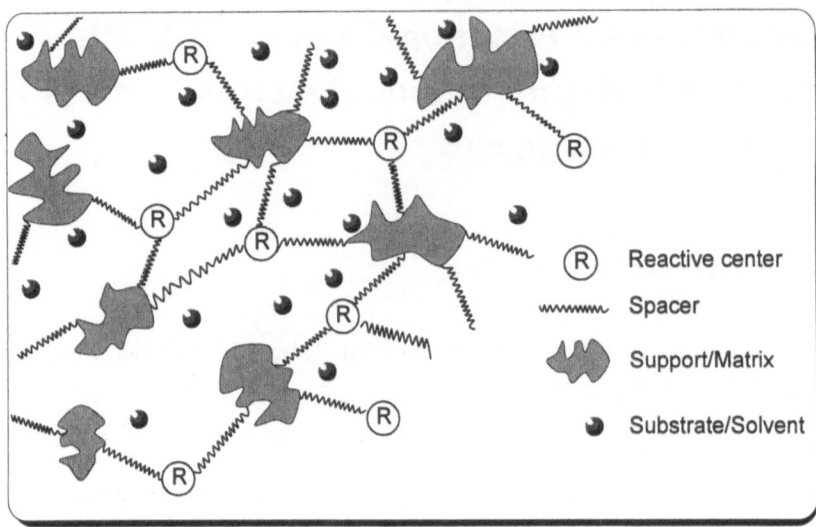

Figure 1

2 Stationary Phase

In the present work the stationary phases consist of metalla-phospha-adamantanes as reactive centers, polyethylene glycol as flexible spacers, and silica as carrier matrix (Scheme 1). The material is synthesized in several steps (Scheme 1). The thermally stable tris(hydroxymethyl) complex **1** is functionalized with polyethylene glycol groups in a base catalyzed oxirane polyaddition reaction. The degree of polymerization was adjusted by the stoichiometric ratio of **1** and oxirane to 50, 145 and 285. The polyethylene glycol functionalized {[M]-tripod} templates **2** exhibit excellent solubility behavior in polar solvents like THF and water, and swelling abilities in ether. The hydroxylic end groups of the homologous polymers of **2** react with $Cl(CH_2)_3Si(OEt)_3$ under Williamson ether synthesis conditions to the corresponding tris[(triethoxy)silyl] functionalized complexes **3**. The simultaneous co-condensation of **3** with various amounts (0, 6, 30 equivalents) of tetraethoxysilane results in organometallic-inorganic hybride polymers. Since the polyethylene glycol spacers already offer a high flexibility, $Si(OEt)_4$ was used as co-condensating agent to achieve a high cross-linking of the polymer. This is important in order to obtain reduced leaching of the reactive centers.

3 Characterization of the Materials by Solid State NMR Spectroscopy

Multinuclear CP/MAS solid-state NMR spectroscopy is an excellent tool to obtain detailed information about the structure and dynamic properties of these amorphous materials. ^{29}Si CP/MAS NMR spectroscopy shows that the degree of condensation is about 80%. The analysis of the relaxation times T_{SiH} and $T_{1\rho H}$ indicate that the reactive centers are distributed homogeneously across the siloxane framework when small amounts of cocondensation agent are applied. When 30 equivalents of Si(OEt)$_4$ are used in the sol-gel process inhomogeneous regions are formed which are larger than 1-2 nm in diameter. Information about the dynamic of the {[M]-tripod} template can be extracted from the line width of the ^{31}P CP/MAS NMR spectra. It was found that the flexibility of the immobilized complexes increases when the degree of polymerization of the spacer lengths is expanded from 50 to 145. If the degree of polymerization is enlarged to 285, the mobility is reduced caused by a stronger interaction among the polyethylene glycol chains. A dependence of the line width on the amount of co-condensate used was also observed. Wide-line-separation NMR experiments demonstrate that the mobility of the spacer is independent from the siloxane frame work and from the metal complexes.

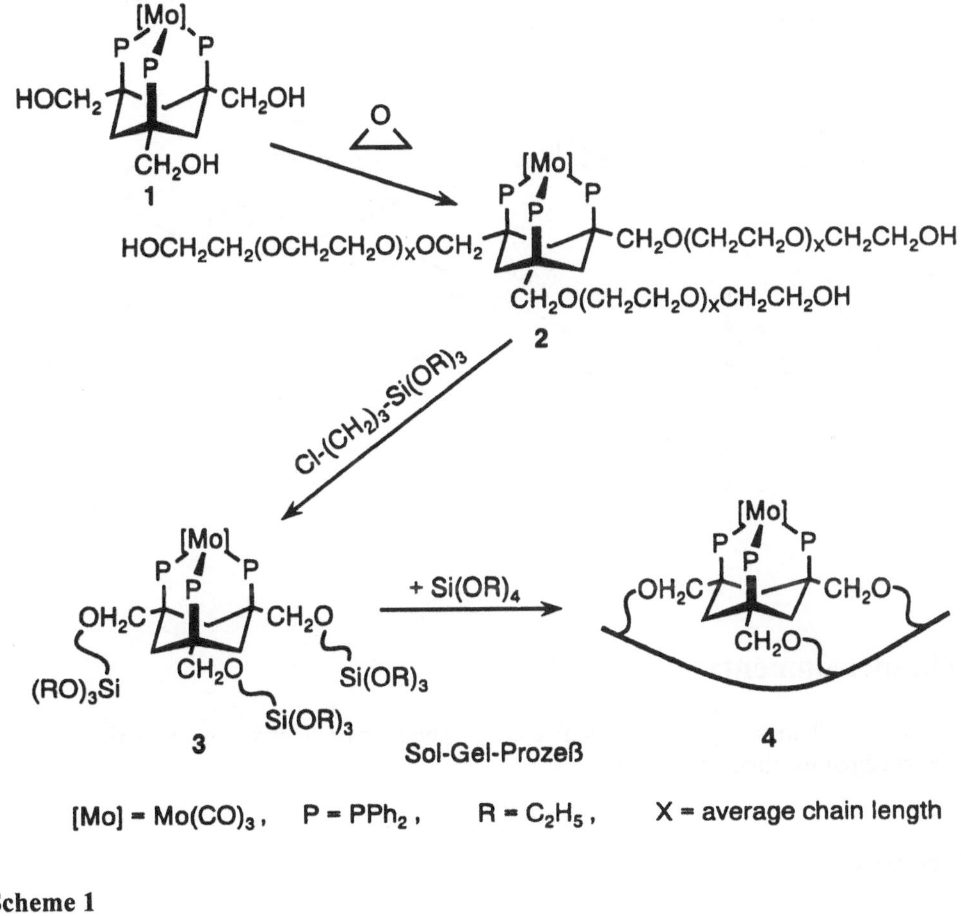

Scheme 1

4 Chemistry in the Interphase

The liberation of the immobilized phosphine ligand from the $Mo(CO)_3$ fragment in **4** is the first example to demonstrate the good accessibility of the reactive centers (Scheme 2). The oxygen transferring agent N_2O is able to selectively oxidizes the $Mo(CO)_3$ fragment to MoO_3 and generate **5** quantitatively upon irradiation of **4** with ultra violet light in the presence of N_2O. This was shown by the absence of the $\nu(CO)$ absorptions of the $Mo(CO)_3$ fragment in the IR spectrum and a shift of the ^{31}P resonance to higher field (δ 29) in the ^{31}P CP/MAS spectrum of **5**. The MoO_3 can be washed out after conversion into MoO_4^- with K_2CO_3. The chemical accessibility of the immobilized tripodal phosphine ligand depends on the size of the metal fragment. While Vaska's complex is too large to enter the interphase and react with the tripodal phosphine the dimer $[Ir[COD]Cl]_2$ is readily split by **5** to form the new immobilized iridum complex **6**. The characteristic ^{31}P NMR signal is shifted to δ -7. The change of the line width from 430 Hz (**4**) to 170 Hz (**5**) and 930 Hz for **6** indicates a highly mobile free phosphine ligand and shows that the mobility of the supported complexes depend on the size of the metal fragment.

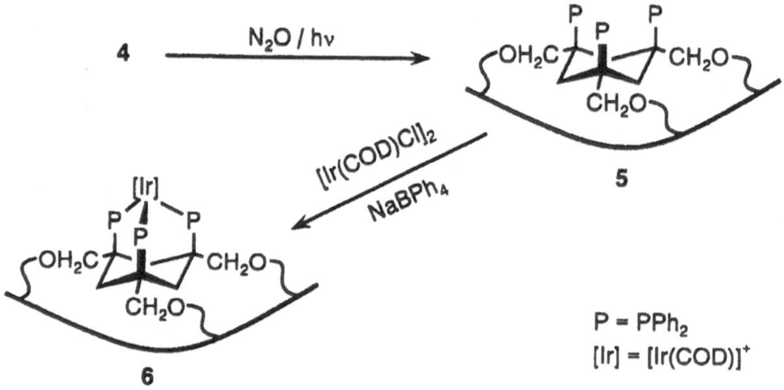

Scheme 2

Acknowledgments

The Deutsche Forschungsgemeinschaft and the Fonds der Chemischen Industrie is greatly acknowledged for financial support.

References

E. Lindner, M. Kemmler, Th. Schneller, H. A. Mayer, *Inorg. Chem.* **1995**, *34*, 5489-5495.
J. Büchele, Diplomarbeit, Tübingen 1996

Calculation of Structures, Activities and Spectroscopic Properties of Metal-Activated Molecules

G. Bringmann*[a], M. Breuning[a], S. Busemann[a], J. Kraus[a], C. Rummey[a], R. Stowasser[a], D. Vitt[a], W. Kiefer*[b], C. Fickert[b], T. Linker*[a], F. Rebien[a], W. Malisch*[c], S. Möller[c], J. Sundermeyer*[d], G. Wahl[d]

[a] Institute of Organic Chemistry, Am Hubland, D-97074 Würzburg, Germany
[b] Institute of Physical Chemistry, Am Hubland, D-97074 Würzburg, Germany
[c] Institute of Inorganic Chemistry, Am Hubland, D-97074 Würzburg, Germany
[d] Department of Chemistry, Hans-Meerwein-Straße, D-35042 Marburg, Germany

1   Introduction

Modern quantumchemical methods are powerful tools for the investigation of metal-activated compounds. The efficiency of interdisciplinary collaboration between theoretical and experimental scientists is demonstrated by several joint SFB 347 projects.

2   Calculation of Activities, Dynamics, and Structures

Mo-complexes of type **1** efficiently catalyze epoxidations of olefins. The ligand **L** has a great influence on the solubility, stability and activity of the catalyst. BLYP/DZVP calculations show that the dimeric structure **2** is by 66 kJ/mol more stable than the monomer **1** (L = OPMe$_3$). This is corroborated experimentally by Raman spectroscopy of this catalyst in the absence of a coordinating solvent. Among various physical properties calculated (PW91/TZVP), the proton affinity showed a good correlation with the activity of the Mo-catalysts [1].

Furthermore, the regio- and stereochemistry of the Cr(CO)$_3$ metal fragment on a helically twisted biaryl lactone was elucidated through DF-calculations. The calculated (BP/DZVP) minimum structures of the different regioisomers of **3** can be subdivided into structures with sterically (the tripodal rotation is blocked) and electronically (the metal fragment rotates rapidly) controlled positions of the Cr(CO)$_3$ rotor [2,3].

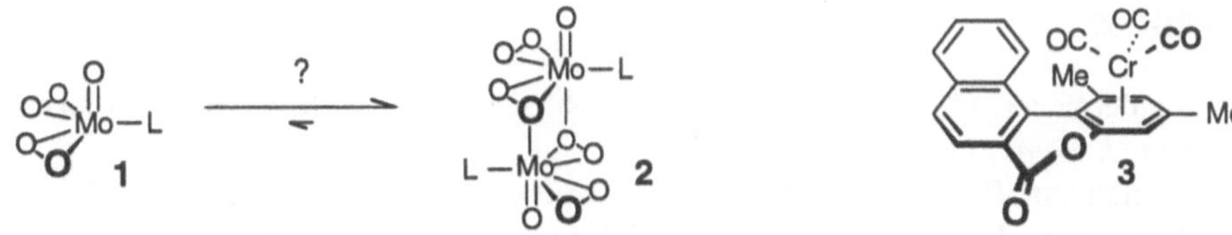

Figure 1   Mono- and dimeric [Mo]-catalysts **1** and **2**; one of the possible regioisomers **3** of a biaryl lactone η^6-chromium-tricarbonyl complex

3 Calculation of Spectroscopic Properties

By comparison of calculated and experimental vibrational data from Raman spectra, the photolysis product of the metallosilane **4** was identified as **5**, a silylene complex without base stabilization [4]. In addition, the absolute configurations of the Jacobsen epoxidation products of dihydro-naphthalenes, *e.g.* **6**, were elucidated by CD-calculations. Conformational analyses were performed by semiempirical NDDO-methods (AM1, PM3), while the CD properties were calculated by the CNDO/S-CI method.

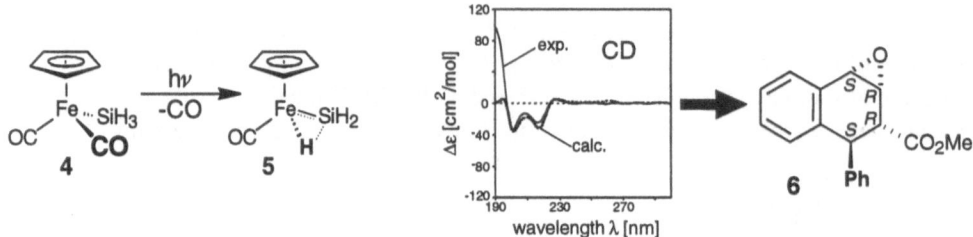

Figure 2   Photolysis of **4** and the stereostructure of the expoxide **6** by CD investigations

4 Mechanistic Studies

The complex mechanistic course of the stereoselective ring opening of configuratively labile biaryl lactones with BH_3 and oxazaborolidines (see previous paper) was studied by semiempirical AM1 calculations [5], showing the *second* hydride attack to be responsible for the overall stereoselectivity due to a dynamic kinetic resolution of the intermediate aldehyde, which was then verified experimentally.

The equilibrium between the intermediate aldehydes and their lactol isomers plays an important role within the stereoselective ring opening. *Ab initio* methods describe this equilibration process better than AM1, due to problems with the AM1-parametrization.

Ackowledgements

This work was supported by the Deutsche Forschungsgemeinschaft (SFB 347, projects B-1, B-2, and C-2) and by the Fonds der Chemischen Industrie.

References

[1]   J. Sundermeyer, G. Wahl, G. Bringmann, R. Stowasser, C. Rummey, C. Fickert, W. Kiefer, *J. Catal.*, in preparation.
[2]   G. Bringmann, R. Stowasser, D. Vitt, *J. Organomet. Chem.* **1996**, *520*, 261-264.
[3]   G. Bringmann, R. Stowasser, L. Göbel, *J. Organomet. Chem.* **1997**, in press.
[4]   G. Bringmann, R. Stowasser, W. Malisch, S. Möller, C. Fickert, W. Kiefer, in preparation.
[5]   G. Bringmann, D. Vitt, *J. Org. Chem.* **1996**, *60*, 7674-7681.

Relativistic Density Functional Calculations on Transition-Metal and Lanthanide Compounds

Christoph van Wüllen

Lehrstuhl für Theoretische Chemie, Ruhr-Universität, D-44780 Bochum, Germany

There is much interest in transition metal complexes, mainly because of the role these compounds play in homogeneous catalysis. Therefore, the availability of structural and energetic data such as bond lengths, force constants, and bond dissociation energies of transition metal complexes is quite important. However, the amount of reliable experimental data is still limited. Even for transition metal carbonyls and their derivatives, which are a well-known class of compounds, only few metal-ligand binding energies have been obtained experimentally, and especially for bond dissociation energies drawn from kinetic studies, the error bars are sometimes not small.

Some standard methods from ab initio quantum chemistry, for example the Hartree-Fock and MP2 (Møller-Plesset perturbation theory to second order) methods, are not very accurate when applied to transition metal complexes, especially to those where the metal is in a low oxidation state. There are more accurate methods like the coupled cluster approach, but such calculations are rather expensive in terms of computer time. Fortunately, calculations based on density functional theory (DFT) are a cost-effective alternative to the ab initio methods. If gradient-corrected ('nonlocal') functionals are used, DFT calculations can provide accurate molecular structures and predict binding energies at least semiquantitatively [1].

Another problem emerges if transition metals from the second or third row are present in the complex. In the cores of the heavy atoms, the kinetic energy of the electrons is so large that relativistic effects cannot be neglected: there are cases where relativistic effects [2] change metal-ligand bond lengths by more than 10 pm (0.1 Å) and contribute more than 50% to the binding energy (for the d elements, relativistic effects generally shorten and strengthen metal-ligand bonds). The present investigator has implemented a Kohn-Sham density functional program [3] and developed a method for the calculation of relativistic corrections by relativistic perturbation theory [4]. Results of such calculations will be shown in the present contribution.

The metal-phosphorus bond in tungsten pentacarbonyl phosphines

The neutral, coordinatively saturated octahedral hexacarbonyls of the group VI metals chromium, molybdenum, and tungsten may be regarded as prototypes of low-valent donor-acceptor complexes formed by these metals. Likewise, the substitution reaction

$$M(CO)_6 + L \rightarrow M(CO)_5L + CO$$

is a model for a large class of chemical reactions. M-L bond lengths and bond dissociation energies have been obtained for a variety of ligands L [5]. It is also interesting to look at the metal-carbon bond lengths and at the C=O force constants in the monosubstituted hexacarbonyls since they depend on the electronic properties of the substituting ligand L. This has been done for various phosphine ligands (PH_3, PMe_3, PF_3, PCl_3, PBr_3, and $P(CF_3)_3$) as to elucidate the contributions of both σ basicity and π acidity to the trends in the M-P binding energy. In experimental studies [6,7] usually only the vibrational frequencies in the CO stretching region are analyzed, while binding energies are not known and the error bars of the bond lengths are too large to allow a comparison.

If the ligand L is a good π acceptor, it will reduce the backbonding to the other (i.e. the carbonyl) ligands. This leads to an increase in the metal-carbon bond lengths, while the C=O bond is strengthened. This effect is more pronounced on the trans (axial) CO ligand ('trans effect') since a pair of trans ligands shares two metal d orbitals of appropriate symmetry, while only one such orbital is shared by a pair of cis ligands. If the ligand L is a good σ donor, backbonding to the carbonyl ligands might be enhanced. This enhancement is not much different for the trans and cis carbonyl groups.

Table 1 Bond lengths, bond dissociation energies, and C=O force constants in octahedral $W(CO)_5$ complexes

L	$W-P$[a]	$W-C_{ax}$[b]	$D_e(W-P)$[c]	$D_e(W-C_{ax})$[c]	k_1[d]	k_2[d]
PH_3	251.8	201.5	138	233	1650	1593
PMe_3	254.2	202.0	182	227	1629	1584
$P(^iPr)_3$	260.7	201.0	148	234		
PF_3	239.9	202.9	147	205	1683	1638
PCl_3	244.2	202.7	122	209	1681	1629
PBr_3	245.7	202.6	107	210	1677	1620
$P(CF_3)_3$	245.2	202.7	121	214	1673	1622
NMe_3		199.1		265	1626	1559
N_2		202.2		233	1678	1605
CO		206.1		193	1685	1645
CH_2		210.0		159	1678	1645

a) Tungsten-phosphorus bond length in pm. c) axial tungsten-carbon bond length in pm. b) Bond dissociation energy in kJmol^{-1}. c) Diagonal force constant of symmetric cis CO stretching mode in d) Diagonal force constant of trans CO stretching mode in Nm^{-1}.

In table 1, results of relativistic density functional calculations are compiled for several pentacarbonyl phosphine tungsten complexes. Data for some other pentacarbonyl tungsten complexes are given for comparison. The metal-phosphorus binding energy increases considerably when replacing the phosphine by the trimethyl phosphine ligand. This substitution also lowers the cis CO force constant considerably. These findings are consistent with the interpretation that the π acidity of PH_3 and PMe_3 are similar, while the latter ligand is a better σ donor, and the enhanced σ basicity of PMe_3 is the reason for the increased metal-ligand bond strength. When comparing PH_3 with PF_3 on the other hand, the larger trans metal-carbonyl bond length and its decreased bond dissociation energy indicate that PF_3 is a much better π acceptor than PH_3, a fact which is also demonstrated by the much higher value of the trans CO stretching force constant k_2.. The cis CO stretching constant k_1 is also much higher than in the phosphine complex, which can not only be explained by the π acidity of PF_3. In addition, this ligand is a much weaker σ donor than PH_3 or PMe_3. The lower σ basicity is also the reason why PCl_3, PBr_3 and $P(CF_3)_3$ (which are all better π acceptors than PH_3) are more weakly bound to the metal.

So far we have only considered electronic effects and ignored steric repulsion. Steric effects become more important if the ligands are larger, and are responsible for the increase in the metal-phosphorus bond length (and the decrease of its bond energy) in the triisopropylphosphine complex compared to $W(CO)_5PMe_3$. The data on the other complexes are given for comparison. It is no surprise that NMe_3 is found to be a good σ donor, but no π acceptor, while dinitrogen is both a weak σ donor and a weak π acceptor. Of course, one finds that CO is a good and methylene is an excellent π acceptor.

Density functional calculation of Lanthanide compounds

The one-particle equations for the Kohn-Sham orbitals of a molecule all contain the same effective potential. This is different from, say, Hartree-Fock calculations where the effective potential is orbital-dependent. As a consequence, the aufbau principle strictly holds in Kohn-Sham calculations. However, our view of Lanthanide systems is that they have partially filled atom-like 4f orbitals whose orbital energy is below the valence orbital energies. If one performs a Kohn-Sham calculation on a diatomic lanthanide oxide molecule like SmO, one encounters a mixing of the 4f open shell orbitals of the Sm atom with closed shell valence orbitals such as the oxygen lone pairs, and this mixing must be considered unphysical.

Some possible ways out of this problem are (i) Kohn-Sham variants with orbital-dependent effective potentials as introduced by e.g. a self-interaction correction or the use of Hartree-Fock type exchange terms in the exchange-correlation energy, (ii) using frozen atomic cores including the partially filled 4f shell or, better, (iii) formulate a restricted Kohn-Sham variational procedure in which the open shell orbitals may only be expanded in the f-type atomic basis functions of the lanthanide atoms. The last alternative looks promising, but remains to be worked out.

In table 2, spectroscopic constants (equilibrium distance, dissociation energy, and harmonic vibrational frequency) of some lanthanide oxides, calculated both at the nonrelativistic and relativistic level, are given together with experimental data. Since only f^0, f^7 and f^{14} configurations have been considered, the problem with partially filled f shells, as mentioned before, is not present in these cases. It can be seen that the relativistic effect on the bond length and on the vibrational frequency is relatively small for these compounds, i.e. the relativistic correction to the total energy does not depend much on the bond length.

However the relativistic effect on the dissociation energy is substantial, and opposite to what is found for the d block elements, relativity decreases the dissociation energy: before a lanthanide atom can form a bond, it must promote to an excited valence state and transfer one or two electrons from the 6s to the 5d shell. Since the 6s orbital is stabilized by relativistic effects, these increase the excitation energy to the valence state and thus decrease the overall dissociation energy. Note that the experimental bond length of YbO is probably wrong. The configuration of YbO is so close in energy that the vibrational bands overlap, and this makes the analysis of the spectra difficult.

Table 2 Spectroscopic constants of some lanthanide oxides

Compound	$R_e[\text{Å}]$	$D_e[\text{eV}]$	$\omega_e\left[cm^{-1}\right]$
LaO $(f^0\sigma^1)$			
nonrel.	1.86	9.5	793
rel.	1.89	8.9	783
expt.	1.83	8.3	813
EuO $(f^7\sigma^0)$			
nonrel.	1.88	6.8	693
rel.	1.90	5.8	686
expt.	1.89	5.0	688
YbO $(f^{14}\sigma^0)$			
nonrel.	1.89	5.6	677
rel.	1.88	4.7	671
expt.	1.81	4.1	683
LuO $(f^{14}\sigma^1)$			
nonrel.	1.80	8.5	841
rel.	1.80	8.0	846
expt.	1.79	7.0	842

References

[1]   T. Ziegler, Can. *J. Chem.* **1995**, 73, 743.
[2]   P. Pyykkö, *Chem. Rev.* **1988**, 88, 563.
[3]   Ch. van Wüllen, *Chem. Phys. Lett.* **1994**, 219, 8.
[4]   Ch. van Wüllen, *J. Chem. Phys.* 1995, 103, 3589; *J. Chem. Phys.* **1996**, 105, 5485.
[5]   Ch. van Wüllen, *J. Comput. Chem.*, in press.
[6]   M.S. Davies, R.S. Armstrong, and M.J. Aroney, *Chim. Chron.* **1995**, 24, 233.
[7]   J. Apel, R. Bacher, J. Grobe, and D. LeVan, *Z. anorg. allg. Chemie* **1979**, 39, 453.

Photodissociation Dynamics of Transition Metal Complexes: Quantum Simulation Supported by Wavepacket Propagations

M. C. Heitz[1], K. Finger[2], D. Guillaumont[1], C. Daniel[1] *

[1] Laboratoire de Chimie Quantique, UPR 139 CNRS, Université Louis Pasteur, Strasbourg, France
[2] Institut für Physikalische und Theoretische Chemie, Freie Universität, Berlin, Germany

Summary

The photodissociation dynamics of three model systems, each being representative of a class of molecules, has been studied through wavepacket propagations on CASSCF/CCI potentials, calculated for the electronic ground and excited states, as a function of the reaction coordinates.

1 Introduction

The photochemistry of transition metal complexes has a wide variety of applications such as formation of unsaturated intermediates or catalytically active radicals, synthesis of new substituted products, formation of long life-time excited states able to promote electron or energy transfer processes among others. Most of the time the UV-visible absorption spectra are poorly resolved and many questions still remain regarding the early events (in the first picoseconds) contributing to the photochemical behavior of the molecule under irradiation.

For example a variety of organometallics undergoes concurrent photoinduced primary reactions. This general behavior has been observed for transition metal hydrides for which two primary reactions, namely the departure of a carbonyl ligand and the homolysis of the metal-hydrogen bonds can occur either at a unique wavalength ($HCo(CO)_4$) or at different wavelengths ($HMn(CO)_5$) [1]. The complex photochemical behavior of transition metal dihydrides is illustrated by the photochemistry of $H_2Os(CO)_4$, for which molecular hydrogen elimination and carbonyl loss have been observed in low temperature matrices, whereas the elimination of H_2 is the only primary reaction observed after irradiation of $H_2Fe(CO)_4$ in the same experimental conditions [2].

Another aspect of the photochemistry of organometallics is the role of the triplet states and the spin-orbit coupling effects. An important question concerns the relative time scales of the direct dissociation from the singlet states vs. the indirect dissociation via singlet to triplet intersystem crossing.

301

Finally, the mixing between different kinds of excited states can complicate the photochemistry. An illustration is given by a series of RM(CO)$_3$(α-diimine) complexes which can either dissociate, leading to the formation of highly reactive intermediates, or manifest the photophysics of Metal-to-Ligand-Charge-Tranfer (MLCT) complexes (luminescence, energy/electrons transfer processes) [3]. This class of molecules is characterized by a high density and a large variety of excited states in the UV-visible energy domain: MLCT, SBLCT (Sigma-Bond-to-Ligand-Charge-Transfer), MC (Metal Centered).

We present here numerical simulations of the photodissociation dynamics of HCo(CO)$_4$, H$_2$Fe(CO)$_4$ and HMn(CO)$_3$(DAB) based on the knowledge of the excited potential energy surfaces (PES) and on the time evolution of wavepackets on these PES calculated for the following dissociation pathways:

i) the photodissociation of CO vs. the homolysis of the Co-H bond in HCo(CO)$_4$:

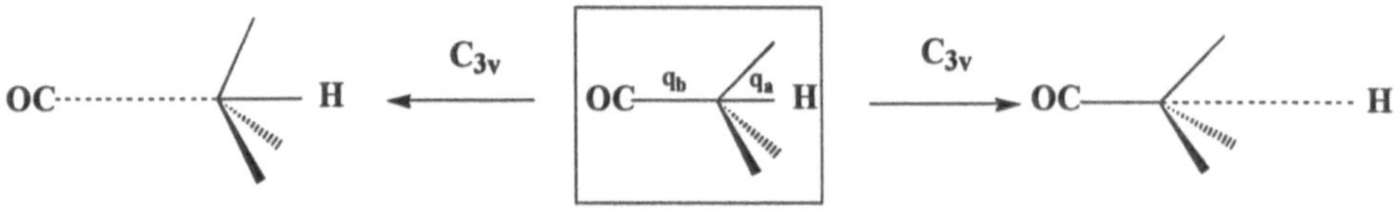

ii) the photodissociation of CO vs. the molecular hydrogen elimination of H$_2$ in H$_2$Fe(CO)$_4$:

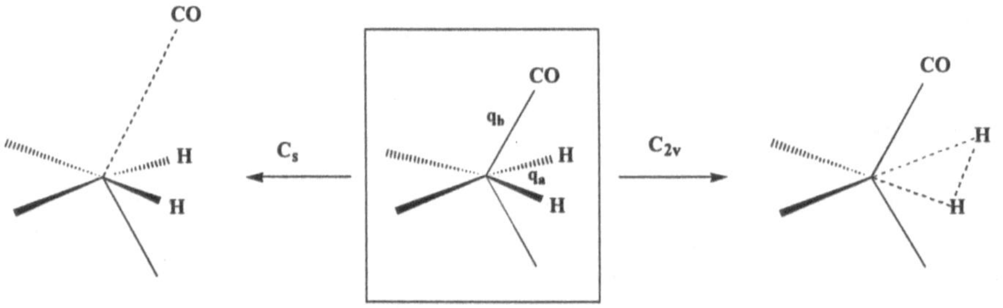

iii) the photodissociation of CO vs. the homolysis of the Mn-H bond in HMn(CO)$_3$(DAB) (DAB=1-4-diaza-1,3-butadiene):

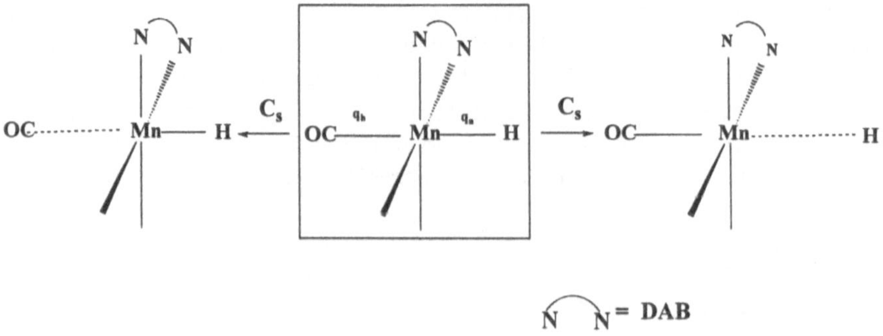

N N = DAB

2 Methods

Two-dimensional potential energy surfaces (PES) for the ground and excited electronic states were evaluated as a function of the bond elongations q_a and q_b, using CASSCF/CCI ab initio methods [4]. The corresponding vibrational wavefunctions were evaluated either by means of the Chebychev relaxation method or through the Fourier Grid Hamiltonian method [5]. The photoabsorption and the subsequent bond breaking are simulated by propagation of selected wavepackets on the non-adiabatically and spin-orbit coupled PES. The time evolution of the wavepackets is obtained by solving the time-dependent Schrödinger equation.

3 Results

3.1 Photodissociation Dynamics of HCo(CO)$_4$

The molecular reaction dynamics of competing direct dissociation on the ^1E ($3d_d \rightarrow \sigma^*_{Co-H}$) state of HCo(CO)$_4$ vs. indirect spin-orbit induced dissociation on the $^3A_1$ ($\sigma_{Co-H} \rightarrow \sigma^*_{Co-H}$) state is illustrated by the evolution of wavepackets Ψ_{1E} (q_a,q_b,t) and Ψ_{A1} (q_a,q_b,t) on the coupled PES, left side of figure 1 and right side of figure 1, respectively.

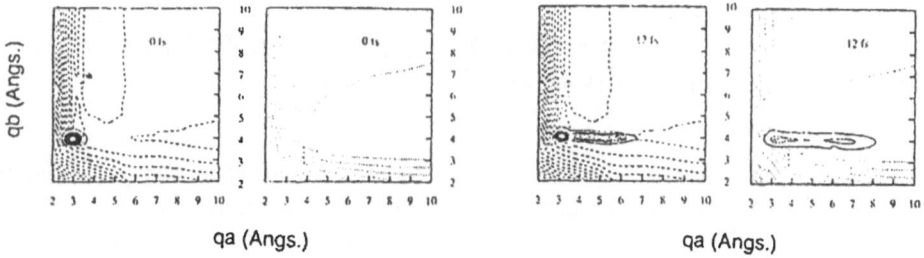

Figure 1 Time evolution of the wavepackets Ψ_{1E} (q_a,q_b,t) (solid line) on the V(^1E) potential (dashed line) (left) and Ψ_{A1} (q_a,q_b,t) (solid line) on the V($^3A_1$) potential (dotted line) (right) (q_a=CO-H, q_b=Co-CO$_{ax}$).

According to the initial conditions, the ^1E state is the only initially populated excited state. The wavepacket breaks into two parts: one part has enough energy to overcome the barrier at the entrance of the valley leading to the metal-hydrogen bond breaking, the other fraction represents intramolecular vibrational energy redistribution. In 20 fs, 35% of the system dissociates to the products H+Co(CO)$_4$(^1E), whereas 100 fs later another 5% leads to the same reaction. Immediately after the initial excitation to the ^1E state, the $^3A_1$ state is

303

weakly populated through ISC (figure 1, right side). The small fraction of the wavepacket coming from the singlet potential runs out exclusively toward the products $H+Co(CO)_4(^3A_1)$. The $^1E \rightarrow {}^3A_1$ transition occurs within 32 ps [6] and is not competitive with the ultra fast direct dissociation.

3.2 Photodissociation Dynamics of $H_2Fe(CO)_4$

The two low-lying a^1B_1 $(3d_{yz} \rightarrow \sigma_{g*})$ and b^1A_1 $(3d_{x^2-y^2}^2 \rightarrow \sigma_g^*)$ excited states, directly accessible under UV irradiation and contributing mainly to the absorption spectrum of $H_2Fe(CO)_4$ [7] have been used in the two-dimensional simulation. After the initial $a^1A_1 \rightarrow a^1B_1$ transition, in a very short time scale (15fs) the wavepacket evoluates to the dissociation channel corresponding to H_2 elimination (figure 2).

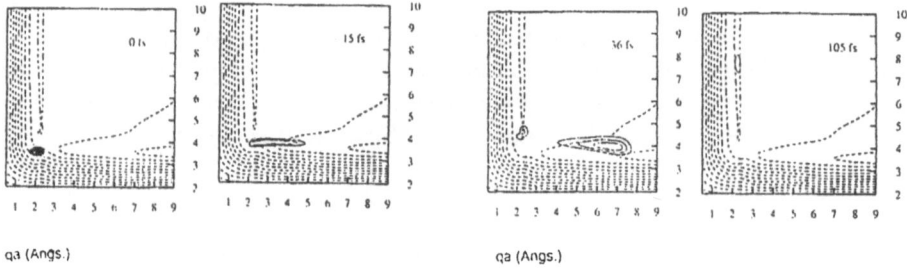

Figure 2 Time evolution of the wavepacket Ψ_{a^1B1} (q_a,q_b,t) (solid line) on the $V(a^1B_1)$ potential (dashed line) $(q_a= Fe-H_2)$, $q_b= Fe-CO_{ax})$.

After 30 fs one observes a splitting of the initial wavepacket in two parts: the main fraction leads to the primary products $H_2+Fe(CO)_4(a^1B_1)$ in 40 fs, the remainning part dissociates to the carbonyl loss primary products in a time scale of the order of 100 fs with a probability of 4%. The elimination of molecular hydrogen is the major process overall the UV absorption domain and its probability increases with the wavelength of irradiation, this reaction being total and ultra-fast (less than 40 fs) after excitation to the b^1A_1 state.

3.3 Photodissociation Dynamics of $HMn(CO)_3(DAB)$

The photodissociation dynamics of $HMn(CO)_3(DAB)$, along the two reaction coordinates $q_a=Mn-H$ and $q_b=Mn-CO_{ax}$, has been followed by propagation of selected wavepackets on the set of non-adiabatically and spin-orbit coupled PES corresponding to the low-lying $^{1,3}MLCT$ $(d \rightarrow \pi^*_{DAB})$ and ^{1,3} SBLCT $(\sigma_{Mn-H} \rightarrow \pi^*_{DAB})$ excited states after the electronic ground state transition either to one of the ^1MLCT states (visible irradiation) (figure 3) or to the ^1SBLCT state (UV irradiation).

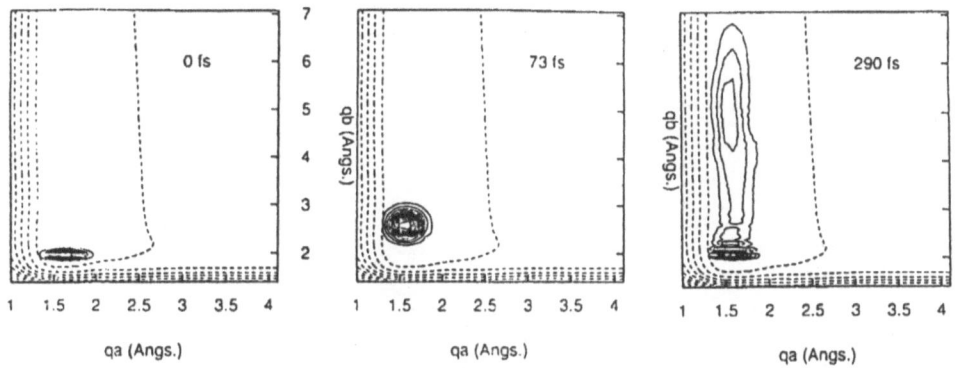

Figure 3 Time evolution of the wavepacket $\Psi_{^1MLCT}(q_a,q_b,t)$ (solid line) on the V(^{1}MLCT) potential (dashed line) (q_a=Mn-H, q_b=Mn-CO$_{ax}$).

The simulation of the visible photochemistry does not show any efficient homolysis. In contrast, a significant fraction of the wavepacket dissociates to the primary products HMn(CO)$_2$(DAB)+CO in a few hundred of fs (figure 3). The remaining part of the system gets trapped into the low-lying ^{1}MLCT potential wells. In the ps time scale only 2% of this part dissociates to the homolysis primary products H + Mn(CO)$_3$(DAB) after ^{1}MLCT $\rightarrow$ ^{3}MLCT intersystem crossing.

References

[1] Sweany, R. L. *Inorg. Chem.*, **1980**, 19, 3512; **1982**, 21, 752; *J. Am. Chem. Soc.*, **1982**, 104, 3739.; Church, S. P.; Poliakoff, M.; Timney, J. A.; Turner, J. J. *Inorg. Chem.* **1983,** 22, 3259.

[2] Sweany, R. L. *J. Am. Chem. Soc.* **1981**, 103, 2410; Swenay, R. L. submitted for publication.

[3] Stufkens, D. J. *Comments Inorg. Chem.*, **1992**, 13, 359; Rossenaar, B. D.; George, M. W.; Jonhson, F. P. A.; Stufkens, D. J.; Turner, J. J.; Vlcek, A. Jr. *J. Am. Chem. Soc.* **1995**, 117, 11582; Rossenaar, B. D.; Kleverlaan, C. J.; Van de Ven, M. C. E.; Stufkens, D. J.; Vlcek, A. Jr. *Chem. Eur. J.* **1996**, 2, 228; Rossenaar, B. D.; Stufkens, D. J.; Oskam, A.; Fraanje, J.; Goubitz, K. *Inorg. Chim. Acta*, **1996**, 247.

[4] Finger, K. Daniel, C. *J. Am. Chem. Soc..*, **1995**, 117, 12322; Heitz, M. C.; Finger, K.; Daniel, C. *Coord. Chem. Rev.,* **1997**, 159, 171.

[5] Heitz, M. C.; Daniel, C. *J. Am. Chem. Soc.* **1997**, in press.

[6] Kosloff, R.; Tal-Hezer, H. *Chem. Phys. Letters,* **1986**, 127, 223; Morston, C. C.; Balint-Kurti, C. G. **1989**, 91, 3571.

[7] Daniel, C.; Heitz, M. C.; Manz, J.; Ribbing, C. *J. Chem. Phys.* **1995,** 102, 905; Heitz, M. C.; Daniel, C.; Ribbing, C. *J. Chem. Phys.* **1997**, 106, 1421.

Time-dependent quantum calculations on the nuclear dynamics of laser excited organometallic molecules

M. Braun[#], O. Rubner[#], B. T. Markert[#], W. Malisch[†], and V. Engel[#]*

[#] Institut für Physikalische Chemie, [†]Institut für Anorganische Chemie, Universität Würzburg, Am Hubland, D-97074 Würzburg, Germany

1 Introduction

Spectroscopy is an indispensable tool to characterize the properties of molecules. In a traditional spectroscopic experiment a laser pulse interacts with a sample of molecules, thus inducing transitions via one- or more-photon processes. In the language of quantum mechanics the interaction between the electromagnetic field and the molecules causes a transition from an initial state ψ_i to a final state ψ_f and the transition rates are proportional to certain matrix elements containing these states. A closer look at the formuli shows that time as a variable has vanished although we like to think of processes as evolving in time. In fact, the time is only absent because nanosecond laser pulses, as usually used in rotational, vibrational and electronic spectroscopy, have a much larger temporal width than all relevant timescales for molecular motion.

This is not the case for femtosecond pulses which are by now available from commercially produced laser sources. Such pulses are shorter than e. g. vibrational periods, as will be discussed below. Since the spectral width of an ultrashort pulse is broad it is possible to excite several molecular eigenstates simultaneously just like in a pulsed NMR-experiment many spin states are excited. The quantum mechanical object created by such a superposition of states is a wave packet which shows a time-evolution to be detected in specially designed experiments (1).

Several experiments on the femtosecond spectroscopy of organometallic compounds have been performed (2,3,4) and it is our aim to simulate the possible outcome of future measurements. Here we treat some simple model systems to arrive at a basic understanding of the femtosecond spectroscopy of such molecules.

2 FeCO-NENEPO spectroscopy

"Negative-ion-to-neutral-to-positive-ion" (NENEPO) time-resolved spectroscopy has been recently introduced as a method to perform experiments on neutral size selected clusters (5). In our example the processes induced by two femtosecond pulses are photodetachment from $FeCO^-$ and time-delayed photoionization $FeCO \rightarrow FeCO^+ + e^-$.

The theory of this double ionization process and an efficient numerical method for a simulation was formulated by us (6). We calculate the kinetic energy distribution of the photoelectrons which are ejected as a function of the delay-time between the detachment and ionization pulse. As done in other studies, only the Fe-CO distance is treated. The interaction of the first pulse with the anion prepares a non-stationary wave packet in the FeCO molecule. The corresponding vibrational dynamics is reflected in the photoelectron spectrum obtained by the ionization process taking place at different delay times.

Figure 1 shows the kinetic energy distribution of the emitted electrons which are obtained in the transition from the ground state of $FeCO^-$ via the $^5\Sigma^-$ FeCO state to the ground state of $FeCO^+$ as a function of the pulse delay. The calculations were performed with pulses of central energies of $\omega_1 = 2.54$ eV and $\omega_2 = 9.5$ eV for the first and second pulse. A temporal width of 25 fs was used. The constant peak at low energies belongs to electrons produced by the detachment pulse. At larger energies the spectrum shows a periodic time-dependence which reflects the vibrational motion in the Fe-CO co-ordinate. The distribution at particular delay-times can be directly correlated to the probability-density distribution in the intermediate electronic state. Thus, we are able to watch the iron dynamics relative to the carbonyl group, i. e. it is possible to "see" quantum dynamics by detecting time-resolved photoelectron spectra (7, 8).

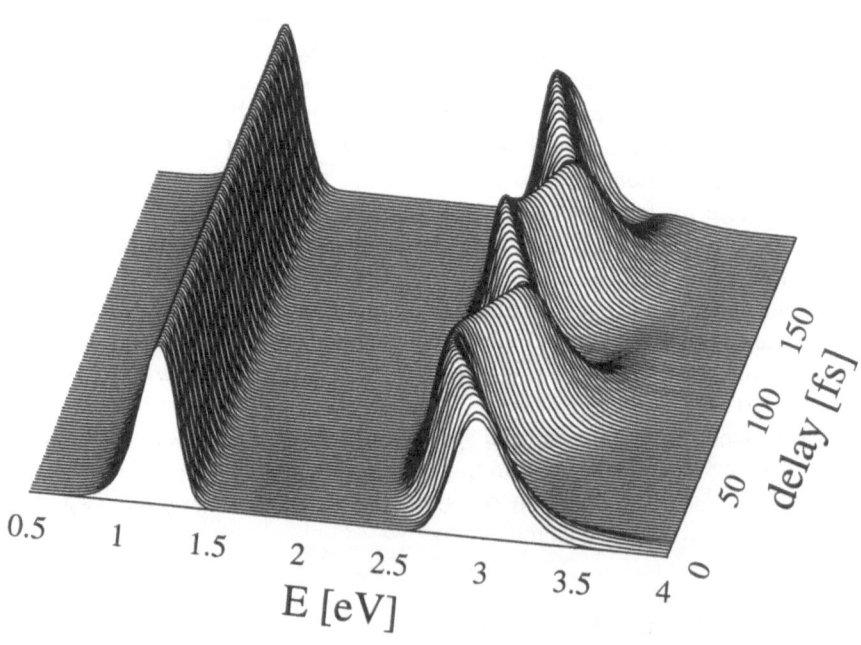

Figure 1 Time-resolved photoelectron spectra obtained by a NENEPO process via the $^5\Sigma^-$ electronic state of FeCO.

3 Infrared excitation of metallotrihydridosilane

Here we treat femtosecond excitation within the electronic ground state of a trihydridosilyl-complex (9). To understand the basic dynamics we perform quantum mechanical calculations on a simplified model of the R-Fe-Si-H_3 unit, where R = C_5H_5 and H_3 are treated as point masses and the R-Fe, Fe-Si and Si-H_3 distances are the internal degrees of freedom. A model potential was constructed which the help of Raman data obtained in the Kiefer group (10). We solve the time-dependent Schrödinger equation for this three-mode system interacting with an infrared pulse (ω = 0.034 eV) of 300 fs width and different peak frequencies. Figure 2 (right) shows energy expectation values obtained for a laser pulse intensity of 10^{12} W/cm^2. The upper panel displays the total energy before, during and after the interaction of the pulse with the molecules. An energy amount of about 0.1 eV is transfered from the field to the system. The middle panel contains the energies which are associated with the Hamiltonians belonging to the single internal modes and can be chosen as a measure of the energy in the corresponding bonds. The absorbed energy is transfered into the R-Fe and Fe-Si internal vibration but not into the Si-H_3 vibrational motion. This suggests that it is possible to excite the complex selectively: the absorbed energy is stored in specific bonds while others are decoupled although the total energy of the system allows for excitation of these modes.

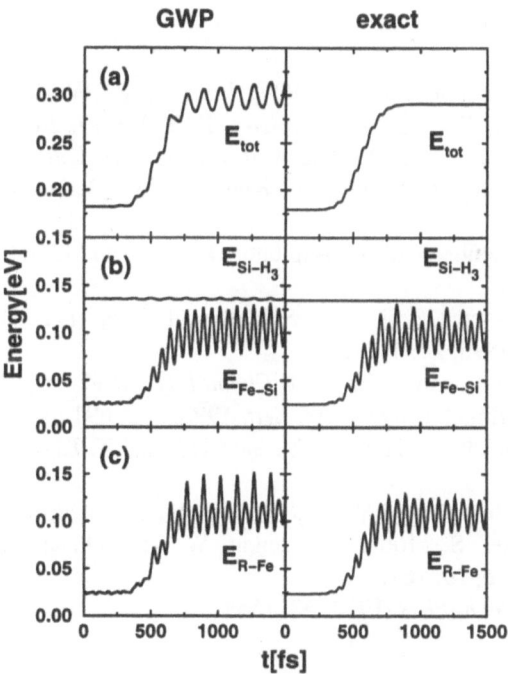

Figure 2 (a): expectation value of the total molecular energy. (b), (c): energy expectation values for the single mode Hamiltonians, as indicated. Results from exact and approximative calculations are compared.

309

To make realistic predictions about the excitation processes the theoretical treatment has to be extended to include all vibrational modes of the complex. This is, however, not possible within a purely quantum calculation and appropriate approximations have to be incorporated. The Gaussian wave-packet approximation (GWP) introduced by Heller (11) was adapted by us and modified for the description of coherent excitation. The central idea here is that a Gaussian moving in a harmonic potential remains of Gaussian shape. The expansion of the potential surface around the center of the wave packet at each time-step yields a set of differential equations for the parameters of the Gaussian. This amounts to a computation of the quantum dynamics within a numerical effort which resembles the one needed to perform a classical calculation. The above described model system was treated within the GWP approximation and the results of the calculation are compared to the exact ones in Figure 2. The agreement is astonishing good which gives us confidence for further studies. The method will allow us to incorporate more internal degrees of freedom which is important for a complete description of the dynamics induced by femtosecond excitation processes.

Acknowledgements

Financial support from the Deutsche Forschungsgemeinschaft (SFB 347, Teilprojekte C-5 and B-2) as well as from the Fonds der Chemischen Industrie is gratefully acknowledged.

References

1. A. H. Zewail, *Femtochemistry,* **1994,** *Vols. 1, 2* (World Scientific, Singapore); J. Manz, L. Wöste, *Femtosecnd Chemistry,* **1995,** *Vols. 1, 2* (VCH, Weinheim)
2. S. K. Kim, S. Pedersen, and A. H. Zewail, *Chem. Phys. Lett.* **1995,** *233,* 500.
3. L. Banares, T. Baumert, M. Bergt, B. Kiefer, and G. Gerber, *Chem. Phys. Lett.* **1996,** *267,* 149.
4. T. Lian, S. E. Bromberg, M. C. Asplund, H. Yang, and C. B. Harris, *J. Chem. Phys.* **1996,** *100,* 11994.
5. S. Wolf, G. Sommerer, S. Rutz, E. Schreiber, T. Leisner, E. Wöste, and R. S. Berry, *Phys. Rev. Lett.* **1995,** *74,* 4177.
6. O. Rubner, C. Meier, and V. Engel, *J. Chem. Phys.* **1997,** *107,* 1066.
7. C. Meier and V. Engel, *Chem. Phys. Lett.* **1993,** *212,* 691.
8. A. Assion, M. Geisler, J. Helbing, V. Seyfried, and T. Baumert, *Phys. Rev. A,* **1996,** *54,* R4605.
9. T. Markert, W. Malisch, and V. Engel, *Chem. Phys. Lett.* **1997,** *270,* 222.
10. R. Pikl, U. Posset, S. Möller, R. Lankat, W. Malisch, and W. Kiefer; *Vibrational Spectroscopy* **1996,** *10,* 161.
11. E. J. Heller, *J. Chem. Phys.* **1975,** *62,* 1544.

Ultrafast Photodissociation Dynamics of Isolated Iron-pentacarbonyl

L. Bañares, T. Baumert, M. Bergt, B. Kiefer, and G. Gerber*

Physikalisches Institut, Experimentalphysik I, Universität Würzburg, Am Hubland, D-97074 Würzburg, Germany.

Introduction

The photochemistry of metal carbonyls has received a wealth of interest in the last decades [1]. Metal carbonyls are important photocatalysts of many organic reactions [2]. Therefore, it has become important in recent years to understand the mechanisms by which these complex molecules photodissociate loosing one or several CO groups.

Fe(CO)$_5$ is a prototype molecule whose study can provide a good understanding of the photodissociation mechanisms of metal carbonyls. For this molecule, concerning its molecular and electronic structure, plenty of experimental and theoretical information is available (see Refs. in [3,4]).

The energetics of this molecule is presented in Fig. 1. The absorption spectrum is rather structureless and exhibits a strong band centered at about 50,000 cm^{-1} with a shoulder at 41,600 cm^{-1} (indicated by arrows in the figure). These two bands have been assigned to metal to ligand charge-transfer (MLCT) d $\rightarrow$ π^* transitions.

From the experimental point of view, many studies have been reported on the photodissociation of Fe(CO)$_5$ in the gas phase using nanosecond lasers in combination with other techniques (see Refs. in [3,4]). In all experiments, where multiphoton ionization (MPI) with nanosecond lasers was applied, the main photodissociation product was Fe$^+$, but in none of these experiments, the parent Fe(CO)$_5^+$ ions have been observed. It has been a subject of controversy, whether the loss of ligands by the metal atom after laser excitation occurs stepwise or in a concerted way.

Several theoretical calculations suggested mechanisms for the photodissociation and gave estimates for the involved time scales. Recent theoretical studies by Daniel *et al.* on metal carbonyl hydrides [5,6] have proposed that the time scale for the cleavage of the metal-CO bond is as short as 100 fs.

Since no technique applied thus far has been fast enough to measure the primary photodissociation of Fe(CO)$_5$ in real time, the application of femtosecond laser studies is very timely. The femtosecond time scale is the ultimate time scale for the nuclear motion in molecules. Femtosecond techniques are therefore an ideal tool for real-time observations and control of chemical reactions[7].

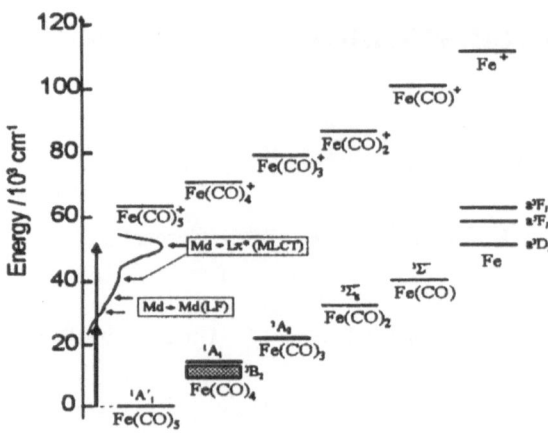

Figure 1 Energetics of Fe(CO)$_5$ and excitation scheme

The first femtosecond time resolved gas phase experiment on metal carbonyls was recently reported by Zewail and coworkers [8]. In that cornerstone experiment, they studied the ultrafast fragmentation dynamics of Mn$_2$(CO)$_{10}$.

We report on experiments where the photodissociation dynamics of Fe(CO)$_5$ in a molecular beam has been studied with femtosecond time resolution . The parent molecule and all the fragments were detected after femtosecond laser ionization in a linear time-of-flight (TOF) spectrometer. Transient ionization spectra of the parent molecule and of every fragment were measured by using the femtosecond pump-probe technique. From the results obtained, it can be concluded that the photodissociation of Fe(CO)$_5$ up to Fe(CO) occurs in about 100 fs. The subsequent dissociation of Fe(CO) into Fe + CO occurs on a longer time scale of 230 fs [3,4].

Experimental Setup

The experimental setup has been described in detail elsewhere [9] and only a brief description will be given here. An amplified Ti:Sapphire laser system yields pulsed femtosecond laser radiation centered at 800 nm with a duration of 70 fs and a pulse energy of 1mJ at a repetition rate of 1 kHz. After second harmonic generation, the 400 nm and 800 nm radiation is separated by means of a dichroic mirror into two beams. The pump laser pulses (400 nm) are delayed with respect to the probe laser pulses (800 nm) using a computer controlled Michelson-type interferometer. Both laser beams, appropriately attenuated, are recombined using a dichroic mirror and focused into the molecular beam chamber.

The molecular beam apparatus consists of two differentially pumped chambers one of them containing the Fe(CO)$_5$ molecular beam source and the other one a linear time-of-flight (TOF) spectrometer with a microchannel plate (MCP) detector. The Fe(CO)$_5$ sample, used without further purification (98%; Stream Chemicals), is taken directly from the cylinder at room temperature (vapor pressure of $\approx 3 \cdot 10^3$ Pa) and expanded through a nozzle of 50 μm into high vacuum (10^{-5} Pa).

The transient ionization spectra, *i.e.* the ion signal for a given mass in the TOF spectra dependent upon the time delay between the pump and the probe lasers, are measured by means of boxcar integrators. The transients were fitted using a non-linear least square method based on a Marquardt-Levenberg algorithm where the corresponding molecular response function (single or multiple-exponential with rise and decay components) was convoluted with a gaussian whose full-width-half-maximum (FWHM) corresponds to the cross-correlation of the pump and probe laser pulses.

Results and Discussion

Nanosecond and femtosecond multiphoton ionization-TOF spectra

Figure 2 shows time-of-flight (TOF) spectra obtained by MPI with nanosecond (ns) and femtosecond (fs) laser pulses. When using nanosecond laser pulses (337 nm, $\approx 10^9$ Wcm^{-2}), the only mass peak observed in the TOF spectra is the final photoproduct Fe$^+$.

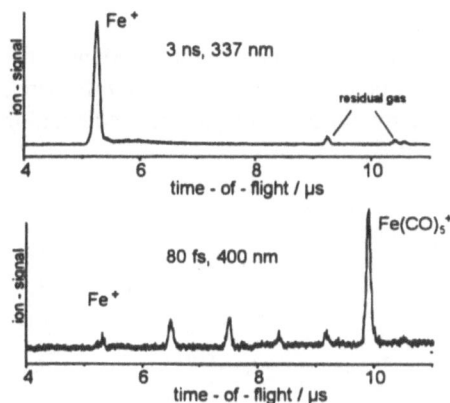

Figure 2 Comparison of TOF spectra obtained by MPI with ns laser (upper part) and fs laser (lower part).

No evidence of larger fragment ions Fe(CO)$_{4-1}^+$, or the parent ion Fe(CO)$_5^+$ was found which is in agreement with earlier nanosecond laser studies (see Refs. in [3,4]). Fe(CO)$_5$ absorbs a photon already within the rising edge of the laser pulse and the molecule dissociates. The resulting fragment molecules absorb additional photons of the same laser pulse and can dissociate further. The ultimate photoproduct is Fe$^+$.

The use of femtosecond laser pulses leads to a completely different picture. Mainly the parent ion Fe(CO)$_5^+$ appears in the fs-TOF spectrum obtained at 400 nm ($\approx 5 \cdot 10^{12}$ Wcm^{-2}). Ladder switching, as observed with ns laser excaitation is efficiently supressed. For laser intensities exceeding $\approx 5 \cdot 10^{13}$ Wcm^{-2} all the fragment ions Fe(CO)$_{4-0}^+$ and even CO$^+$ start to appear. Increasing of the pulse durations from 80 fs to 100 ps enlarges the amount of fragmentation. For laser pulswidths of more than 70 ps, Fe^+ is the only ion observed. These observations already show that the photodissociation of $Fe(CO)_5$ is ultrafast and one has to use ultrashort laserpulses to examine the fragmentation dynamics of this molecule in detail. The fragmentation pattern strongly depends on the fs-laser wavelength, the intensity and the pulse duration, which is discussed in detail in [4].

Transient ionization spectra

One and two color pump-probe experiments were performed using the fundamental wavelength (800 nm) and the second harmonic (400 nm) of the Ti:Sapphire femtosecond laser system. The 400 nm (pump) laser pulse excites $Fe(CO)_5$ by two photon excitation to the MLCT band (see Figure 1) and the molecules start to dissociate. After a variable delay time, the 800 nm (probe) laser pulse takes a snapshot of the evolving system. The probe laser produced ions are detected by a TOF mass spectrometer. The transient for a given mass is the ion signal *vs.* delay time.

Figure 3 shows the measured transients of the parent molecule and of every fragment. For positive delay times, the 400 nm laser was the pump and the 800 nm was the probe. All the transients were measured with attenuated pump and probe laser beams, where each of the lasers alone produces a negligible amount of ions. Under these experimental conditions, the observed dynamics has to occur in the neutral molecule where the probe laser induces the ionization.

The ion signal before time zero, when pump (400 nm) and probe (800 nm) lasers interchange their role, will not be discussed here. In the following, that part of the transients is analyzed for which the 400 nm pulse excites the molecule and the subsequent dynamics is probed by 800 nm photons.

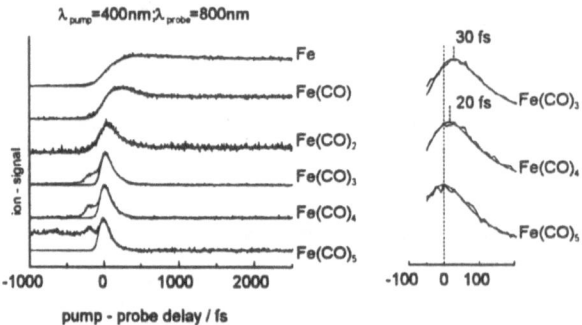

Figure 3 Two color (400 nm / 800 nm) femtosecond transient ionization spectra of $Fe(CO)_5$ and of the fragments. The solid lines represent fits to the data based on exponential rise and decay constants. The inset shows the expanded transients around time delay zero. The observed time shifts are indicated.

All the measured tran-sients were fitted to single or multiple-exponentials with rise and decay times taking into account the pump and probe laser pulse durations as indicated in the experi-mental section. The results of the fits for the parent molecule and all fragments are shown in Figure 3 as solid lines. The transients of the parent and the fragments $Fe(CO)_{4,3,2}$ were fitted to single exponentials with decay times of 100±5 fs, 105±5 fs, 115±5 fs and 150±20 fs, respectively. Note that $Fe(CO)_{4,3,2}$ transients show time shifts of 20±5 fs, 30±5 fs and 60±15 fs respectively, with respect to the maximum of the $Fe(CO)_5$ signal. The Fe(CO) ionization transient shows a slower rise than that of the cross correlation of the pump and probe laser, and was fitted with a rise time of 120±20 fs and a longer decay time of 230±20 fs. The transient of the atomic Fe fragment was fitted to a single-exponential rise with a time constant of 260±20 fs, and a longer decay time of 490±50 fs.

One color (400 nm / 400 nm) pump-probe experiments have also been performed (see Figure 4). Due to the fact, that pump and probe wavelength as well as intensities of both lasers are the same, the transients are symmetric with respect to time delay zero. The observed time constants are very similar to the one of the two color experiments and the width of the transients $Fe(CO)_{5-2}$ rises with the number of ligands lost, which corresponds to the increasing time shift of the 400/800 nm transients.

From all these results we conclude that the photo-dissociation of $Fe(CO)_5$ in the gas phase occurs on an ultrafast time scale (few hundreds of femtoseconds). This ultrafast time scale for the photodissociation process should exclude the possi-bility of statistical energy transfer within the molecule prior to fragmentation (IVR) or internal conversion, as it has been pointed out for the related $Mn_2(CO)_{10}$ molecule [8].

On the basis of the observed transients the following dissociation model can be proposed. $Fe(CO)_5$ is excited to a state with a total energy of 50,000 cm^{-1}. This state will evolve in a structural rearrangement. During the first 20±5 fs (the time shift observed in the $Fe(CO)_4$ transient) the $[Fe(CO)_5]^{\ddagger}$ complex reaches a geometry which after ionization prefers to decay to a $Fe(CO)_5^+$ ion. The evolution of the multidimensional $[Fe(CO)_5]^{\ddagger}$ complex continues until another configuration is reached after 30±5 fs. This new geometry of the complex now favors the production of $Fe(CO)_5^+$ ions, which undergo fragmentation into the measured $Fe(CO)_3^+$ ions. This is the origin of the $Fe(CO)_3$ transient. With a very similar interpretation, we can account for the observed $Fe(CO)_2$ transient. Within the model, the decay times of about 100 fs correspond to the dissociation time of $Fe(CO)_5$ into Fe(CO) + 4 CO. The measured transients for $Fe(CO)_{4-2}$ represent snapshots of the evolution of the multidimensional $[Fe(CO)_5]^{\ddagger}$ transition state towards the loss of four CO-ligands yielding the Fe(CO) fragment.

Furthermore, the ionization transient of Fe(CO) shows a rise time of 120±20 fs. This rise time matches the $Fe(CO)_5$ dissociation time of 100-150 fs discussed above. In addition, a decay to a time independent level is observed in the Fe(CO) with a time constant of 230±20 fs. This is an indication that at least two different, maybe excited, Fe(CO) fragments are formed in the dissociation, and only part of the Fe(CO) fragments decay further. This decay time is close to the observed rise time of 260±20 fs for the Fe fragment.

The ultrafast photodissociation of $Fe(CO)_5$ after fs excitation of the MLCT band with two 400 nm photons can be summarized as a two step process. The first step is a concerted loss of four CO ligands in about 100 fs. In the second step the formed Fe(CO) looses the last CO ligand within 230 fs.

$$
\begin{aligned}
&1. \quad [Fe(CO)_5]^{\ddagger} \quad \rightarrow \quad Fe(CO) + 4\ CO \\
&2. \quad Fe(CO) \quad \rightarrow \quad Fe + CO
\end{aligned}
$$

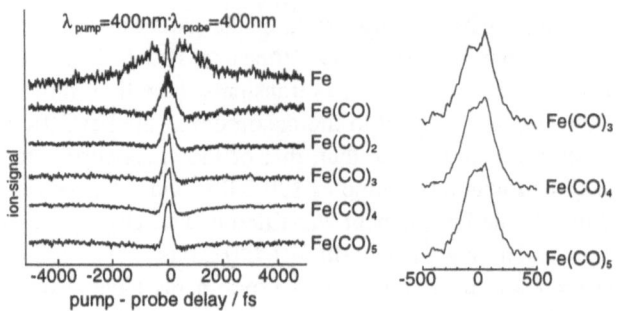

Figure 4 One color (400 nm / 400 nm) femtosecond transient ionization spectra of $Fe(CO)_5$ and of the fragments.

Conclusions

The photodissociation of $Fe(CO)_5$ has been investigated in the gas phase using nanosecond and femtosecond laser pulses. Due to the ultrafast dissociation process, nanosecond laser pulses fail to ionize the parent molecule $Fe(CO)_5$. By the use of femtosecond laser pulses, ionization preceeds fragmentation and the ionized parent molecule is detected.

In addition, one and two color femtosecond pump-probe experiments have been performed by two photon excitation of the MLCT band with 400 nm femtosecond pump pulses and probe pulses of 400 nm and 800 nm. From the analysis of the transients in terms of rise and decay times, a model for the photodissociation of $Fe(CO)_5$ is proposed. In this model, $Fe(CO)_5$ dissociates up to $Fe(CO)$ in a concerted fashion, *i.e.* the parent molecule looses after absorption of two 400 nm photons four CO-ligands on the time scale of a vibrational period of the Fe-CO bond. The measured $Fe(CO)_{4,3,2}$ ionization transients represent snapshots of the evolution of the multidimensional $[Fe(CO)_5]^{\ddagger}$ transition state on its way to $Fe(CO)$. The excess energy is sufficient for a further dissociation of $Fe(CO)$ into Fe and CO on a longer time scale.

More experiments using femtosecond lasers at different wavelengths and even shorter pulses are in preparation to provide additional information about the photodissociation dynamics of $Fe(CO)_5$ and of more complex metalorganic molecules in the gas phase.

Acknowledgments

Financial support from the DFG (SFB 347, Teilprojekt C-4) is gratefully acknowledged.

References

[1] G. I. Geoffrey and M. S. Wrighton, Organometallic Photochemistry, Academic Press, New York, **1979**.
[2] See for examle Organic Synthesis via Metal Carbonyls,Vols. 1 and 2, eds. I. Wender and P. Pino, Wiley, New York, **1977**.

[3] L. Bañares, T. Baumert, M. Bergt, B. Kiefer, and G. Gerber, *Chem. Phys. Lett.* **1997** *267*, 141.

[4] L. Bañares, T. Baumert, M. Bergt, B. Kiefer, and G. Gerber, submitted to *J. Chem. Phys.* **1997**.

[5] C. Daniel, M. C. Heitz, L. Lehr, T. Schröder, and B. Warmuth, *Int. J. Quan. Chem.* **1994** *52* 71.

[6] C. Daniel, M.-C. Heitz, J. Manz, and C. Ribbing, *J. Chem. Phys.* **1995** *102* 905.

[7] T. Baumert, J. Helbing, and G. Gerber, in *Advances in Chemical Physics-Photochemistry: Chemical Reactions and their Control on the Femtosecond Time Scale*, edited by I. Prigogine and S. Rice, vol. 101, pages 47-77, John Wiley and Sons, Inc, New York, **1997**.

[8] S. K. Kim, S. Pedersen, and A. H. Zewail, *Chem. Phys. Lett.* **1995** *233* 500.

[9] T. Baumert and G. Gerber, *Adv. At. Mol. Opt. Phys.* **1995** *35* 163.

Organic Synthesis via Organometallics (OSM5)

Proceedings of the Fifth Symposium in Heidelberg,
September 26 to 28, 1996

Edited by Günter Helmchen
with Jörg Dibo, Dietmar Flubacher, and Burkhard Wiese.

1997. viii, 376 pp. Hardcover. DM 168,–
ISBN 3-528-06905-8

Organometallic chemistry and homogeneous catalysis
with transition metal complexes are of increasing importance
in organic synthesis. Methods from these areas allow the control of
regio- and stereoselectivity with ever increasing efficiency. Recent
progress and state of the art in the field are documented in this volume
which covers the lectures delivered at the 5th symposium "Organic Synthesis
via Organometallics" held in Heidelberg. A selection of 13 plenary and 10
invited lectures present a broad survey on new avenues in organic synthesis including timely topics such as polymer synthesis via organometallics,
synthesis of new organometallics, mechanism of catalytic processes,
asymmetric catalysis of carbon bond forming reactions and stereoselective
synthesis of biologically active compounds.

Abraham-Lincoln-Str. 46, Postfach 1547, 65005 Wiesbaden
Fax: (06 11) 78 78-4 00, http://www.vieweg.de

Stand 1.2.98.
Änderungen vorbehalten.
Erhältlich im Buchhandel
oder beim Verlag.

vieweg